Encyclopaedia of Mathematical Sciences

Volume 58

Editor-in-Chief: R. V. Gamkrelidze

A. N. Parshin I. R. Shafarevich (Eds.)

Algebra VII

Combinatorial Group Theory
Applications to Geometry

With 39 Figures

Springer-Verlag
Berlin Heidelberg New York
London Paris Tokyo
Hong Kong Barcelona
Budapest

Title of the Russian edition:
Itogi nauki i tekhniki, Sovremennye problemy matematiki,
Fundamental'nye napravleniya, Vol. 58, Algebra 7
Publisher VINITI, Moscow 1990

Mathematics Subject Classification (1991):
08A50, 20Exx, 20Fxx, 20H10, 20J05, 57Mxx, 57N10, 68Q68

ISBN 3-540-54700-2 Springer-Verlag Berlin Heidelberg New York
ISBN 0-387-54700-2 Springer-Verlag New York Berlin Heidelberg

Library of Congress Cataloging-in-Publication Data
Algebra VII. Combinatorial group theory. Applications to geometry /
A. N. Parshin, I. R. Shafarevich, eds.
p. cm. - (Encyclopaedia of mathematical sciences; v. 58)
Includes bibliographical references and index.
ISBN 0-387-54700-2
1. Combinatorial group theory. 2. Geometric group theory. I. Parshin, A. N.
II. Shafarevich, I. R. (Igor Rostislavovich), 1923- . III. Series.
QA182.5A43 1993 512'.2--dc20 92-13652

Printed in the United States of America
Typesetting: Springer TEX in-house system
41/3140-543210 - Printed on acid-free paper

List of Editors, Authors and Translators

Editor-in-Chief

R. V. Gamkrelidze, Russian Academy of Sciences, Steklov Mathematical Institute, ul. Vavilova 42, 117966 Moscow, Institute for Scientific Information (VINITI), ul. Usievicha 20a, 125219 Moscow, Russia

Consulting Editors

A. N. Parshin, I. R. Shafarevich, Steklov Mathematical Institute, ul. Vavilova 42, 117966 Moscow, Russia

Authors

D. J. Collins, School of Mathematical Sciences, Queen Mary and Westfield College, University of London, Mile End Road, London E1 4NS, England
R. I. Grigorchuk, MIIT, Moscow Institute for Railway Engineers, ul. Obrastsova 15, Moscow, Russia
P. F. Kurchanov, MIIT, Moscow Institute for Railway Engineers, ul. Obrastsova 15, Moscow, Russia
H. Zieschang, Fakultät und Institut für Mathematik, Ruhr-Universität Bochum, Universitätsstraße 150, 4630 Bochum 1, FRG

Translator

P. M. Cohn, Department of Mathematics, University College London, Gower Street, London WC1E 6BT, England

Contents

I. Combinatorial Group Theory and Fundamental Groups
D. J. Collins, H. Zieschang
1

II. Some Questions of Group Theory Related to Geometry
R. I. Grigorchuk, P. F. Kurchanov
167

Author Index
233

Subject Index
236

I. Combinatorial Group Theory and Fundamental Groups

D.J. Collins, H. Zieschang

Contents

Introduction 3

Chapter 1. Group Presentations and 2-Complexes 6

§ 1.1. Presentations of Groups 6
§ 1.2. Complexes and Fundamental Groups 13
§ 1.3. Subgroups and Coverings 23

Chapter 2. Free Groups and Free Products 31

§ 2.1. Free Groups 31
§ 2.2. Amalgamated Free Products and Graphs of Groups 34
§ 2.3. Automorphisms of Free Groups 47
§ 2.4. One-Relator Groups 55

Chapter 3. Surfaces and Planar Discontinuous Groups 61

§ 3.1. Surfaces 61
§ 3.2. Planar Discontinuous Groups 66
§ 3.3. Subgroups of Planar Groups 75
§ 3.4. Automorphisms of Fuchsian Groups 79
§ 3.5. Relations to Other Theories of Surfaces 86

Chapter 4. Cancellation Diagrams and Equations Over Groups 90

§ 4.1. Cancellation Diagrams 90
§ 4.2. Locally Indicable Groups and Equations Over Groups 100

Chapter 5. 3-Manifolds and Knots 105

§ 5.1. Fundamental Groups of 3-Manifolds 105
§ 5.2. Haken Manifolds 108
§ 5.3. On Knots and Their Groups 114

Chapter 6. Cohomological Methods and Ends 127

§ 6.1. Group Extensions and Cohomology 127
§ 6.2. Ends of Groups 135

Chapter 7. Decision Problems 143

§ 7.1. Decision Problems and Algorithms 143
§ 7.2. Unsolvable Decision Problems 146
§ 7.3. Automata and Groups 151

Bibliography 157

Index of Notation 165

Introduction

Combinatorial group theory has its roots in topology and in particular in the theory of the fundamental group. Introduced by Poincaré in giving an example of a 3-manifold which has trivial homology but is not homeomorphic to the 3-sphere, the fundamental group is a powerful if sometimes intractable invariant of a topological space. The aim of this essay is to describe the complex interaction between the algebra and the geometry that is transmitted through the medium of the fundamental group.

In its broadest sense we understand combinatorial group theory to refer to the theory of group presentations, that is of groups specified by a set of generators and corresponding defining relations. The theory begins in about 1880 as part of complex analysis with the work of Klein, Poincaré and others on Fuchsian groups. Of particular importance from the standpoint of combinatorial group theory was the work of Dyck, who was the first to isolate the notion of a free group and whose work marks the start of combinatorial group theory as an independent discipline. Some thirty years later, around 1910, the topic came of age with the work of Dehn on decision problems for fundamental groups of closed surfaces. Above all the word problem — when do two words in a system of generators represent the same group element — and the isomorphism problem — when do two group presentations define the same group — have been the motors which have driven combinatorial group theory.

Fuchsian groups and fundamental groups in general were introduced to provide tools to deal with problems in analysis and topology and it was the need to sharpen these tools into effectiveness that led to the development of the concepts and techniques of combinatorial group theory. Of prime importance among these are free groups and their generalisation to amalgamated free products of groups. To begin with the methods used were comparatively algebraic in character: the cancellation method of Nielsen and the method of Schreier transversals used to show that subgroups of free groups are free, and the elaboration of the latter by Kurosh to describe subgroups of a free product, are typical examples. Part of our aim here is to emphasise that there is two-way traffic between combinatorial group theory and topology and we shall prove these theorems by using the relationship between coverings of a space and subgroups of its fundamental group. In practice the full analytic machinery of topology is not necessary for this and we shall follow Reidemeister in working almost entirely with combinatorial cell complexes. Indeed the theory of 2-dimensional complexes is to a large extent synonomous with combinatorial group theory as we understand it and it is striking how frequently difficult algebraic arguments have later been replaced by more elegant arguments of a geometric or toplogical nature.

In Chapter 1 we begin with a description of the two notions of group presentation and combinatorial 2-complex and a discussion of the basic relationship between the two. The main topics are the Seifert-van Kampen Theorem and the link between subgroups and coverings. In Chapter 2, the basic group theoretic tools are developed and we start with the Nielsen cancellation method, since this technique still has many applications, some of which are illustrated in Chapter 5 for Heegaard splittings of 3-manifolds. Chapter 2 continues by picking up the theme of the Seifert-van Kampen Theorem from Chapter 1 with a description of amalgamated free products and their generalisation to the Bass-Serre theory of groups acting on trees. Then we return to free groups and in particular to their group of automorphisms, and finish Chapter 2 with a description of that standard testbed for methods of combinatorial group theory, the theory of one-relator groups.

Our third chapter goes back to the very start of the subject and deals with Fuchsian groups and fundamental groups of surfaces via their action on the plane. Here the rich interplay between geometric and algebraic methods is especially striking. In Chapter 4 the main focus is on the application of topological and geometric techniques to problems in combinatorial group theory. The main technique here, the method of cancellation diagrams, is an outstanding example of how the use of geometric ideas can simplifiy and clarify algebraic arguments and then lead to further developments. Chapter 5 deals with 3-manifolds and concentrates on those aspects of the subject where significant results can be obtained primarily through the use of the fundamental group. In contrast to the case of surfaces where the fundamental groups are relatively easy to determine and contain practically everything one would ever wish to know about a surface, fundamental groups of 3-manifolds are difficult to deal with and do not always provide answers to the questions at hand. Nonetheless in many interesting cases such as those of Seifert fibre spaces, where the fundamental group is closely related to a surface group, and knots, the techniques of combinatorial group theory can be applied.

In Chapter 6 we return to the theme of the influence of topological ideas on group theory. The theory of ends, originally developed as a way to compactify spaces, was used by Stallings as a way of proving a theorem on 3-manifolds and resulted in a proof of the purely group-theoretical result that a torsion-free group containing a free subgroup of finite index must itself be free. The argument proceeds via the notion of cohomological dimension and a brief account is given of this. Finally in Chapter 7, we examine the limitations that nature has imposed on solving the word problem by giving an example of a group presentation for which no algorithm to solve the word problem can exist. We also give a characterisation of finitely presented groups with a free subgroup of finite index in terms of automata theory — the interesting aspect of this is that the proof rests heavily on the theory of ends developed in Chapter 6.

We conclude by giving a list of books which contain full details and proofs of the basic theorems in the text. Concerning group theory: [Coxeter-Moser 1972], [Lyndon-Schupp 1977], [Magnus-Karrass-Solitar 1966], [Zieschang-Vogt-Coldewey 1980,1988]; concerning topology: [Massey 1967], [Reidemeister 1932],

[Seifert-Threlfall 1934], [Spanier 1966], [Zieschang-Vogt-Coldewey 1980,1988] – in view of the large number of citations in the text, these last two sources will be referred to as [ZVC 1980] and [ZVC 1988].

Added for the English edition. The present text is essentially unchanged from the Russian text but the opportunity has been taken to correct a few minor errors and misprints. The authors would like to add that while a number of proofs are given, in some cases what follows the word *Proof* is a sketch and the reader should refer to more detailed sources if the argument appears to omit some steps. While no systematic attempt has been made to update the material in the interval between the preparation of the text for the Russian edition and the publication of the English edition, it seems worth mentioning that the existence of infinite Burnside groups of even exponent (cf. Theorem 4.1.12) has been announced by G. Lysionok and a negative solution to the problem of accessibility of finitely generated groups (cf. Theorem 6.2.14) has been announced by M.J. Dunwoody.

Chapter 1
Group Presentations and 2-Complexes

§ 1.1. Presentations of Groups

In combinatorial group theory groups are presented using generating sets of elements and systems of defining relations. More precisely:

1.1.1. Definitions and Simple Properties. Let G be a group. A system $\mathcal{X}$ of elements of G is called a *generating system of* G or a *system of generators of* G if the smallest subgroup of G containing $\mathcal{X}$ is equal to G, i.e. every element of G is expressible as a product of the elements of $\mathcal{X}$ and their inverses. The least number of elements needed to generate G is sometimes called the *rank of* G and is denoted by $d(G)$. (Note that for an abelian group G the minimal number of elements needed to generate the quotient of G by the torsion subgroup is often called the rank of G; however we will call this the *Betti number of* G, see 1.1.12.)

Let $\mathcal{X}$ be a generating system for G and let $\hat{\mathcal{X}}$ be a system of letters such that there is a bijection $\hat{\mathcal{X}} \to \mathcal{X}$. We use capital letters $X, Y, Z, A, B, C, \ldots$ for elements of $\hat{\mathcal{X}}$ and the corresponding small letters $x, y, z, a, b, c, \ldots$ for the corresponding elements of $\mathcal{X}$. A *word (over* $\hat{\mathcal{X}}$*)* is a formal expression

$$W \equiv W(\hat{\mathcal{X}}) \equiv X_1^{\varepsilon_1} X_2^{\varepsilon_2} \ldots X_k^{\varepsilon_k} \equiv \prod_{j=1}^{k} X_j^{\varepsilon_j}$$

where $X_1, \ldots, X_k \in \hat{\mathcal{X}}$, $\varepsilon_1, \ldots, \varepsilon_k \in \{1, -1\}$. The number k is the *length* $|W|$ of the word W. The word $W(\hat{\mathcal{X}})$ *represents* or *defines* the element $g \in G$ if $g = W(\mathcal{X}) = \prod_{j=1}^{k} x_j^{\varepsilon_j}$. If $V(\hat{\mathcal{X}})$ is another word, representing $h \in G$, then the *product* $W(\hat{\mathcal{X}})V(\hat{\mathcal{X}})$ of the words $W(\hat{\mathcal{X}})$ and $V(\hat{\mathcal{X}})$ is the word obtained by first writing $W(\hat{\mathcal{X}})$ and then $V(\hat{\mathcal{X}})$. Clearly, the product $W(\hat{\mathcal{X}})V(\hat{\mathcal{X}})$ of words represents the product gh of elements. The *inverse word* of $W(\hat{\mathcal{X}})$ is the word $W(\hat{\mathcal{X}})^{-1} \equiv X_k^{-\varepsilon_k} \ldots X_2^{-\varepsilon_2} X_1^{-\varepsilon_1}$; of course, it defines the inverse element g^{-1} of g. We also introduce the *trivial* or *empty word* consisting of no letters and denote it by 1; it has length 0 and defines the neutral element of G. Two words V and W are called *freely equivalent*, also written $V \equiv W$, if one can be transformed into the other by inserting and deleting *peaks* $X^\varepsilon X^{-\varepsilon}$, $X \in \hat{\mathcal{X}}$, $\varepsilon = \pm 1$. A word $R(\hat{\mathcal{X}}) = X_1^{\varepsilon_1} \cdot \ldots \cdot X_k^{\varepsilon_k}$ is called a *relator (relative to* $\mathcal{X}$ *and* G*)* if $x_1^{\varepsilon_1} \cdot \ldots \cdot x_k^{\varepsilon_k} = 1$ in G. A system $\mathcal{R}$ of relators is called a *system of defining relators* if every relator is a *consequence* of those of $\mathcal{R}$, that is, is freely equivalent to a word

$$L_1(\hat{\mathcal{X}})R_1(\hat{\mathcal{X}})^{\eta_1}L_1(\hat{\mathcal{X}})^{-1} \cdot \ldots \cdot L_k(\hat{\mathcal{X}})R_k(\hat{\mathcal{X}})^{\eta_k}L_k(\hat{\mathcal{X}})^{-1}$$

where $R_j(\hat{\mathcal{X}}) \in \mathcal{R}$, $\eta_j \in \{1, -1\}$ and the $L_j(\hat{\mathcal{X}})$ are words. A *trivial relator* is freely equivalent to the empty word. As indicated, the notion of a relator depends on the group G *and* the system of generators.

Given a generating system $\mathcal{X}$ for G, it is often convenient to identify $\hat{\mathcal{X}}$ with $\mathcal{X}$. Care must then be taken to distinguish a word $W(\hat{\mathcal{X}})$ from the element $W(\mathcal{X})$ it represents — the intended meaning should always be clear from the context.

If $\mathcal{X}$ is a generating system of the group G and $\mathcal{R}$ a corresponding system of defining relators then $\langle \mathcal{X} \mid \mathcal{R} \rangle$ is called a *presentation of* G and we shall indicate this by writing $G = \langle \mathcal{X} \mid \mathcal{R} \rangle$. Other forms are also used, for instance, $G = \langle (x_j)_{j\in J} \mid (R_k(\hat{\mathcal{X}}))_{k\in K} \rangle$ or $G = \langle S_1, \ldots, S_n \mid R_1, \ldots, R_q \rangle$ or $G = \langle s_1, \ldots, s_n \mid - \rangle$, where the last form indicates that the set of relators is empty. A group G is called *finitely generated* if it has a finite system of generators and *finitely presentable* (or *presented*) if it has a presentation with a finite number of generators and defining relators.

1.1.2. Examples

(a) Let G be the group $\mathbb{Z}$ of integers, relative to addition, with the single generator 1. Let $\mathcal{X}$ consist of one letter A. Then the possible words are of the form

$$W \equiv A^{\varepsilon_1} \cdot A^{\varepsilon_2} \cdot \ldots \cdot A^{\varepsilon_k} \equiv A^{n_1} \cdot \ldots \cdot A^{n_q} \quad \text{where } \varepsilon_j = \pm 1,\ n_j \in \mathbb{Z};$$

in the latter form we collect together all consecutive letters with the same exponent ε, that is $n_j \cdot n_{j+1} < 0$. Henceforth we shall use this *power* notation. Clearly, W is freely equivalent to the word A^n where $n = \sum_{j=1}^{q} n_j$. Two words A^n and A^m, where $n, m \in \mathbb{Z}$, are freely equivalent if and only if $m = n$. Now A^n represents the integer $n \in \mathbb{Z}$, and so a word A^n is a relator if and only if $n = 0$; hence, there are only trivial relators and $\mathbb{Z} = \langle A \mid - \rangle$.

(b) Let $G = \mathbb{Z}_n$, the group of integers modulo n, $n \in \mathbb{Z}$, $n \geq 2$. If we associate to A the class 1 mod n we obtain a generator for $\mathbb{Z}_n$, but now A^n is a relator. If a word A^k is a relator then $k \equiv 0 \bmod n$ and hence A^k is a power of A^n. This shows that $\mathbb{Z}_n = \langle A \mid A^n \rangle$. Obviously, the finite or infinite cyclic groups are the only groups of rank 1.

(c) Let $G = \mathbb{Z} \oplus \mathbb{Z}$, let $\mathcal{X} = \{A, B\}$ and associate to A, B the elements $(1,0), (0,1) \in \mathbb{Z}\oplus\mathbb{Z}$, respectively. Then $ABA^{-1}B^{-1}$ is a non-trivial relator. A word $A^{n_1}B^{m_1} \ldots A^{n_k}B^{m_k}$ is a relator if and only if $\sum_{i=1}^{k} n_i = \sum_{i=1}^{k} m_i = 0$. Moreover $\mathbb{Z} \oplus \mathbb{Z} = \langle A, B \mid ABA^{-1}B^{-1} \rangle$. The proof is not difficult and we illustrate it with an example. Write $[A, B] = ABA^{-1}B^{-1}$. Now $A^2B^2A^{-2}B^{-2}$ is a relator since a little messy calculation gives

$$A^2B^2A^{-2}B^{-2} \equiv$$
$$[A,B] \cdot BAB^{-1}[A,B]BA^{-1}B^{-1} \cdot B[A,B]B^{-1} \cdot B^2AB^{-1}[A,B]BA^{-1}B^{-2}.$$

In example (c) AB and BA represent the same element $(1,1) \in \mathbb{Z} \oplus \mathbb{Z}$ and this correponds to the fact that $AB(BA)^{-1}$ is a relator. More generally, if $G = \langle \mathcal{X} \mid \mathcal{R} \rangle$ and W and W' are words then W and W' represent the same element of G if and only if WW'^{-1} is a relator. When this occurs we write $W = W'$ and call this formal equality a *relation of* G. Note that if $W = W'$ is a relation then so is $UWV = UW'V$ for any U, V. A *system of defining relations* $\mathcal{R}$ is a system of relations of the form $R = R'$ such that the corresponding system of words RR'^{-1} is a system of defining relators. We shall also write $G = \langle \mathcal{X} \mid \mathcal{R} \rangle$ when $\mathcal{R}$ is a system of defining relations. It is usually simpler to work with relations than with relators.

Before we continue with examples we formulate and prove the fundamental theorem of Dyck [Dyck 1882]. Let $G = \langle \mathcal{X} \mid \mathcal{R} \rangle$ and let H be a group. Let $\psi \colon \mathcal{X} \to H$ be a mapping with the property that for every defining relation $R(\hat{\mathcal{X}}) \equiv X_1^{\varepsilon_1} \dots X_k^{\varepsilon_k} \in \mathcal{R}$ (here all $X_j \in \mathcal{X}$)

$$R(\psi(\mathcal{X})) = \psi(X_1)^{\varepsilon_1} \dots \psi(X_k)^{\varepsilon_k} = 1$$

where this last equation holds in H. Then we say that *the system* $\mathcal{X}^\psi = (\psi(X))_{X \in \mathcal{X}}$ *satisfies the relations* $\mathcal{R}$.

1.1.3. Dyck's Theorem. *Let* $\langle \mathcal{X} \mid \mathcal{R} \rangle$ *be a presentation of the group* G. *Let* H *be a group and* $\psi \colon \mathcal{X} \to H$ *be a mapping. If the system* $\mathcal{X}^\psi$ *satisfies the relations* $\mathcal{R}$ *then there is a uniquely determined homomorphism* $\Psi \colon G \to H$ *with* $\Psi(x) = \psi(X)$, $X \in \mathcal{X}$.

Proof. For any $g \in G$ there is a word $W(\mathcal{X}) = X_1^{\varepsilon_1} \dots X_k^{\varepsilon_k}$ such that $g = x_1^{\varepsilon_1} \cdot \dots \cdot x_k^{\varepsilon_k}$. Define $\Psi(g) = \psi(X_1)^{\varepsilon_1} \cdot \dots \cdot \psi(X_k)^{\varepsilon_k} = W(\psi(\mathcal{X}))$. Since X^ψ satisfies the relations $\mathcal{R}$, it follows that Ψ is well-defined and a homomorphism. The uniqueness follows from the fact that G is generated by $\mathcal{X}$. □

1.1.4. Examples

(d) As in 1.1.2 (c) one concludes that $\mathbb{Z}^n = \langle A_1, \dots, A_n \mid \{[A_i, A_j] : 1 \le i < j \le n\} \rangle$. This presentation has n generators and $\frac{1}{2}(n-1)n$ relations. Linear algebra or elementary group theory shows that n is the minimal number of elements needed to generate $\mathbb{Z}^n$, that is, the rank of $\mathbb{Z}^n$ is $d(\mathbb{Z}^n) = n$. Does there exist a presentation with less than $\frac{1}{2}(n-1)n$ relations? Using homological algebra, the answer to this will be shown to be negative. Hence, see 5.1.2, for $n > 3$, $\mathbb{Z}^n$ cannot be the fundamental group of a 3-dimensional manifold.

(e) Clearly, every group G has a presentation, for instance: $G = \langle G \mid \{x \cdot y \cdot (xy)^{-1} : x, y \in G\} \rangle$. Mostly this presentation is not helpful and, in general, one tries to minimize the numbers of generators and relations.

(f) Consider $G = \langle X, Y \mid X^2Y^3, X^3Y^4 \rangle$. If $x, y \in G$ correspond to X, Y, we have the equations $x^2 = y^{-3}$, $x^3 = y^{-4}$ whence $x = y^{-1}$ and hence $y = 1$, $x = 1$. Thus G is the trivial group. In the same way one can show that $\mathbb{Z}_\Delta \cong \langle X, Y \mid X^aY^b, X^cY^d \rangle$ where $\Delta = |ad - bc| > 0$.

(g) Consider the *dihedral group* D_n, $n > 1$, the symmetry group of the regular n-polygon P_n, see [Shafarevich 1986, p. 125]. It is generated by the rotation r with angle $\frac{2\pi}{n}$ and a reflection s in the line through the centre and one of the vertices. Then $r^n = s^2 = \mathrm{id}$, $srs = r^{-1}$. Via these relations every element of D_n can be expressed in the form $r^k s^l, 0 \leq k < n,\ l \in \{0,1\}$. These expressions give different motions and hence $D_n = \langle r, s \mid r^n, s^2, (sr)^2 \rangle$. Among these groups is the symmetric group $\mathrm{Sym}(3) = D_3$. Similarly let D_∞ be the infinite dihedral group consisting of the motions of $\mathbb{R}$ which map the integers to integers, i.e. the transformations $\mathbb{R} \to \mathbb{R},\ x \mapsto \pm x + k$, with $k \in \mathbb{Z}$. It follows that $D_\infty = \langle s, t \mid s^2, stst \rangle$. The groups D_n are not abelian if $n > 2$ and have rank 2.

(h) Consider the group $SL(2, \mathbb{Z})$ of 2-by-2 integer matrices with determinant 1. Let $A = \left(\begin{smallmatrix} 0 & -1 \\ 1 & 0 \end{smallmatrix}\right)$ and $B = \left(\begin{smallmatrix} 0 & 1 \\ -1 & 1 \end{smallmatrix}\right)$. Using the matrices $AB = \left(\begin{smallmatrix} 1 & -1 \\ 0 & 1 \end{smallmatrix}\right)$ and $BA = \left(\begin{smallmatrix} 1 & 0 \\ 1 & 1 \end{smallmatrix}\right)$ it is easy to see that A and B generate $SL(2, \mathbb{Z})$. Moreover $A^2 = B^3$ and $A^4 = \left(\begin{smallmatrix} 1 & 0 \\ 0 & 1 \end{smallmatrix}\right)$ and these are defining relations, see [Zieschang 1981, 23.1], [ZVC 1988, 8.3.1]. Thus $SL(2, \mathbb{Z}) = \langle A, B \mid A^2 B^{-3}, A^4 \rangle$. The next example is the group $GL(2, \mathbb{Z})$ of all invertible integer matrices. To the previous generating matrices A, B we add the matrix $R = \left(\begin{smallmatrix} 0 & 1 \\ 1 & 0 \end{smallmatrix}\right)$; the corresponding presentation is

$$GL(2, \mathbb{Z}) = \mathrm{Aut}(\mathbb{Z} \oplus \mathbb{Z}) = \langle A, B, R \mid A^2 B^{-3}, A^4, R^2, (RA)^2, (RB)^2 \rangle.$$

(i) Consider the *modular group* $PSL(2, \mathbb{Z})$ of linear fractional transformations $z \mapsto \frac{az+b}{cz+d}$ with $a, b, c, d \in \mathbb{Z}$ and $ad - bc = 1$. The mapping which sends the matrix $\left(\begin{smallmatrix} a & b \\ c & d \end{smallmatrix}\right)$ to the above linear fractional transformation is an epimorphism $SL(2, \mathbb{Z}) \to PSL(2, \mathbb{Z})$ with kernel $\{\left(\begin{smallmatrix} 1 & 0 \\ 0 & 1 \end{smallmatrix}\right), A^2\}$ and we get the presentation $PSL(2, \mathbb{Z}) = \langle a, b \mid a^2, b^3 \rangle$; here a, b are the images of A, B, see [Shafarevich 1986, p. 150]. (The term modular group stems from the classification of closed Riemann surfaces of genus 1, see 3.5.6.)

(j) Consider $Q = \langle \{x_n : n \geq 1\} \mid \{x_n = x_{nk}^k : n, k \geq 1\} \rangle$. The mapping $x_n \mapsto \frac{1}{n}$ induces an isomorphism from Q to the group $\mathbb{Q}$ of additive rationals. This is an example of a presentation of a group which is not finitely generated and needs infinitely many defining relations. There also exist groups which are finitely generated but not finitely presentable, see 2.2.7 (e).

(k) The *n-th braid group* (see 5.3.12, [Shafarevich 1986, p. 157]) is

$$\begin{aligned} B_n = &\langle \sigma_1, \ldots, \sigma_{n-1} \mid \{\sigma_i \sigma_j = \sigma_j \sigma_i : 1 \leq i < j - 1 \leq n - 2\} \\ &\cup \{\sigma_i \sigma_{i+1} \sigma_i = \sigma_{i+1} \sigma_i \sigma_{i+1} : 1 \leq i < n\} \rangle \ . \end{aligned}$$

By 1.1.3 there is an epimorphism $B_n \to \mathrm{Sym}(n)$ defined by $\sigma_i \mapsto (i, i+1)$.

(l) The following group will be of some interest in 3-dimensional topology (see 5.2): $P = \langle a, b \mid (ab)^2 = b^3 = a^5 \rangle$. The element $a^5 = b^3$ lies in the centre of P and generates a normal subgroup of order 2, see [Fuks 1986, p. 308], [Magnus 1974, p. 78], [Seifert-Threlfall, p. 218]. The factor group has order 60, see 1.1.14 (d), 3.2.11.

We see from (f) that a group can have infinitely many different presentations and so there arises the problem of finding the connection between different presentations of the same group. Another question is whether to a given $\langle \hat{\mathcal{X}} \mid \mathcal{R} \rangle$ where $\hat{\mathcal{X}}$ is some system of arbitrary elements and $\mathcal{R}$ an arbitrary collection of words over $\hat{\mathcal{X}}$ there exists a group presented by $\langle \hat{\mathcal{X}} \mid \mathcal{R} \rangle$.

1.1.5. Proposition. *For an arbitrary presentation expression $\langle \hat{\mathcal{X}} \mid R \rangle$ there exists a group presented by $\langle \hat{\mathcal{X}} \mid R \rangle$.*

Proof. Consider the set $\mathcal{W}$ of all words with letters from $\hat{\mathcal{X}}$, including the empty word. In 1.1.1 we have explained products and inverses of words. Two words $W(\hat{\mathcal{X}})$, $W'(\hat{\mathcal{X}})$ are called *equivalent* if $W'(\hat{\mathcal{X}})W(\hat{\mathcal{X}})^{-1}$ is a consequence of $\mathcal{R}$, see 1.1.1. This is an equivalence relation; the equivalence classes with multiplication defined by juxtaposition of representatives form the required group. □

1.1.6. Proposition and Definition

(a) *A group F is called a* free group of rank n *if F has a presentation free of defining relators with n generators: $F = \langle X_1, \dots, X_n \mid - \rangle$. The corresponding system of generators is called a free basis of F. Here n may also be infinite. If n is finite we write F_n for F.*

(b) *If F is a free group with free basis $\mathcal{X}$, G an arbitrary group and $\varphi : \mathcal{X} \to G$ an arbitrary mapping then there is a uniquely determined homomorphism $\Phi : F \to G$ with $\Phi(x) = \varphi(X)$ for $X \in \mathcal{X}$.*

(c) *Every group is a homomorphic image of a free group. More precisely: For $G = \langle \mathcal{X} \mid \mathcal{R} \rangle$ take $F = \langle \mathcal{X} \mid - \rangle$ and define a homomorphism $F \to G$ by sending the element of F represented by $X \in \mathcal{X}$ to the element of G represented by X. This is an epimorphism whose kernel consists of the elements of F represented by the relators of G.* □

This is a direct consequence of Dyck's theorem and demonstrates the importance of free groups. We will see below, see 1.1.13, that $d(F_n) = n$ and so the two notions of rank coincide.

It is clear that a group G admits different presentations. We describe next a set of procedures allowing one to move from any presentation of G to any other.

1.1.7. Tietze Operations [Tietze 1908]. Let $G = \langle \mathcal{X} \mid \mathcal{R} \rangle$.

(a) Let $\mathcal{U}$ be a system of symbols disjoint from $\mathcal{X}$ and $\{W_U(\mathcal{X}) : U \in \mathcal{U}\}$ a system of words over $\mathcal{X}$. Define $\mathcal{X}' = \mathcal{X} \cup \mathcal{U}$ and $\mathcal{R}' = \mathcal{R} \cup (U \cdot (W_U(\mathcal{X})^{-1})_{U \in \mathcal{U}}$. The procedure $\langle \mathcal{X} \mid \mathcal{R} \rangle \Rightarrow \langle \mathcal{X}' \mid \mathcal{R}' \rangle$ is called *adding (the) new generators $\mathcal{U}$*. The inverse operation is called *deleting (the) generators $\mathcal{U}$*.

(b) Let $\mathcal{Q}$ be a system of relations which are consequences of the defining relations $\mathcal{R}$, possibly including trivial relations. Define $\mathcal{X}' = \mathcal{X}$ and $\mathcal{R}' = \mathcal{R} \cup \mathcal{Q}$. The step $\langle \mathcal{X} \mid \mathcal{R} \rangle \Rightarrow \langle \mathcal{X}' \mid \mathcal{R}' \rangle$ is called *adding consequences* and the inverse process is called *deleting redundant relations*.

These procedures are called *Tietze operations*. If the sets $\mathcal{U}$ and/or $\mathcal{Q}$ are finite then we call the operation a *finite Tietze operation*.

It is clear that presentations which differ only by a Tietze operation define isomorphic groups, but the converse statement is also true. For, consider two presentations $\langle \mathcal{Y} \mid \mathcal{Q} \rangle$ and $\langle \mathcal{X} \mid \mathcal{R} \rangle$ of the group G. Assume that $W_X(\mathcal{Y})$, $X \in \mathcal{X}$ and $W_Y(\mathcal{X})$, $Y \in \mathcal{Y}$ are words such that $W_X(\mathcal{Y})$ and X represent the same element of G and similarly for $W_Y(\mathcal{X})$ and Y. We assume that the formal generator and relator sets are pairwise disjoint. By adding first generators and then consequence relations we obtain

$$\langle \mathcal{X} \mid \mathcal{R}\rangle \Rightarrow \langle \mathcal{X} \cup \mathcal{Y} \mid \mathcal{R} \cup \{Y(W_Y(\mathcal{X}))^{-1} : Y \in \mathcal{Y}\}\rangle$$
$$\Rightarrow \langle \mathcal{X} \cup \mathcal{Y} \mid \mathcal{R} \cup \{Y(W_Y(\mathcal{X}))^{-1} : Y \in \mathcal{Y}\} \cup \{X(W_X(\mathcal{Y}))^{-1} : X \in \mathcal{X}\} \cup \mathcal{Q}\rangle.$$

Since this last presentation can also be obtained by adding generators and consequence relations starting from $\langle \mathcal{Y} \mid \mathcal{Q}\rangle$, we have proved:

1.1.8. Tietze Theorem. *Two presentations define isomorphic groups if and only if one can be transformed into the other by a sequence of Tietze transformations. If both presentations are finite then only finitely many finite Tietze transformations are needed.* □

Let $\langle \mathcal{X} \mid \mathcal{R}\rangle$ be a finite presentation and $\langle \mathcal{Y} \mid \mathcal{Q}\rangle$ a presentation with a finite number of generators of the same group. An easy argument shows that already some finite subset of $\mathcal{Q}$ constitutes a set of defining relators.

1.1.9. On Decision Problems. The theory of group presentations brings with it important examples of *decision problems.* In a decision problem, a class of mathematical objects is partitioned into two subclasses by some determining property and the problem is to provide an algorithm which specifies in a finite number of steps which of the two subclasses an arbitrary object of the class belongs to. A decision problem is called *solvable* if such an algorithm exists.

In the case of a group presentation $\langle \mathcal{X} \mid \mathcal{R}\rangle$ one considers the class of all words over X and asks whether or not a word is a consequence of $\mathcal{R}$. This is the *word problem for* $\langle \mathcal{X} \mid \mathcal{R}\rangle$, first formulated in [Dehn 1910], who also introduced the *conjugacy* or *transformation problem*: Given $G = \langle \mathcal{X} \mid \mathcal{R}\rangle$, determine of any two words U and V over $\mathcal{X}$ whether there is a word W over $\mathcal{X}$ such that $W^{-1}UW$ and V represent the same element of G. Moreover there is the *isomorphism problem*: determine of any two presentations $G = \langle \mathcal{X} \mid \mathcal{R}\rangle$ and $G^* = \langle \mathcal{X}^* \mid \mathcal{R}^*\rangle$ whether they present isomorphic groups.

These questions are discussed in greater detail in Chapter 7. Here we make only two observations.

1.1.10. Corollary. *If* $\langle \mathcal{X} \mid \mathcal{R}\rangle$ *and* $\langle \mathcal{Y} \mid \mathcal{Q}\rangle$ *are finite presentations of the same group* G *then the word problem for* $\langle \mathcal{X} \mid \mathcal{R}\rangle$ *is solvable if and only if the word problem for* $\langle \mathcal{Y} \mid \mathcal{Q}\rangle$ *is solvable.*

Proof. This follows from Tietze's theorem since adding generators or consequences of the defining relations — or the inverse operations — preserves solvability of the word problem. Furthermore, since $\langle \mathcal{X} \mid \mathcal{R}\rangle$ and $\langle \mathcal{Y} \mid \mathcal{Q}\rangle$ present the same group, suitable words $U_X(\mathcal{Y})$ and $V_Y(\mathcal{X})$ exist and are, by hypothesis, assumed to be known. □

Note that Tietze's theorem does not lead to a solution of the isomorphism problem, even for the case of finite presentations since there are infinitely many possible ways to add finitely many new generators and consequence relations to a given presentation.

1.1.11. Commutator Subgroup and Abelianizing. One way to try to deal with the word problem is to map the group $\langle \mathcal{X} \mid \mathcal{R} \rangle$ to some known group by a homomorphism, for instance, by abelianizing.

If G is a group then the smallest normal subgroup of G containing the commutators $[x,y] = xyx^{-1}y^{-1}$, $x,y \in G$ is called the *commutator subgroup of* G and is denoted by $[G,G]$ or G'. The group G is abelian if and only if $[G,G] = 1$. If $\varphi : G \to H$ is a homomorphism then, obviously, $\varphi([G,G]) \leq [H,H]$ and, thus, an automorphism of G maps $[G,G]$ onto itself. The image $\varphi(G)$ is abelian if and only if $[G,G] < \ker \varphi$.

The *abelianization of* G is the group $G^{ab} = G/[G,G]$ and the projection is written $ab : G \to G^{ab}$. The abelianization has the following universal property: *If $\varphi : G \to A$ is a homomorphism of G to an abelian group A then there is a uniquely determined homomorphism $\varphi^{ab} : G^{ab} \to A$ such $\varphi = \varphi^{ab} \circ ab$. Hence, if G and H are isomorphic then $G/[G,G]$ and $H/[H,H]$ are isomorphic, too.* Now we can use the Classification Theorem of Finitely Generated Abelian Groups [Kurosh 1967, Sec. 20].

1.1.12. Classification Theorem of Finitely Generated Abelian Groups. *Let A be a finitely generated abelian group.*

(a) *Then A has a presentation of the following type:*

$$A = \langle a_1, \dots, a_n \mid [a_i, a_j],\ 1 \leq i < j \leq n, a_1^{t_1}, \dots, a_r^{t_r} \rangle \cong \mathbb{Z}_{t_1} \oplus \mathbb{Z}_{t_2} \oplus \dots \oplus \mathbb{Z}_{t_r} \oplus \mathbb{Z}^p$$

where $1 < t_1 | t_2 | \dots | t_r$. The numbers $t_1, \dots, t_r$ are called the torsion coefficients *and p is called the* Betti number *of A. The rank of A is $d(A) = p + r = n$.*

(b) *Two finitely generated abelian groups are isomorphic if and only if their Betti numbers and torsion coefficients coincide.*

(c) *The elements of finite order of A form the subgroup* Tor A *generated by $a_1, \dots, a_r$, and hence $A =$* Tor $A \oplus \mathbb{Z}^p$. *Any subgroup $U < A$ needs at most as many generators as A, i.e. $d(U) \leq d(A)$.* □

This important theorem allows us to classify free groups.

1.1.13. Theorem. *Let $F_n = \langle x_1, \dots, x_n \mid - \rangle$. Then $F_n^{ab} \cong \mathbb{Z}^n$; hence $d(F_n) = n$ and free groups of different rank are not isomorphic.* □

1.1.14. More Examples

(a) The following groups will be considered later as fundamental groups of surfaces: $\pi_1(S_g) = \langle t_1, u_1, \dots, t_g, u_g \mid \prod_{i=1}^{g} [t_i, u_i] \rangle$ and $\pi_1(N_g) = \langle v_1, \dots, v_g \mid v_1^2 \dots v_g^2 \rangle$. Then

$$H_1(S_g) := \pi_1(S_g)^{ab} \cong \mathbb{Z}^{2g},\ H_1(N_g) := \pi_1(N_g)^{ab} \cong \mathbb{Z}_2 \oplus \mathbb{Z}^{g-1}$$

and this shows that different values of g give non-isomorphic groups and that no $\pi_1(S_g)$ can ever be isomorphic to some $\pi_1(N_h)$.

(b) Consider the dihedral groups of 1.1.4 (g). Then

$$(D_n)^{ab} \cong \begin{cases} \mathbb{Z}_2 \oplus \mathbb{Z}_2 & \text{if } 2|n, \\ \mathbb{Z}_2 & \text{otherwise.} \end{cases}$$

(c) $PSL(2,\mathbb{Z})^{ab} \cong \mathbb{Z}_2 \oplus \mathbb{Z}_3 \cong \mathbb{Z}_6$, $SL(2,\mathbb{Z})^{ab} \cong \mathbb{Z}_{12}$, and $GL(2,\mathbb{Z})^{ab} \cong \mathbb{Z}_2 \oplus \mathbb{Z}_2$, see 1.1.4 (h, i).

(d) Consider the group $\Delta = \langle s_1, s_2, s_3 \mid s_1^2, s_2^3, s_3^5, s_1 s_2 s_3 \rangle$. Then $\Delta^{ab} = 0$ but this group is not trivial since it has the group of orientation preserving symmetries of the regular icosahedron as factor group; in fact, they are isomorphic, comp. [ZVC 1980; ZVC 1988, 4.7.1] and $|\Delta| = 60$. The group P from 1.1.4 (l) is mapped onto Δ by $a \mapsto s_3$, $b \mapsto s_2$. Since $a^5 = b^3$ belongs to the centre of P it follows that the kernel consists of 1, a^5 and that $|P| = 120$. However, the abelianized group $P^{ab} = 0$.

(e) For the braid group, see 1.1.4 (k), $B_n^{ab} \cong \mathbb{Z}$.

§ 1.2. Complexes and Fundamental Groups

An important invariant of a topological space X is its fundamental group $\pi_1(X, x_0)$, $x_0 \in X$, consisting of homotopy classes of closed paths starting at x_0, see [Massey 1967], [Novikov 1986, p. 14]. Fundamental groups play an important role in several fields of mathematics and are closely related to combinatorial group theory. For calculations of fundamental groups two methods are developed: one uses the so-called Seifert-van Kampen Theorem which relates the fundamental group of $X = X_1 \cup X_2$, X_1, X_2 and $X_1 \cap X_2$ where X_1, X_2 are open subspaces of X and $X_1 \cap X_2$ is pathwise connected, comp. [Massey 1967, Chap. 4]. In the other approach one restricts to the case of polyhedra or CW-complexes since these spaces cover virtually all topological spaces for which the fundamental group is in any way useful. In these circumstances the simplicial (or cellular) approximation theorem allows one to obtain $\pi_1(X, x_0)$ from just the 2-skeleton. We introduce these ideas in a combinatorial form.

1.2.1. Definition. A *graph* C, or 1-*dimensional complex*, consists of two disjoint (countable) sets $V = V(C)$ and $E = E(C)$ together with mappings $s, t\colon E \to V$, ${}^{-1}\colon E \to E$, $\sigma \mapsto \sigma^{-1}$, satisfying

(a) $$\sigma^{-1} \neq \sigma, \ (\sigma^{-1})^{-1} = \sigma,$$

(b) $$s(\sigma^{-1}) = t(\sigma), \ t(\sigma^{-1}) = s(\sigma).$$

The elements of V are *points* or *vertices*, the elements of E are *edges*; if $\sigma \in E$ then σ^{-1} is the *inverse* of σ, $s(\sigma)$ the *initial vertex* (or *source*) and $t(\sigma)$ the *final vertex* (or *target*) of σ (allowing $s(\sigma) = t(\sigma)$). The *degree* of a vertex v is the number of edges with initial vertex v.

A *subgraph* C' of C consists of a subset V' of $V(C)$ and a subset E' of $E(C)$ such that the initial and terminal vertices of any edge in E' lie in V' and the inverse edge of any edge in E' also lies in E'.

1.2.2. Definition. A *path* ω of *length* n is either just a single vertex v (a *trivial* or *constant path* of length 0) or a finite sequence $\sigma_1, \sigma_2, \dots, \sigma_n$ of edges such that $t(\sigma_i) = s(\sigma_{i+1})$ for $1 \le i < n$. We write, respectively, $\omega = v$ or $\omega = \sigma_1\sigma_2\dots\sigma_n$ and say in the latter case that ω *joins* $s(\omega) = s(\sigma_1)$ *to* $t(\omega) = t(\sigma_n)$. A path ω is *closed* if $s(\omega) = t(\omega)$ and *reduced* if it contains no *spur* $\sigma\sigma^{-1}$. A graph C is *connected* if any two vertices can be joined by a path and *finite* if $V(C)$ and $E(C)$ are finite. If ω is a closed path and $0 < n \in \mathbb{Z}$ then the path obtained by repeating ω n-times is written as the power ω^n; furthermore, if $\omega = \sigma_1 \dots \sigma_k$ then $\omega^{-1} = \sigma_k^{-1} \dots \sigma_1^{-1}$, $\omega^{-n} = (\omega^{-1})^n$ and ω^0 is the constant path $s(\omega)$.

1.2.3. Definition and Proposition. *A graph that contains no non-trivial reduced closed path is called* acyclic *and a connected acyclic graph is called a* tree.

(a) *In every connected graph C there is a tree called a* spanning tree *which contains all vertices of C. Furthermore, given any acyclic subgraph B_0 of C, there is a spanning tree of C which contains B_0.*

(b) *In any tree there is a unique reduced path joining any two vertices.*

Proof. Take some vertex $v_0 \in C$. Let the subgraph B_0 consist of v_0. Next take a maximal system of edges with initial vertex v_0, whose terminal vertices are distinct and different from v_0. Let B_1 be the complex consisting of these edges, their inverses and their endpoints. Clearly, B_1 is a tree. In general, the complex B_i is formed from B_{i-1} by taking all vertices which are one edge distant from B_{i-1}, together with a connecting edge and its inverse. The union $B = \bigcup_{i=0}^{\infty} B_i$ is a subgraph of C which is a tree and contains all vertices of C. A similar argument can be applied to an arbitrary acyclic B_0. □

The graph just constructed has the following minimality property:

1.2.4. Corollary. *Given a vertex v_0, there is a tree B such that any vertex is joined to v_0 by a path in B which is of minimal length among such paths in C.* □

1.2.5. Definition. A *2-complex* C consists of a vertex set $V(C)$, an edge set $E(C)$, and a *face* set $F(C)$ together with mappings $s, t, ^{-1}, \partial$ such that:

(a) $V(C)$ and $E(C)$ together with $s, t, ^{-1}$ form a graph C^1, called the *1-skeleton of C*.

(b) For $\varphi \in F(C)$, the boundary $\partial\varphi$ is a set of paths in C^1 consisting of all cyclic rearrangements of some given path. Elements of $\partial\varphi$ are the *positive boundary paths* of φ. We often specify $\partial\varphi$ by choosing some representative path ω and writing $\partial\varphi = \omega$.

(c) For each $\varphi \in F(C)$ there is a face $\varphi^{-1} \in F(C)$ with $\varphi^{-1} \neq \varphi$ and $(\varphi^{-1})^{-1} = \varphi$. Moreover, $\partial\varphi^{-1} = (\partial\varphi)^{-1}$ in the obvious sense, namely, that $\omega \in \partial\varphi^{-1}$ if and only if $\omega^{-1} \in \partial\varphi$.

(d) If the complex C is finite, let a_0 be the number of vertices, a_1 the number of pairs of inverse edges, and a_2 the number of pairs of inverse faces of C. Then $\chi(C) = a_0 - a_1 + a_2$ is the *Euler characteristic of* C.

1.2.6. Definition and Examples. Let C be a complex and let there be defined an equivalence relation $\sim$ on C, that is, equivalence relations on the systems of vertices, edges and faces which are compatible with the boundary relations: if, for example, $\sigma \sim \tau$ then $\sigma^{-1} \sim \tau^{-1}$ and $t(\sigma) \sim t(\tau)$. Then the equivalence classes also form a 2-complex $C/\sim$ called the *quotient complex*. One often speaks of *identifying* the different members of an equivalence class.

(a) A *circle* or 1-*sphere* S^1 is a graph with one vertex and one edge pair. Let the graph C be the disjoint union of n circles S_i^1, $i = 1, 2, \ldots, n$. The quotient complex obtained by identifying all the vertices is denoted by $\bigvee_{i=1}^n S_i^1$ and consists of one vertex v and n pairs of inverse edges, see Fig. 1.2.1. The Euler characteristic is $\chi(\bigvee_{i=1}^n S_i^1) = 1 - n$.

(b) The complex P^2 consists of one vertex v, one pair of edges σ, σ^{-1} and one pair of faces ψ, ψ^{-1} such that $\partial\psi = \sigma^2$, see Fig. 1.2.2. This complex is called *the projective plane* and $\chi(P^2) = 1$.

(c) Let S_g consist of one vertex v, $2g$ pairs of edges, namely $\tau_1, \mu_1, \ldots, \tau_g, \mu_g$ and their inverses, and a face ψ and its inverse. The boundary is $\partial\psi = \prod_{i=1}^g [\tau_i, \mu_i]$; here $[\tau_i, \mu_i] = \tau_i\mu_i\tau_i^{-1}\mu_i^{-1}$. This complex is called an *orientable closed surface of genus* g, see Fig. 1.2.3, and $\chi(S_g) = 2 - 2g$.

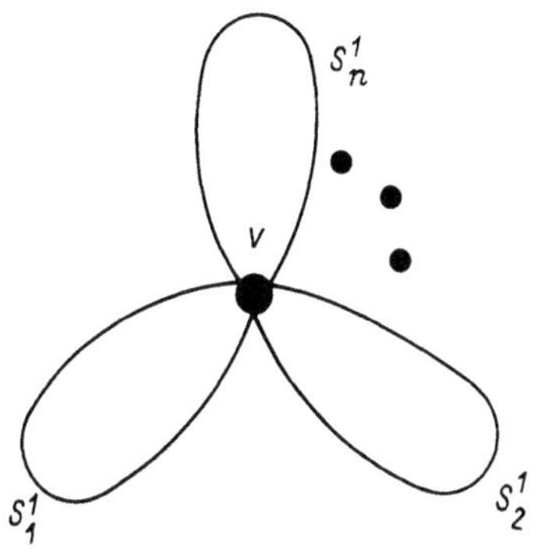

Fig. 1.2.1

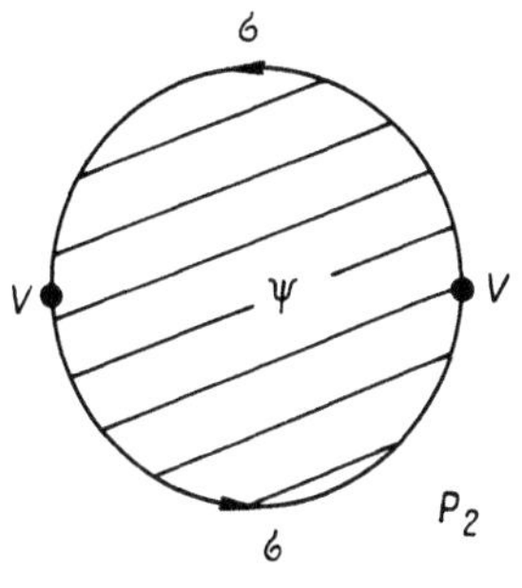

Fig. 1.2.2

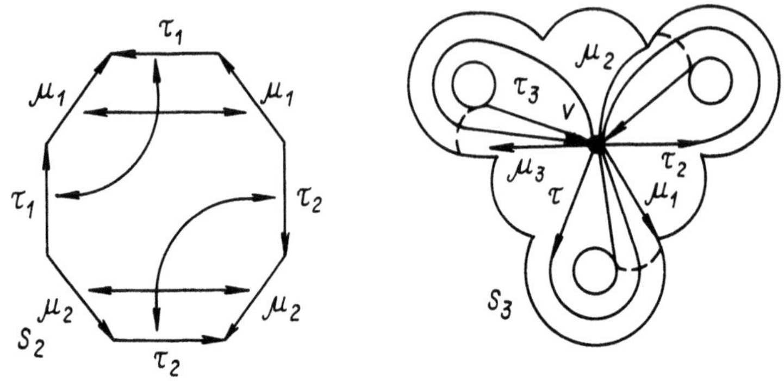

Fig. 1.2.3

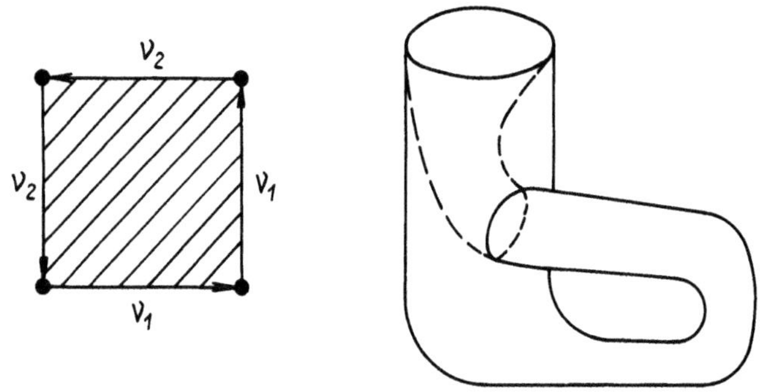

Fig. 1.2.4

(d) Let N_g consist of one vertex v, g edges ν_i and their inverses and a face ψ and its inverse. The boundary is $\partial\psi = \prod_{i=1}^{g} \nu_i^2$. This complex is called a *non-orientable closed surface of genus* g, see Fig. 1.2.4. Here $\chi(N_g) = 2 - g$.

(e) The complex L with $V(L) = \{v_i : i \in \mathbb{Z}\}$ and $E(L) = \{\sigma_i, \sigma_i^{-1} : i \in \mathbb{Z}\}$ where $t(\sigma_i) = v_{i+1},\ s(\sigma_i) = v_i$ is a *line*.

(f) The following complex L^2 consists of all points in $\mathbb{R}^2$ with integer coordinates as vertices, the horizontal and vertical segments $\tau_{i,j}, \mu_{i,j}$, respectively, of length 1 starting at the point (i,j) and their inverses. The faces are the squares $\psi_{i,j}$ with corners (i,j), $(i+1,j)$, $(i+1,j+1)$, $(i,j+1)$ and their inverses. The boundary relations are $s(\tau_{i,j}) = (i,j)$, $t(\tau_{i,j}) = (i+1,j)$, $s(\mu_{i,j}) = (i,j)$, $t(\mu_{i,j}) = (i,j+1)$, $\partial\psi_{i,j} = \tau_{i,j}\mu_{i+1,j}\tau_{i,j+1}^{-1}\mu_{i,j}^{-1}$.

(g) Consider the complex consisting of one pair of faces $\varphi^{\pm 1}$, four pairs of edges $\sigma^{\pm 1}, \tau^{\pm 1}, \sigma_0^{\pm 1}, \tau_0^{\pm 1}$ and four vertices v_1, v_2, v_3, v_4 arranged in a square such that $\partial\varphi = \sigma\tau\sigma_0^{-1}\tau_0^{-1}$. Define $\sigma \sim \sigma_0$, $\tau \sim \tau_0$ and correspondingly $v_1 \sim v_4$, $v_2 \sim v_3$, $v_2 \sim v_1$, $v_3 \sim v_4$. Then $C/\sim$ is the complex S_1 in (c). The other surfaces S_g, N_g can also be obtained from a complex with one pair of inverse faces by a similar identification. If, in the complex L in (e), one identifies all edges σ_i then all the vertices have to be identified and one gets the circle S^1 as quotient space. Similarly, by identifying all faces $\psi_{i,j}$ of L^2 in (f), and all vertices (i,j), all edges $\tau_{i,j}$ and all edges $\mu_{i,j}$ one obtains the complex S_1 as quotient.

1.2.7. Construction and Definition. A *path in a 2-complex* C is a path in the 1-skeleton. We assume that C^1 is connected and that v_0 is some vertex of C. Two paths ω_1, ω_2 are *homotopic* if one can be transformed into the other by a finite number of the following *elementary homotopies*:

(i) insertion or removal of a *spur*, i.e. $\omega' = \nu_1\nu_2 \iff \nu_1\sigma\sigma^{-1}\nu_2 = \omega''$ where ν_i are subpaths of ω' with with $s(\nu_2) = t(\nu_1) = s(\sigma)$, σ an edge;

(ii) elementary deformations over a face, i.e. if $\omega = \omega_1\omega_2\omega_3$ and if $\omega_2^{-1}\omega_2'$ is the boundary of a 2-cell of C^2 then ω may be replaced by $\bar{\omega} = \omega_1\omega_2'\omega_3$.

This defines an equivalence relation $\simeq$ for paths in which equivalent path have the same initial and final vertices. An equivalence class $[\omega]$ is called a *homotopy class of paths.* A closed path with initial vertex v is called *nullhomotopic in* C if it is homotopic to the constant path v.

1.2.8. Definition. A *mapping* or *homomorphism* $f : C \to D$ between two 2-complexes C, D assigns to each vertex of C a vertex of D, to each edge of C an edge or vertex of D, and to each face of C a face or path of D preserving the boundary behavior:

— if $v \in C$ is the initial vertex of the oriented edge $\sigma \in C$ then $f(v)$ is the initial vertex of $f(\sigma)$ if this is an edge, and otherwise $f(\sigma) = f(v)$;

— if $\prod_{i=1}^{k} \sigma_i$ is the boundary of the disc $\psi \in C$ then $\prod_{i=1}^{k} f(\sigma_i)$ bounds $f(\psi)$ if this is a face; otherwise $f(\psi)$ must be a path nullhomotopic in D^1;

— $f(\sigma^{-1}) = (f(\sigma))^{-1}$ and $f(\psi^{-1}) = (f(\psi))^{-1}$ for every edge σ and face ψ of C.

The mapping $f : C \to D$ *preserves dimension* if the image of an edge or face is an edge or face, respectively. A homomorphism between two complexes is called a *isomorphism* if it is injective and surjective. In this case we call the image and preimage *isomorphic.*

Under an isomorphism the images of nullhomotopic paths are nullhomotopic. The inclusion i: $C^1 \hookrightarrow C^2$ is a homomorphism preserving dimension. If C is a complex and $\sim$ an equivalence relation on C, then the identification mapping $C \to C/\sim$ is a homomorphism between complexes. In particular, when a group G of automorphisms of a complex C is given, in other words, when G acts on C then the orbits under G form equivalence classes and the quotient complex is denoted by C/G. The projection $C \to C/G$ is a surjective homomorphism of complexes.

If ω_1, ω_2 are paths and $t(\omega_1) = s(\omega_2)$ then there is the *product path* $\omega = \omega_1\omega_2$ which traverses first ω_1, then ω_2. From the definitions it follows immediately that if the factors are homotopic then so are the products and that the inverses of homotopic paths are homotopic: $\omega_i \simeq \omega_i'$, $i = 1,2 \Rightarrow \omega_1\omega_2 \simeq \omega_1'\omega_2'$, $\omega_1^{-1} \simeq (\omega_1')^{-1}$. Thus one can define a product and inverse for homotopy classes. Moreover, a mapping $f : C \to D$ maps paths to paths and homotopic paths to homotopic paths.

1.2.9. Proposition and Definition

(a) *The equivalence classes of closed paths with fixed initial vertex v_0 constitute a group, the* fundamental group of C *with basepoint* v_0. *This group is also called the* first homotopy group *and is denoted by* $\pi_1(C, v_0)$ *or, if the basepoint is immaterial, by* $\pi_1(C)$.

(b) *A mapping $f : C \to D$ between 2-complexes induces a homomorphism* $f_\# : \pi_1(C, v_0) \to \pi_1(D, f(v_0))$. *If $g : D \to E$ is another mapping then* $(g \circ f)_\# = g_\# \circ f_\#$. *Moreover,* $(\mathrm{id}_C)_\# = \mathrm{id}_{\pi_1(C)}$.

(c) *A connected complex C is called* simply connected *if* $\pi_1(C) = 1$. □

Standard categorial arguments show that an isomorphism between complexes induces an isomorphism between their fundamental groups. Of course, the fundamental group is also defined for complexes without 2-cells, that is for graphs, and there is a homomorphism $i_\# : \pi_1(C^1) \to \pi_1(C)$, where i: $C^1 \hookrightarrow C$, and this is surjective.

1.2.10. Finding a Presentation of the Fundamental Group $\pi_1(C)$

(a) **Generators.** Take a spanning tree $B \in C^1$. From each pair σ, σ^{-1} of edges from $C^1 \setminus B$ select one to obtain a system $(\sigma_i)_{i\in I}$ of (oriented) edges. There are unique reduced paths μ_i, ν_i in B going from the basepoint v_0 to, respectively, the initial and terminal vertex of σ_i. Let $\omega_i = \mu_i\sigma_i\nu_i^{-1}$ and let s_i be the homotopy class of ω_i with respect to the complex C^1. Then $(s_i)_{i\in I}$ *is a generating system for* $\pi_1(C^1, v_0)$. To see this consider some closed path $\omega = \tau_1 \dots \tau_k$ starting at v_0. Let ξ_i be the reduced path in B from v_0 to the final vertex of $\prod_{j=1}^{i} \tau_j$, and ξ_0 the constant path v_0. Then $\prod_{j=1}^{k} \tau_i \simeq \prod_{j=1}^{k} \xi_{j-1}\tau_j\xi_j^{-1} \simeq \prod_{l=1}^{m} \omega_{i(l)}^{\varepsilon_l}$ where the third term is obtained by deleting all factors $\xi_{j-1}\tau_j\xi_j^{-1}$ of the second product for which τ_j is an edge of the tree B. If τ_j does not belong to B then $\tau_j = \sigma_{i(l)}^{\varepsilon_l}$ for some $i(l) \in I$, $\varepsilon_l \in \{1, -1\}$ and $\xi_{j-1} = \mu_{i(l)}$, $\xi_j = \nu_{i(l)}$ if $\varepsilon_l = 1$ and $\xi_{j-1} = \nu_{i(l)}$, $\xi_j = \mu_{i(l)}$ if $\varepsilon_l = -1$. This proves that $[\omega] = \prod_{l=1}^{m} s_{i(l)}^{\varepsilon_l}$. Since insertion or deletion of a spur either leaves the representation of $[\omega]$ unchanged or gives rise to the insertion or deletion of a pair $s_i^{\varepsilon}s_i^{-\varepsilon}$, $\pi_1(C^1, v_0) = \langle (s_i)_{i\in I} \mid - \rangle$, and the images of the s_i under $i_\# : \pi_1(C^1, v_0) \to \pi_1(C, v_0)$ generate $\pi_1(C, v_0)$.

(b) **Defining Relations.** From each inverse pair of faces we choose one and obtain a system $(\psi_j)_{j\in J}$. Moreover, we fix a positive boundary path ϑ_j of ψ_j, take a path λ_j from v_0 to the starting point of ϑ_j, say in the tree B, and form the path $\rho_j = \lambda_j\vartheta_j\lambda_j^{-1}$. Then the homotopy class of ρ_j in C^1 is a product $r_j(s)$ in the generators from (a). From the definition of homotopy it follows easily that the $(r_j)_{j\in J}$ form a system of defining relations for $\pi_1(C, v_0)$; thus $\pi_1(C, v_0) = \langle (s_i)_{i\in I} \mid (r_j)_{j\in J} \rangle$.

1.2.11. Homology. Let K be a 2-complex. Select in every pair of inverse edges or faces one edge or face, respectively. Denote by $C_0(K)$ the free abelian group with free generators the vertices of K, by $C_1(K)$ the free abelian group with basis the chosen edges, and by $C_2(K)$ the free abelian group generated by the chosen faces. The elements of these groups are called $0-$, $1-$ or $2-$*chains*. Now $\sigma \mapsto t(\sigma) - s(\sigma)$ defines a homomorphism $\partial_1 : C_1(K) \to C_0(K)$. If ψ is a selected face with boundary $\sigma_1^{\varepsilon_1} \dots \sigma_k^{\varepsilon_k}$, where the σ_j are selected edges, then define $\partial_2(\psi) = \varepsilon_1\sigma_1 + \dots + \varepsilon_k\sigma_k$ and extend this to a homomorphism $\partial_2 : C_2(K) \to C_1(K)$. Now $\partial_1\partial_2 = 0$, hence, the group $Z_1(K) = \ker\, \partial_1$ of $1-$*cycles* contains the group $B_1(K) = \partial_2(C_2(K))$ of $1-$*boundaries*. Then $H_1(K) = Z_1(K)/B_1(K)$ is the *first homology group of* K and $z + B_1(K)$ is called the *homology class of* z. Moreover $H_0(K) = C_0(K)/\partial_1(C_1(K))$ and $H_2(K) = \ker\, \partial_2$ are the $0-$*th* and *second homology groups* and terms similar to the above are used for elements.

A path ω in K defines a 1-chain w and ω is closed if $\partial_1 w = 0$. If K is connected then every 1-cycle is in this sense the image of a path. An elementary homotopy, see 1.2.7, either has no effect on the image in $C_1(K)$ of a path (case (i)) or it induces the addition of the boundary of a face (case (ii)) and, thus, does not alter the homology class. This gives

1.2.12. Proposition. *Let K be a $2-$complex. Then there is a homomorphism $h : \pi_1(K) \to H_1(K)$. If K is connected then the homomorphism is the abelianization and $H_1(K) \cong \pi_1(K)^{ab}$.* □

Now a direct consequence of the Definition 1.1.6 (a) of a free group is

1.2.13. Corollary. *The fundamental group of a graph is a free group; the first homology group is free abelian. If C^1 is a finite graph then $\pi_1(C^1)$ is a free group of rank $(1 - \chi(C^1))$.* □

1.2.14. Examples. We use the notation of 1.2.6.

(a) $\pi_1(\bigvee_{i=1}^{k} S_i^1) = \langle \{\sigma_1\}, \dots, \{\sigma_k\} \mid - \rangle$, $\qquad H_1(\bigvee_{i=1}^{k} S_i^1) \cong \mathbb{Z}^k$;

(b) $\pi_1(P^2) = \langle s \mid s^2 \rangle$, $\qquad H_1(P^2) \cong \mathbb{Z}_2$;

(c) $\pi_1(S_g) = \langle t_1, u_1, \dots, t_g, u_g \mid \prod_{j=1}^{g} [t_j, u_j] \rangle$, $\qquad H_1(S_g) \cong \mathbb{Z}^{2g}$;

(d) $\pi_1(N_g) = \langle v_1, \dots, v_g \mid v_1^2 \cdot \dots \cdot v_g^2 \rangle$, $\qquad H_1(N_g) \cong \mathbb{Z}_2 \oplus \mathbb{Z}^{g-1}$;

(e) $\pi_1(L) = 1$, $\qquad H_1(L) = 0$;

(f) $\pi_1(L^2) = 1$, $\qquad H_1(L^2) = 0$.

1.2.15. Cut and Paste. Let us call two 2-complexes *homeomorphic* if one may be converted into the other by a finite sequence of the following elementary processes (this amounts to a special case of the usual topological notion):

(a) Let σ be an edge leading from v_1 to v_2. The modified complex is obtained by deleting σ and σ^{-1} and introducing a new vertex v, and new edges σ_1, σ_2 and their inverses, where σ_1 runs from v_1 to v and σ_2 from v to v_2. In boundary paths σ is replaced by $\sigma_1\sigma_2$ and additional paths beginning at v are adjoined to the set of boundary paths of a face. This process is called a *subdivision of an edge.*

(b) Let φ be a face of the complex and $\partial\varphi = \omega_1\omega_2$. The new complex contains all terms of the original one except $\varphi^{\pm 1}$. In addition it contains a new edge σ which runs from $s(\omega_1)$ to $t(\omega_1)$, with its inverse σ^{-1}, new faces φ_1, φ_2 with $\partial\varphi_1^{-1} = \omega_1\sigma$ and $\partial\varphi_2 = \sigma\omega_2$, together with their inverses. (*Subdivision of a face.*)

(c) The paste processes inverse to the cut procedures (a) and (b) whenever they are possible, that is, lead to 2-complexes.

1.2.16. Proposition. *Homeomorphic 2-complexes have isomorphic fundamental groups and homology groups. Finite homeomorphic complexes have the same Euler characteristic.* □

Taking a complex with one vertex, for every generator one pair of edges and for every relation a face (strictly a pair) in the obvious way, we obtain:

1.2.17. Proposition. *Given a presentation $\langle \mathcal{X} \mid \mathcal{R} \rangle$ there exists a 2-complex $C\langle \mathcal{X}|\mathcal{R}\rangle$ which defines this presentation when the construction of 1.2.10 is applied. Let us remark that every finite connected CW-complex of dimension at most 2 is of the homotopy type of some 2-complex $C\langle \mathcal{X} \mid \mathcal{R}\rangle$.*

An important method for calculating fundamental groups deals with the case of a complex consisting of the union of two subcomplexes which have non–trivial intersection:

1.2.18. Seifert-van Kampen Theorem (Amalgamation Form) [Seifert, 1931], [van Kampen, 1933a]. *Let C be a 2-complex, C_1, C_2 connected subcomplexes such that $C_0 = C_1 \cap C_2$ is non-empty and connected and let $i_l\colon C_l \hookrightarrow C$, $j_l\colon C_0 \hookrightarrow C_l$, $l = 1, 2$ be the embeddings. Fix a base point $v \in C_0^0$. Assume that $\pi_1(C_0, v)$ is generated by $(y_k)_{k\in K}$ and that $\pi_1(C_l, v) = \langle \mathcal{X}_l \mid \mathcal{R}_l\rangle$. Let $W_k(\mathcal{X}_1)$ and $V_k(\mathcal{X}_2)$ be words representing, respectively, the elements $j_{1\#}(y_k) \in \pi_1(C_1, v) = \langle \mathcal{X}_1 \mid \mathcal{R}_1\rangle$ and $j_{2\#}(y_k) \in \pi_1(C_2, v) = \langle \mathcal{X}_2 \mid \mathcal{R}_2\rangle$. Then:*

$$\pi_1(C, v) = \langle \mathcal{X}_1 \cup \mathcal{X}_2 \mid \mathcal{R}_1 \cup \mathcal{R}_2 \cup (W_k(\mathcal{X}_1) \cdot V_k(\mathcal{X}_2)^{-1})_{k\in K}\rangle \ .$$

Proof. Form a spanning tree for C by taking a spanning tree in C_0 and extending it to spanning trees for C_1, C_2. The result is a presentation for $\pi_1(C, v)$ of the form $\langle \mathcal{Y}_1 \cup \mathcal{Y}_2 \mid \mathcal{Q}_1 \cup \mathcal{Q}_2\rangle$ where $\langle \mathcal{Y}_1 \mid \mathcal{Q}_1\rangle$ and $\langle \mathcal{Y}_2 \mid \mathcal{Q}_2\rangle$ are presentations of $\pi_1(C_1, v)$ and $\pi_1(C_2, v)$. Tietze transformations now yield the presentation of the above form. □

The Seifert-van Kampen Theorem has a variant form which, in topology, deals with the process of adding a handle to a space. For complexes this is treated as follows. Let the connected 2-complex C_0 contain two disjoint connected isomorphic subcomplexes D_1, D_2 with inclusions $i_j : D_j \hookrightarrow C_0, j = 1, 2$ and let $g\colon D_1 \to D_2$ be an isomorphism. Let the 2-complex C be obtained from C_0 by identifying images and preimages under g: $C = C_0/g$. Take a vertex v not lying in $D_1 \cup D_2$, a vertex $v_1 \in D_1$, and $v_2 = g(v_1) \in D_2$. Let $\pi_1(C_0, v) = \langle \mathcal{X}_0 \mid \mathcal{R}_0 \rangle$ and $(\omega_k)_{k\in K}$ be a system of closed paths in D_1, with $s(\omega_k) = v_1$, the homotopy classes of which generate $\pi_1(D_1, v_1)$. Fix paths ν_1, ν_2 from v to v_1, v_2, respectively, and let the words $W_k(\mathcal{X}_0)$ and $V_k(\mathcal{X}_0)$ represent the homotopy classes in C_0 of, respectively, the paths $\nu_1\omega_k\nu_1^{-1}$ and $\nu_2\omega_k\nu_2^{-1}$. Finally let t denote the homotopy class of the path $\nu_1\nu_2^{-1}$. Then:

1.2.19. Seifert-van Kampen Theorem (HNN-Form). *With the notation from above,* $\pi_1(C, v_0) = \langle \mathcal{X}_0 \cup \{t\} \mid \mathcal{R}_0 \cup (W_k(\mathcal{X}_0) \cdot t \cdot (V_k(\mathcal{X}_0)^{-1} \cdot t^{-1}))_{k\in K} \rangle$. □

The theorem and its proof will become transparent if we look at some examples.

1.2.20. Examples

(a) Let C_0 consist of a pair of edges $\sigma^{\pm 1}$ and the vertices $v_0 = s(\sigma), v_1 = t(\sigma)$, i.e. C_0 is a segment. This is a miniature tree, hence has trivial fundamental group. Let S^1 be obtained by identifying the two vertices. (Of course, this is a 1-sphere.) By 1.2.19, $\pi_1(S^1) = \langle t \mid - \rangle \cong \mathbb{Z}$.

(b) Let $C = C_1 \cup C_2$ and $v = C_1 \cap C_2$ a vertex. If $\pi_1(C_1, v) = \langle \mathcal{X}_1 \mid \mathcal{R}_1 \rangle$, $\pi_1(C_2, v) = \langle \mathcal{X}_2 \mid \mathcal{R}_2 \rangle$ then, by 1.2.18, $\pi_1(C) = \langle \mathcal{X}_1 \cup \mathcal{X}_2 \mid \mathcal{R}_1 \cup \mathcal{R}_2 \rangle = \langle \mathcal{X}_1 \mid \mathcal{R}_1 \rangle * \langle \mathcal{X}_2 \mid \mathcal{R}_2 \rangle$; in this case $\pi_1(C)$ is called the *free product* of the groups $\langle \mathcal{X}_1 \mid \mathcal{R}_1 \rangle$ and $\langle \mathcal{X}_2 \mid \mathcal{R}_2 \rangle$. In particular, by induction $\pi_1(\bigvee_{i=1}^n S_i^1) = \mathbb{Z} * \mathbb{Z} * \ldots * \mathbb{Z}$, n times.

(c) Let $S_{1,1}$ denote a torus with one hole - see Example 3.1.4 for notation. Using the Seifert-van Kampen Theorems, we calculate its fundamental group. We construct it as shown in Fig. 1.2.5, where the v_i denote the vertices, the α, β, γ_i edges. Clearly, $\pi_1(R) = 1$. By 1.2.19, $\pi_1(A, v_1) = \langle b \mid - \rangle$ where b contains the homotopic paths $\gamma_1\gamma_2$ and $\alpha\beta\alpha^{-1}$. For T formed by glueing two copies of A we obtain by 1.2.18 $\pi_1(T) = \pi_1(A') * \pi_1(A'') = \langle b', b'' \mid - \rangle$. Finally we add a handle and obtain $\pi_1(S_{1,1}, v_1) = \langle t, b', b'' \mid tb't^{-1}b''^{-1} \rangle \cong \langle t, b' \mid - \rangle$. The boundary γ^* is from the class $b'^{-1}b''$.

(d) If we take two copies C_1, C_2 of the complex above and identify the boundary curves γ_1^* and γ_2^{*-1} we obtain a complex homeomorphic to S_2 and, by 1.2.18, $\pi_1(S_2) = \langle t_1, b_1', t_2, b_2' \mid [t_1, b_1'] = [t_2, b_2']^{-1} \rangle$. Similar constructions yield presentations for $\pi_1(S_{g,r})$, $\pi_1(N_{g,r})$ (for notation, see Example 3.1.4) and reprove 1.2.6 (c,d).

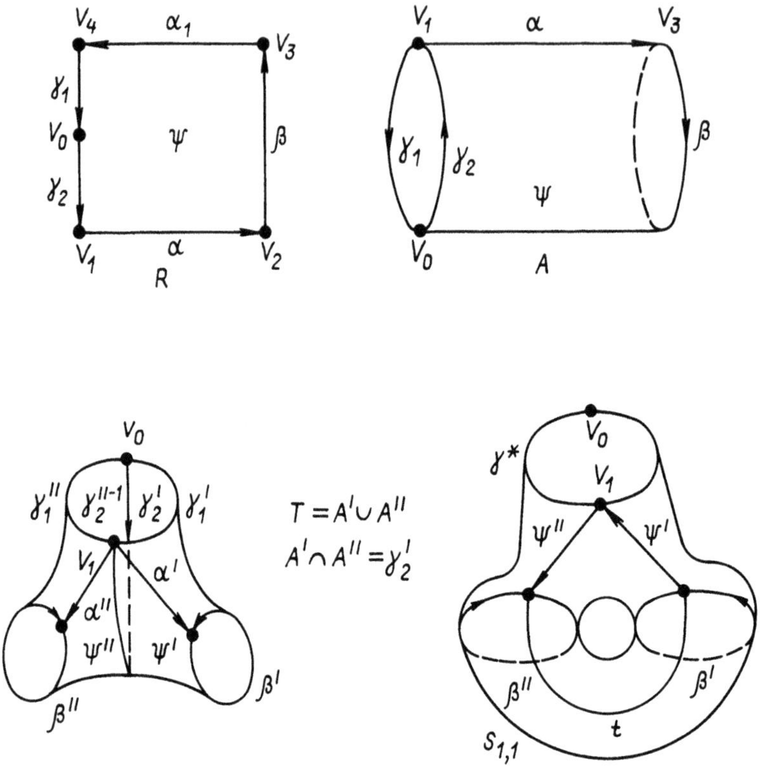

Fig. 1.2.5

The above arguments show that in principle a presentation for the fundamental group of an arbitrary 2-complex can be obtained by iterated use of the Seifert-van Kampen Theorem, and this is the way fundamental groups of topological spaces are usually calculated.

Next we will introduce the group diagram, introduced in [Cayley 1878]and rediscovered in [Dehn 1910], as a tool for the application of complexes to group theory.

1.2.21. Cayley Diagrams. Let $G = \langle \mathcal{X} \mid \mathcal{R} \rangle$. Construct a 2-complex Δ as follows:

(a) Let $V(\Delta) = G$.

(b) Let $E(\Delta) = \{(g, X) : g \in G,\ X \in \mathcal{X} \cup \mathcal{X}^{-1}\}$ and define $s(g, X) = g$, $t(g, X) = gx$, where x is the element of G represented by X, and $(g, X)^{-1} = (gx, X^{-1})$. (It is essential to distinguish betwen X and x here since, if the group element x has order 2, then $x^{-1} = x$ whereas X^{-1} is distinct from X.) Thus far we have obtained a graph which is sometimes denoted by $\Gamma\langle \mathcal{X} \mid \mathcal{R} \rangle$. Call X the *label* on (g, X). Given $g \in G$ and a word W over $\mathcal{X}$ there is a uniquely defined path, which we shall denote by (g, W), with initial vertex g whose edge labels, in sequence, give the word W.

(c) For any $g \in G$ and $R^\varepsilon, \varepsilon = \pm 1$ there is an oriented face $\psi(g, R^\varepsilon)$ whose boundary $\partial\psi(g, R^\varepsilon)$ is the path (g, R^ε), which is closed since R^ε is a relator. Furthermore $\psi(g, R^\varepsilon)^{-1} = \psi(g, R^{-\varepsilon})$. The complex $\Delta = \Delta\langle \mathcal{X} \mid \mathcal{R} \rangle$ defined above is called the *Cayley* or *group diagram* of $\langle \mathcal{X} \mid \mathcal{R} \rangle$.

1.2.22. Remarks

(a) Originally only the graph $\Gamma\langle \mathcal{X} \mid \mathcal{R} \rangle$was regarded as the Cayley diagram (or, in the German literature, (Dehnsches) Gruppenbild).

(b) The group G acts on $\Delta\langle \mathcal{X} \mid \mathcal{R} \rangle$ as a group of automorphisms in the obvious way: the transformation corresponding to $g \in G$ maps the vertex g' to the vertex gg', the edge (g', X^ε) to (gg', X^ε) and the face $\psi(g', R^\varepsilon)$ to $\psi(gg', R^\varepsilon)$. Clearly, the boundary of a face is mapped to boundary of the image face etc. This action of G on $\Delta\langle \mathcal{X} \mid \mathcal{R} \rangle$ is *without inversion*, that is, if c is an edge or face and $g \in G$, then $gc \neq c^{-1}$. Moreover the action of G is also *free* in that if a transformation g maps some vertex, edge or face to itself then g is the identity. For examples see 3.2.

(c) $\pi_1(\Delta\langle \mathcal{X} \mid \mathcal{R} \rangle) = 1$.

(d) If $G = \langle a, b \mid [a, b] \rangle$ then the Cayley diagram is the complex L^2 of 1.2.6 (f), with $a^k b^\ell$ corresponding to the point (k, ℓ). For another example see 3.2.13.

§ 1.3. Subgroups and Coverings

For the study of subgroups of the fundamental group of a complex it is convenient to use a new geometric concept, namely that of a covering. Two remarks are appropriate here. Firstly the absence of any concept paralleling the topological notion of neighbourhood means that the definition of covering is slightly awkward and requires careful formulation. In compensation, however, a universal covering of a complex always exists, in contrast to the situation for a topological space, where a condition such as being semilocally simply connected must be imposed.

1.3.1. Definition. A homomorphism $p : C' \to C$ between 2-complexes preserving the dimension is called a *covering (with base C and total space C')* if it has the following two properties.

(a) If v' is an arbitrary vertex of C' and $p(v') = v$ then p maps the set of edges $\sigma_1', \sigma_2', \ldots$ with initial vertex v' bijectively to the set of edges $\sigma_1, \sigma_2, \ldots$ with initial vertex v. In particular, the number of edges of C' starting at v' is the same as the number of edges in C starting at v.

(b) Let φ be a face of C and v a vertex. Define $m(\varphi, v)$ to be the number of times v appears in a boundary path of φ (counting only one of start and finish). For any two vertices v' of C' and v of C and face φ of C, such that $p(v') = v$, then $\sum_{p(\varphi')=\varphi} m(\varphi', v') = m(\varphi, v)$.

1.3.2. Some Definitions and Remarks. If $p : C' \to C$ is a covering and the vertex v' of C' is mapped to the vertex v of C, then we say v' *lies over* v or *is a lift of* v. Similar phrases apply to edges, faces and paths (of the same length).

(a) A covering $p : C' \to C$ defines a homomorphism $p_{\#} : \pi_1(C', v') \to \pi_1(C, p(v'))$; this is a key observation for what follows.

(b) If one of C or C' is a graph then so is the other.

(c) When C is connected it follows easily that the cardinalities of the inverse images $p^{-1}(v)$, $p^{-1}(\sigma)$ and $p^{-1}(\varphi)$ are the same for all vertices v, edges σ and faces φ of C. It is called the *order of the covering* $p : C' \to C$ (or the *number of sheets of the covering*). If C' is a finite complex then the two Euler characteristics are related by $\chi(C') = c \cdot \chi(C)$, where c is the order of the covering.

(d) Given any path ω in C and a vertex v' of C' lying over the initial vertex of ω, there is a unique path ω' which has initial vertex v' and lies over ω.

(e) Condition (a) in 1.3.1 excludes "branched coverings" as used in the theory of Riemann surfaces.

(f) For examples of coverings see Fig. 1.3.1–4.

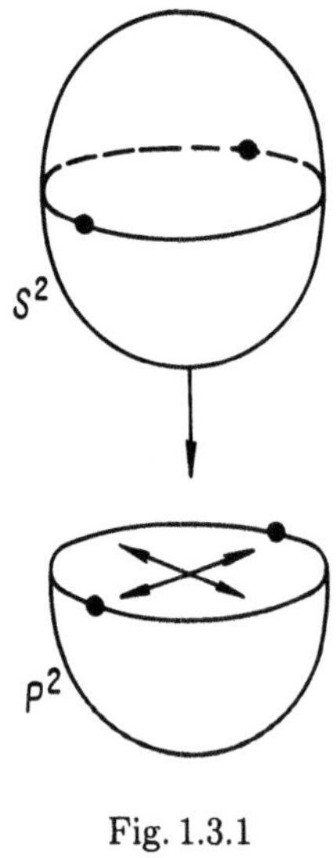

Fig. 1.3.1

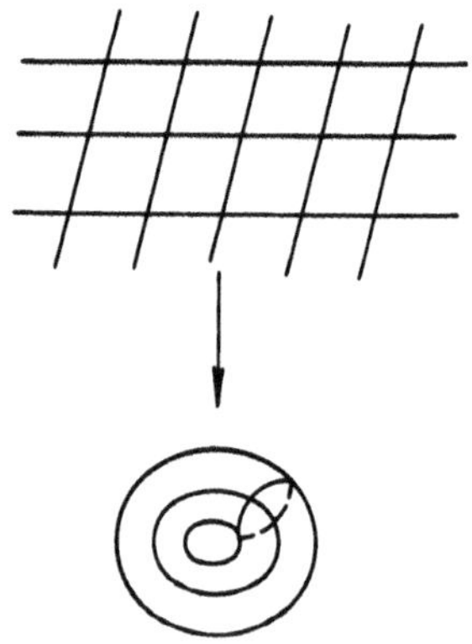
Fig. 1.3.2

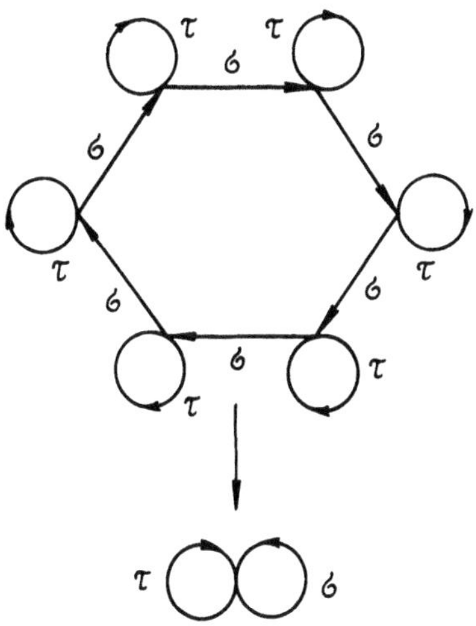

Fig. 1.3.3

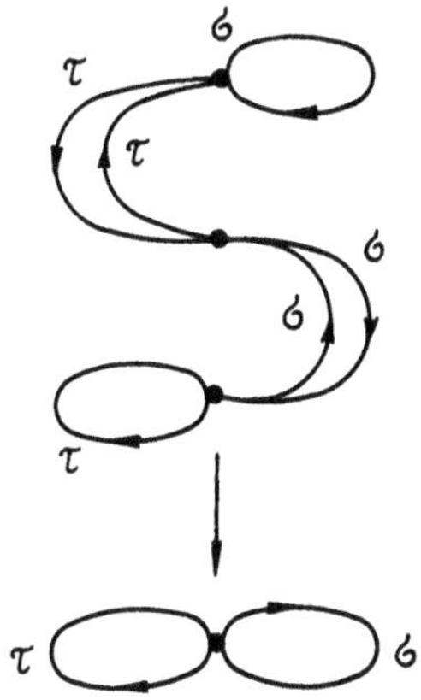

Fig. 1.3.4

Let us describe an important way to construct coverings. Let C' be a 2-complex and G a group of automorphisms of C' acting freely and without inversion. If $C = C'/G$ is the quotient complex, then the identification mapping $p : C' \to C$ is a covering. Clearly, when $g \in G$ then $p \circ g = p$, that is the action of g disappears on the base C of the covering. We formalise this concept:

1.3.3. Definition. Let $p : C' \to C$ be a covering. An automorphism φ' of C' is called a *covering transformation* if $p \circ \varphi' = p$. The covering transformations together form a group $\mathrm{Aut}(p)$, the *group of automorphisms of the covering.* The covering $p : C' \to C$ is called *regular* if to any pair of vertices $v_1', v_2' \in C'$ with $p(v_1') = p(v_2')$ there is one (and only one) covering transformation g with $g(v_1') = v_2'$. The general example of a regular covering is that described prior to the definition.

It follows easily from the definition that a covering transformation fixing any vertex, edge or face is the identity. We consider first some coverings between graphs. In Fig. 1.3.2, restricted to the 1-skeletons, the fundamental group of the covering space is a free group of infinite rank and is mapped to a subgroup of the free group F_2 of rank 2. The image is normal, since it is the commutator subgroup $[F_2, F_2]$, and the group of covering transformations is isomorphic to $\mathbb{Z} \oplus \mathbb{Z} = F_2/[F_2, F_2]$. In Fig. 1.3.3 a spanning tree consists of five edges denoted by τ, and we get, for instance, the free generators t^6, $t^i \cdot s \cdot t^{-i}$, $0 \le i \le 5$. These also generate a normal subgroup of $\langle s, t \mid - \rangle$ and the group of covering transformations consists of the identity and five rotations, i.e., is isomorphic to $\mathbb{Z}_6$. An example of a non-regular covering is that given in Fig. 1.3.4. Here the group of covering transformations is trivial.

If $p : C' \to C$ is a covering and $\sigma'\tau'$ is a path in C' such that $p(\tau') = p(\sigma')^{-1}$ then by 1.3.1 (a): $\tau' = \sigma'^{-1}$. Thus, a path mapped onto a spur is itself a spur, and paths in a graph C' with the same initial point which are mapped to homotopic paths of C are themselves homotopic; hence, $p_\# : \pi_1(C', v') \to \pi_1(C, p(v'))$ is injective. This argument can be generalized to the case of 2-complexes C', C since by 1.3.1 (b) every elementary deformation over a face of C can be lifted to an elementary deformation over a face of C'.

If v'' is another point of the same component of C' over $v = p(v')$ then $p_\#(\pi_1(C', v'))$ and $p_\#(\pi_1(C', v''))$ are conjugate subgroups of $\pi_1(C, v)$ (a conjugating factor comes from the image of a path from v' to v'').

Now assume that C is a connected graph, v a vertex of C and U a subgroup of $\pi_1(C, v)$. We will show that there is a covering $p : C' \to C$ such that $U = p_\#(\pi_1(C', v'))$ where $v' \in p^{-1}(v)$. For convenience let us assume that C has only one vertex. Let g_i, $i \in I$ be a system of representatives of the right cosets Ug of $\pi_1(C)$; here U shall be represented by the trivial element $g_0 = 1$. For each coset Ug_i we take a point v_i'. If σ is an edge of C then it represents an element $[\sigma] \in \pi_1(C, v)$. For $i \in J$ we take an edge σ_i' with initial vertex v_i' and terminal vertex v_k' where $k \in J$ is defined by the equation $U(g_i \cdot [\sigma]) = Ug_k$. The inversely oriented edge starts at v_k', lies over the edge σ^{-1} and goes to v_i'. Then one can check that the system of all these vertices and edges forms a graph C' and that the obvious mapping from C' to C is a covering inducing a mapping from $\pi_1(C', v_0)$ to U. This construction can also be done for a connected 2-complex C and subgroup $U < \pi_1(C, v)$: For $U^* = i_\#^{-1}(U)$, where $i : C^1 \hookrightarrow C$, construct a covering $p : (C^1)' \to C^1$ with $p_\#(\pi_1((C^1)', v')) = U^*$, $v = p(v')$. For a face $\psi \in F(C)$ and some vertex v'' over the initial vertex of $\partial\psi$ the uniquely determined lifting of $\partial\psi$ with initial vertex v'' is a closed path, denoted by $(\partial\psi, v'')$, in $(C^1)'$; introduce a face (ψ, v'') and define $\partial(\psi, v'') = (\partial\psi, v'')$. Let C' be the 2-complex obtained and $p : C' \to C$ the obvious projection. Then this is a covering which induces the previous covering on the 1-skeleton.

The assumption that C contains only one vertex is not essential. In general one starts with a spanning tree B in C, represents each coset by a copy of B and repeats the construction described above to obtain (a) and (b) of the following fundamental result:

1.3.4. Theorem [Reidemeister 1928]. *Let C', C be connected 2-complexes.*

(a) *A covering $p : C' \to C$ induces a monomorphism $p_\# : \pi_1(C', v') \to \pi_1(C, p(v'))$, that is, the fundamental group of the covering space can be considered as a subgroup of the fundamental group of the base space.*

(b) *Conversely, for any subgroup U of $\pi_1(C, v)$ there is a covering $p : C'' \to C$ such that $p_\#\pi_1(C'', v'') = U$, for $v'' \in p^{-1}(v)$.*

(c) *Two coverings $p : \bar{C} \to C$ and $p' : \tilde{C} \to C$ are called* equivalent *if there is an isomorphism $f : \bar{C} \to \tilde{C}$ such that $p = p' \circ f$. Equivalent coverings correspond to the same subgroup.*

(d) *The covering* $p : C' \to C$ *is regular if and only if* $p_\#(\pi_1(C', v'))$ *is a normal subgroup of* $\pi_1(C, p(v'))$. *In this case the group* Aut(p) *of covering transformations is isomorphic to* $\pi_1(C, p(v'))/p_\#(\pi_1(C', v'))$.

Proof of (d) is left as an exercise. When the covering is not regular, then the group of covering transformations is isomorphic to the quotient group of the normalizer of $p_\#(\pi_1(C', v'))$ in $\pi_1(C, p(v'))$ modulo $p_\#(\pi_1(C', v'))$. For details see [ZVC 1980, 1988, 2.5.8]. □

As a first application let us consider a subgroup U of a free group F. Now F can be realized as the fundamental group of a graph. The covering corresponding to U also has a graph as total space, the fundamental group of which is a free group mapped isomorphically to U. This gives the first part of the following theorem.

1.3.5. Theorem. *A subgroup of a free group is free. If the subgroup* U *of the free group* F *has finite index* c *then* rank $U = c \cdot ((\text{rank } F) - 1) + 1$.

The rank equation immediately follows from the equation $\chi(C') = c \cdot \chi(C)$, see 1.3.2 (c). □

It is nice to know that the subgroup U is free and that its rank is determined. But often one is interested in having a more detailed description of U, for instance, by knowing free generators. From the geometric construction above we obtain a simple way to determine a system of free generators of U. Again we consider the covering $p : C' \to C$ and assume that C contains only one vertex v. Let $\{v_i' \mid i \in J\}$ be the vertices of C' where $0 \in J$ and $U = p_\#(\pi_1(C', v_0'))$. To find a generating system of U we choose first a spanning tree $B' \subset C'$. Then the reduced paths starting at v_0' and leading to v_i', $i \in J$, define, after application of p, reduced closed paths in C and, thus, a set $\mathcal{W}$ of reduced words in the generators of $\pi_1(C, v)$ which are defined by the edges of C. Each initial subword of a word in $\mathcal{W}$ also belongs to $\mathcal{W}$, and $\mathcal{W}$ determines a *transversal*, i.e. a system of representatives, of the right cosets of $U = p_\#(\pi_1(C', v_0'))$ which has the following property:

1.3.6. Schreier Property. *Every initial subword of a representative is a representative.*

Now generators of $\pi_1(C', v_0')$ are obtained as follows. Choosing one edge from each pair not in B' we follow the reduced path in B' from v_0' to the initial vertex of the edge, cross the edge and then return to v_0' in B'. Translated to the language of coset representatives and generators, we obtain a system of free generators of U, each of the form $W \cdot X \cdot \overline{WX}^{-1}$, where $\overline{WX}$ denotes the coset representative of the element WX of $\pi_1(C)$.

In the general case when $G = \langle \mathcal{X} \mid \mathcal{R} \rangle$ and $U < G$ we proceed as follows. Consider the free group $F(\mathcal{X})$ and the map $p : F(\mathcal{X}) \to G$. Let $U^* = p^{-1}(U)$ and construct free generators for U^* as above. It should be observed that the images in G of the Schreier transversal $\mathcal{W}$ for U^* in $F(\mathcal{X})$ form a transversal for U in G. The normal subgroup $\mathcal{N}_{\mathcal{F}(\mathcal{X})}(\mathcal{R})$ of $F(\mathcal{X})$ defined by $\mathcal{R}$ is contained in U^* and is generated by the elements $\{Y \cdot R \cdot Y^{-1} \mid Y \in F(\mathcal{X}),\ R \in \mathcal{R}\}$. If we write $Y = ZW$ for suitable $Z \in U^*$ and $W \in \mathcal{W}$, then we see that $\mathcal{N}_{\mathcal{F}(\mathcal{X})}(\mathcal{R})$ is the smallest normal subgroup of U^* containing $\{W \cdot R \cdot W^{-1} \mid R \in \mathcal{R},\ W \in \mathcal{W}\}$. This proves

1.3.7. Reidemeister-Schreier Theorem [Reidemeister 1927], [Schreier 1927]. *Let U be a subgroup of $G = \langle \mathcal{X} \mid \mathcal{R} \rangle$ and let $\mathcal{W}$ be a system of representatives of the right cosets Ug of G modulo U which fulfils the Schreier condition* 1.3.6 *when the elements of $\mathcal{W}$ are written as reduced words in the generators $\mathcal{X}$. Then*

$$U = \langle \{(W,X) = W \cdot X \cdot \overline{WX}^{\,-1} : W \in \mathcal{W},\ X \in \mathcal{X},\ \overline{WX} \neq WX\} \\ \mid \{(W,R) = W \cdot R \cdot W^{-1} : W \in \mathcal{W},\ R \in \mathcal{R}\} \rangle\ .$$

Here the relators (W,R) are written as words the generators (W,X). □

The formulation of the Reidemeister-Schreier Theorem is rather messy. The theorem becomes clearer if one considers examples.

1.3.8. Example. Consider the free group $F = \langle x, y \mid - \rangle$, the homomorphism $\varphi : F \to \mathbb{Z}_n = \{0, 1, \dots, n-1\}$, $n \geq 2$ defined by $x, y \mapsto 1$ and let U be the kernel of φ. As coset representatives we take $1, y, y^2, \dots, y^{n-1}$; clearly, this system fulfils the Schreier condition. Then the Reidemeister-Schreier generators for U are the non-trivial elements of the following set: $\{y^i \cdot x \cdot \overline{y^{i+1}}^{\,-1},\ y^i \cdot y \cdot \overline{y^{i+1}}^{\,-1} \mid i = 0, \dots, n-1\}$. We obtain the non-trivial elements $y_* = y^{n-1} \cdot y = y^n$, $x_i = y^i \cdot x \cdot y^{-(i+1)}$ for $i = 0, \dots, i-2$ and $x_{n-1} = y^{n-1} \cdot x$. The rank of this subgroup is $n+1$. Note that this shows that *the free group of rank* 2 *contains a free group of rank n as a subgroup of index* $n-1$.

1.3.9. Example. Let $G = \langle a, b \mid a^2, b^3 \rangle = \mathrm{PSL}(2, \mathbb{Z})$, see 1.1.4 (i), and $\mathbb{Z}_6 = \{\bar{0}, \bar{1}, \bar{2}, \bar{3}, \bar{4}, \bar{5}\}$ and $\alpha : G \to \mathbb{Z}_6 \cong G^{ab}$ be defined by $a \mapsto \bar{3}$, $b \mapsto \bar{2}$ and let $U = \ker \alpha$. Then $1, a, b, b^2, ab, ab^2$ is a system of coset representatives of G modulo U which satisfies the Schreier condition. It defines the following generators: $1 \cdot a \cdot \overline{1a}^{\,-1} = aa^{-1}$, $a \cdot a \cdot \overline{a^2}^{\,-1} = a \cdot a \cdot 1^{-1} = a^2$, $b \cdot a \cdot \overline{ba}^{\,-1} = bab^{-1}a^{-1}$, ... and after omitting the trivial generators we obtain seven generators, which we expect in view of 1.3.5, namely

$$x_1 = a^2,\ x_2 = bab^{-1}a^{-1},\ x_3 = b^2ab^{-2}a^{-1},\ x_4 = abab^{-1}, \\ x_5 = ab^2ab^{-2},\ y_1 = b^3,\ y_2 = ab^3a^{-1}.$$

To determine the defining relations we have to conjugate each of the defining relations of G by all the representatives and write these elements as words in the generators $x_1, \dots, x_5, y_1, y_2$, for example $b \cdot a^2 \cdot b^{-1} = bab^{-1}a^{-1} \cdot abab^{-1} = x_2x_4$. After calculation we obtain the sequence of relations

$$x_1,\ x_1,\ x_2x_4,\ x_3x_5,\ x_4x_2,\ x_5x_3,\ y_1,\ y_2,\ y_1,\ y_1,\ y_2,\ y_2;$$

hence, we may drop the generators x_1, y_1, y_2 and the corresponding 8 relations and may replace x_4 and x_5 by x_2^{-1} and x_3^{-1}, respectively. Doing so we lose the remaining 4 relations and so deduce that U is the free group with free generators x_2, x_3. This proves that *the modular group contains a free subgroup of rank* 2 *and index* 6.

Examining the geometry connected with the Reidemeister-Schreier Theorem in more detail, see the text before 1.3.4, one finds the reason why the relations x_1, x_2x_4, x_3x_5 appear twice and the relations y_1, y_2 three times, see [ZVC 1980, 1988, 2.2].

1.3.10. Example. Consider $G = \pi_1(N_g) = \langle v_1, \dots, v_g \mid v_1^2 \cdot \dots \cdot v_g^2 \rangle$ and let $\alpha : G \to \mathbb{Z}_2 = \{1, -1\}$ be the homomorphism mapping every v_i to -1; geometrically an element $x \in G$ is represented by a two-sided curve when $\alpha(x) = 1$, and otherwise by a one-sided curve. Let $V = \ker \alpha$. As coset representatives we take $1,\ v_g$. Then we obtain the generators: $x_i = v_iv_g^{-1}$, $y_i = v_gv_i$, $i = 1, \dots, g-1$, and $z = v_g^2$. We have to consider two relations:

$$\begin{aligned} v_1^2 \cdot \dots \cdot v_g^2 &= x_1y_1x_2y_2 \cdot \dots \cdot x_{g-1}y_{g-1} \cdot z \\ v_g \cdot v_1^2 \cdot \dots \cdot v_g^2 \cdot v_g^{-1} &= y_1x_1y_2x_2 \cdot \dots \cdot y_{g-1}x_{g-1} \cdot z\ . \end{aligned}$$

Eliminating z, we obtain the presentation:

$$V = \langle x_1, y_1, \dots, x_{g-1}, y_{g-1} \mid x_1y_1 \dots x_{g-1}y_{g-1} \cdot x_{g-1}^{-1}y_{g-1}^{-1} \dots x_1^{-1}y_1^{-1} \rangle.$$

and this group is isomorphic to the group $\pi_1(S_{g-1})$, see 1.2.14 (c). This result reflects the fact that a closed orientable surface of genus $g-1$ is a twofold cover of a non-orientable closed surface of genus g (see 3.1.9).

The disadvantage of the Reidemeister-Schreier method is that one has to know a system of coset representatives and must be able to determine the coset representative of an element. This is not possible in general. An alternative construction is the Todd-Coxeter method [Todd-Coxeter 1936]; in fact, essentially this method was used by K. Reidemeister in his paper [Reidemeister 1928] to establish the general connection between subgroups and coverings, see Theorem 1.3.4.

1.3.11. Classification of n-Sheeted Coverings. Let G be a group. A homomorphism $f : G \to \mathrm{Sym}(n)$ is called *transitive* if to $i, j \in \{1, \dots, n\}$ there is a $g \in G$ such that $f(g)(i) = j$. Two homomorphisms $f, f' : G \to \mathrm{Sym}(n)$ are *equivalent* if there is a $\sigma \in \mathrm{Sym}(n)$ such that $f'(g) = \sigma^{-1}f(g)\sigma$, for all $g \in G$. Let $\langle f \rangle$ denote the equivalence class of f and $\langle G, \mathrm{Sym}(n) \rangle$ the set of classes $\langle f \rangle$ with transitive f.

Let $p : \tilde{C} \to C$ be an n-sheeted covering, C connected. Take some vertex v of C and enumerate the vertices over v in some way: $\tilde{v}_1, \ldots, \tilde{v}_n$. Given $a = [\alpha] \in \pi_1(C, v)$, the permutation $\bar{a} \in \text{Sym}(n)$ maps i to j if the lifting of the path α starting at $\tilde{v}_i$ ends in $\tilde{v}_j$. Then $a \mapsto \bar{a}$ defines a transitive homomorphism $f : \pi_1(C, v) \to \text{Sym}(n)$. The equivalence class of f does not depend on the chosen enumeration of the vertices over v; it will be denoted by $S(\tilde{C}, p) = \langle f \rangle \in \langle \pi_1(C, v), \text{Sym}(n) \rangle$. Now the following assertions are easy to establish:

(a) *Two n-sheeted connected coverings* $p : \tilde{C} \to C$, $p' : \tilde{C}' \to C$ *are equivalent if and only if* $S(\tilde{C}, p) = S(\tilde{C}', p')$.

(b) *For every transitive homomorphism* $f : \pi_1(C, v) \to \text{Sym}(n)$ *there is an n-sheeted connected covering* $p : \tilde{C} \to C$ *such that* $S(\tilde{C}, p) = \langle f \rangle$.

These results permit the effective determination of the n-sheeted coverings over a given complex C. We will explain it with an example.

1.3.12. Example. Consider the 2-complex N_2 on the Klein bottle consisting of one vertex, two pairs of edges and one pair of faces, see 1.2.6 (d). Then $\pi_1(N_2) = \langle a, b \mid a^2b^2 \rangle$. The symmetric group Sym(3) consists of the six elements 1, $r = (12)$, $s = (13)$, $t = (23)$, $u = (123)$, $v = (132)$, where r, s, t have order 2 and u, v have order 3. Elements $f(a), f(b)$ with $f(a)^2 f(b)^2 = 1$ determine a homomorphism $f : \pi_1(N_2) \to \text{Sym}(3)$. If $f(a) = 1$ or $f(b) = 1$ then f is not transitive. If $f(a)$ has order 2 then, up to equivalence, $f(a) = r$, $f(b) = s$. If the order of $f(a)$ is 3 then $f(b) = f(a)^{-1}$ and, up to equivalence, $f(a) = u$, $f(b) = v$. Hence *there are, up to equivalence, only two* 3*−sheeted coverings of the Klein bottle.*

The method above allows one to determine the subgroups of a given finite index in finitely many steps and can be implemented on computers. It can be modified to a trial-and-error method to determine the index of the subgroup generated by finitely many given elements and to get a presentation — as long as the index is finite. However it is not clear how many steps are needed to determine the index.

For further applications see 5.3.7. An important covering is that defined by the trivial subgroup:

1.3.13. Definition. Let $\hat{C}$ be a connected 2-complex with $\pi_1(\hat{C}) = 1$. Then a covering $p : \hat{C} \to C$ is called *universal.*

A universal covering has the following categorical property: *A covering* $\hat{p} : \hat{C} \to C$ *is universal if and only if for any other covering* $\bar{p} : \bar{C} \to C$ *with connected total space there is a covering* $q : \hat{C} \to \bar{C}$ such that $\hat{p} = \bar{p} \circ q$. By 1.3.4 (b), we see that to every 2-complex there exists an universal covering which is uniquely determined up to isomorphism.

1.3.14. Examples of Universal Coverings

(a) The sphere is the universal cover of the projective plane, see Fig. 1.3.1.

(b) The plane with the lattice of horizontal and vertical lines with integer coefficients is the universal cover of the torus S_1, see Fig. 1.3.2 (considered as 2-complex).

(c) For Cayley diagrams the mapping $p : \Delta\langle \mathcal{X} \mid \mathcal{R}\rangle \to C\langle \mathcal{X} \mid \mathcal{R}\rangle$ identifying all vertices, all edges labeled by the same generator and all faces labeled by the same relation is a universal covering, by 1.2.22 (c).

Chapter 2
Free Groups and Free Products

§ 2.1. Free Groups

Recall from Chapter 1 that a group is *free* if it has a presentation with no non-trivial relations. By Corollary 1.1.6 (c) every group is the homomorphic image of a free group and this indicates the central role that free groups play in combinatorial group theory. We have already encountered free groups in Chapter 1 as fundamental groups of graphs and seen how this kind of representation of a free group enables one to obtain results about subgroups. In this section we work directly with words and derive some elementary properties of free groups.

Let $F = \langle\ \mathcal{X} \mid -\ \rangle$ be a free group. A word W over $\mathcal{X}$ is *reduced* if it contains no *peak* $X^{\varepsilon}X^{-\varepsilon}$, $X \in \mathcal{X}$, $\varepsilon = \pm 1$. Clearly every element of F can be represented by a reduced word. The first principal result in the theory of free groups is the solution to the word problem.

2.1.1. Theorem. *Let $F = \langle\ \mathcal{X} \mid -\ \rangle$. Then each element of F is represented by a unique reduced word.*

Proof. Let $\mathcal{W}_0$ be the set of all reduced words over $\mathcal{X}$. We define an action of F on $\mathcal{W}_0$ — by Theorem 1.1.3 it suffices to define an action for each $X \in \mathcal{X}$. Given $W \equiv X_1^{\varepsilon_1} \cdots X_n^{\varepsilon_n}$ we define

$$X.W = \begin{cases} XX_1^{\varepsilon_1} \cdots X_n^{\varepsilon_n} & \text{if } X \neq X_1^{-\varepsilon_1}; \\ X_2^{\varepsilon_2} \cdots X_n^{\varepsilon_n} & \text{if } X = X_1^{-\varepsilon_1}. \end{cases}$$

It follows that if U is a reduced word representing $g \in F$, then $g.1 = U$ and hence exactly one reduced word represents g. □

Theorem 2.1.1 justifies defining the *length* $|g|$ relative to $\mathcal{X}$ of an element of F as the length of the unique reduced word representing g.

2.1.2. Remark. Several proofs of Theorem 2.1.1 are available – see [Cohen 1989]. The above is the shortest and most elegant.

The conjugacy problem for $F = \langle \mathcal{X} \mid - \rangle$ is also solvable.

2.1.3. Definition and Proposition. *A word* $W \equiv X_1^{\varepsilon_1} \cdots X_n^{\varepsilon_n}$ *is called* cyclically reduced *if it is reduced and* $X_n^{\varepsilon_n} X_1^{\varepsilon_1}$ *is not a peak. Let* $F = \langle \mathcal{X} \mid - \rangle$.

(a) *Every element of* F *is conjugate to an element represented by a cyclically reduced word.*

(b) *Two cyclically reduced words represent conjugate elements if and only if one is a cyclic permutation of the other.* □

We have already seen in 1.1.13 how to solve the isomorphism problem for free groups.

2.1.4. Theorem. *Let* $F = \langle \mathcal{X} \mid - \rangle$ *and let* $F^* = \langle \mathcal{X}^* \mid - \rangle$. *Then* F *and* F^* *are isomorphic if and only if* $\mathcal{X}$ *and* $\mathcal{X}^*$ *have the same cardinality.* □

The solution to the word problem yields, without difficulty, some information about subgroups of a free group.

2.1.5. Proposition

(a) *A free group is torsion-free.*

(b) *An abelian subgroup of a free group is cyclic.*

(c) *A free group has trivial centre — except when it is infinite cyclic.* □

To obtain further information on subgroups of a free group, more powerful methods are needed. Already in 1.3.5 we have used the method of coverings to deduce that a subgroup of a free group is free and to give a formula for the rank of a subgroup of finite index. Here we give another proof of this result, based on methods of Nielsen [Nielsen 1919]. This approach works with systems of generators of a subgroup and can be regarded as more practical since one is often given a subgroup by generators rather than by the transversal needed for the covering space method.

2.1.6. Definition

(a) Let $(W_1, W_2, \ldots, W_n)$ be a system of reduced words of the free group $F = \langle \mathcal{X} \mid - \rangle$. We are interested in the subgroup generated by $(W_1, \ldots, W_n)$. Now, for $i = 1, 2, \ldots, n$ there are unique words P_i and Q_i such that $|P_i| = |Q_i|$ and $W_i \equiv P_i Z_i Q_i^{-1}$ where $Z_i \in \mathcal{X} \cup \mathcal{X}^{-1}$ if $|W_i|$ is odd and Z_i denotes the empty word if $|W_i|$ is even. If $|W_i|$ is odd then Z_i is the *central letter* of W_i and if $|W_i|$ is even then W_i has two *central letters*, namely the last letter of P_i and the first letter of Q_i^{-1}.

(b) A system $(W_1, \ldots, W_n)$ is called *Nielsen reduced* if the following hold, where V_1, V_2 and V_3 are any words of the form $W_i^{\pm 1}$:

(N1) W_i is non-empty, $1 \leq i \leq n$;

(N2) $|W_i^{\varepsilon} W_j^{\eta}| \geq |W_i|, |W_j|$ for any pair W_i^{ε}, W_j^{η}, $1 \leq i, j \leq n$, $\varepsilon, \eta \in \{1, -1\}$ with $i \neq j$ or $i = j$ and $\varepsilon \neq -\eta$;

(N3) if $V_1 V_2 \neq 1$ and $V_2 V_3 \neq 1$ then $|V_1 V_2 V_3| > |V_1| - |V_2| + |V_3|$.

A product $W_{i_1}^{\varepsilon_1} W_{i_2}^{\varepsilon_2} \cdots W_{i_r}^{\varepsilon_r}$, where $\varepsilon_k = \pm 1, k = 1, 2, \ldots, r$, is called *reduced* if $i_k = i_{k+1}$ implies $\varepsilon_k = \varepsilon_{k+1}$.

2.1.7. Proposition

(a) *A reduced product of r elements of a Nielsen reduced system* $(W_1, \ldots, W_n)$ *of elements has length at least r in the free group. Indeed*

$$|W_{i_1}^{\varepsilon_1} W_{i_2}^{\varepsilon_2} \cdots W_{i_r}^{\varepsilon_r}| \geq \frac{1}{2}(|W_{i_1}| + |W_{i_r}|) + r - 2 \qquad \textit{and}$$

$$|W_{i_1}^{\varepsilon_1} W_{i_2}^{\varepsilon_2} \cdots W_{i_r}^{\varepsilon_r}| \geq |W_{i_k}|, \quad 1 \leq k \leq r.$$

(b) *A Nielsen reduced system* $(W_1, \ldots, W_n)$ *of elements of the free group* $F = \langle\, \mathcal{X} \mid - \,\rangle$ *freely generates a free subgroup of* F.

Proof. Consider a reduced product $V_1 V_2 \cdots V_n$ with $V_k \equiv W_{i_k}^{\varepsilon_k}$, and $\varepsilon_k = \pm 1$. Now the Nielsen conditions (N1)-(N3) guarantee that, in cancelling any triple $V_{k-1} V_k V_{k+1}$to reduced form in F, a central letter of V_k survives the cancellation process. Hence, when the words $V_1 V_2$, $V_1 V_2 V_3$, $V_1 V_2 V_3 V_4$, ... are cancelled stepwise to reduced form, a central letter of every V_k survives and therefore no reduced product of elements of the system can give the identity in F. This gives (b) and closer analysis of the argument yields (a). □

To show that any finitely generated subgroup of a free group is free, it suffices, by Proposition 2.1.7, to find a Nielsen reduced system of generators. This is achieved by operations known as Nielsen transformations which seek to eliminate violations of the conditions (N1) — (N3).

2.1.8. Definition. The following operations are called *Nielsen transformations*:

(T1) replace W_i by W_i^{-1};

(T2) replace W_i by $W_i W_j$, $j \neq i$;

(T3) delete W_i if W_i is the empty word.

2.1.9. Proposition. *Any finite system* $(W_1, \ldots, W_n)$ *of reduced words of* $F = \langle\, \mathcal{X} \mid - \,\rangle$ *can be transformed into a Nielsen reduced system by a finite number of Nielsen transformations.*

Proof. In view of the Nielsen transformation (T1) it suffices to work with n-tuples of the form $(\{W_1\}, \ldots, \{W_n\})$ where $\{W\}$ denotes the pair $\{W, W^{-1}\}$. The aim is to use Nielsen transformations in an inductive way with the idea that a system which is minimal relative to the inductive parameter will be Nielsen reduced. Naive considerations suggest that one might attempt to base the inductive argument on the sum of the lengths of the elements in such a n-tuple. However this is an insufficiently delicate measure and the proof actually proceeds by introducing an order relation $\prec$, somewhat like a lexicographical order, on the collection of all such n-tuples. The critical case in which the need for such subtlety is apparent occurs when one considers a violation of condition

(N3) in a product of the form $W_1W_2W_3 \equiv P_1Z_1Q_1^{-1} \cdot P_2Q_2^{-1} \cdot P_3Z_3Q_3^{-1}$ where $Q_1 = P_2, Q_2 = P_3$ and $|W_1W_2| = |W_1|$, $|W_2W_3| = |W_3|$. Now $W_1W_2 \equiv P_1Z_1Q_2^{-1}$ and $W_2W_3 \equiv P_2Z_3Q_3^{-1}$ and no decrease in length occurs as a result of applying a Nielsen transformation (T2). However the properties of $\prec$ ensure that

$$(\{W_1\}, \{W_2\}, \{W_2W_3\}) \prec (\{W_1\}, \{W_2\}, \{W_3\})$$

and the induction hypothesis applies. □

The argument of the theorem in fact establishes:

2.1.10. Proposition. *If the systems $(W_1, \ldots, W_m)$ and $(W_1', \ldots, W_n')$ are minimal relative to the order $\prec$ and generate the same subgroup, then they coincide, i.e. $m = n$ and $W_i = W_i'$, $1 \leq i \leq n$.* □

2.1.11. Corollary. *A finitely generated subgroup of a free group is free.* □

2.1.12. Corollary. *Given a finitely generated subgroup H of a free group F, there is an effective procedure to determine when an element of F lies in H.*

Proof. By using the method of the proof of 2.1.9, one may assume that H is given by a Nielsen reduced system of generators. The argument of Proposition 2.1.7 shows that if a word of length n lies in H then it is a product of at most n occurrences of the generators of H. This can be tested by simply examining all possibilities. □

2.1.13. Remarks

(a) It is also possible to use the method of ordering to derive the full subgroup Theorem 1.3.5 without the restriction of finiteness (see [Lyndon-Schupp 1977]).

(b) A Nielsen reduced system of generators for a subgroup can be obtained from the covering space method of 1.3.6-1.3.7 by choosing a *minimal* spanning tree B' of C', that is a tree in which the distance from the basepoint v_0' to any vertex of C' is the least possible.

§ 2.2. Amalgamated Free Products and Graphs of Groups

In 1.2.18–19 we have seen ways to construct the fundamental group of a complex from the fundamental groups of subcomplexes. The group-theoretic analogue of this procedure consists of certain ways of building up groups from subgroups. This ultimately leads to the notion of the *fundamental group of a graph of groups* which we shall describe below. We begin, however, with a discussion of some important special cases.

2.2.1. Definition. Let $(G_i)_{i\in I}$ be a family of groups and $(\kappa_i : A \to G_i)_{i\in I}$ a family of embeddings of the fixed group A in the various groups G_i. A group G is called the *free product* of the family $(G_i)_{i\in I}$, *amalgamating* the group A (*via the embeddings* $(\kappa_i)_{i\in I}$) if there exists a family $(\mu_i : G_i \to G)_{i\in I}$ of homomorphisms such that for any $i, j \in I$, $\mu_i \circ \kappa_i = \mu_j \circ \kappa_j$ and, for any family $(\varphi_i : G_i \to H)_{i\in I}$ of homomorphisms to a fixed group H satisfying $\varphi_i \circ \kappa_i = \varphi_j \circ \kappa_j$, there is a unique homomorphism $\varphi : G \to H$ such that $\varphi_i = \varphi \circ \mu_i$. The groups G_i are called the *factors* of G and A the *amalgamated subgroup*. When A is trivial G is called *the free product*, written $G = \ast_{i\in I} G_i$.

2.2.2. Proposition. *For any family $(G_i)_{i\in I}$ of groups and any family $(\kappa_i : A \to G_i)_{i\in I}$ of embeddings there exists a group G which is the free product of the family $(G_i)_{i\in I}$ amalgamating A via $(\kappa_i)_{i\in I}$. Furthermore G is unique up to isomorphism.*

Proof. Existence is demonstrated by choosing, for $i \in I$, disjoint presentations $\langle \mathcal{X}_i \mid \mathcal{R}_i \rangle$ of the groups G_i and systems $\mathcal{Y}_i$ of words which define generators for the groups $\kappa_i(A)$. Then a presentation $\langle \mathcal{X} \mid \mathcal{R} \rangle$ for the desired group G is obtained by setting $\mathcal{X} = \bigcup_{i\in I} \mathcal{X}_i$ and $\mathcal{R} = \left(\bigcup_{i\in I} \mathcal{R}_i\right) \cup \mathcal{R}'$ where $\mathcal{R}'$ consists of relations which identify corresponding elements of the systems $\mathcal{Y}_i$.

The uniqueness of G up to isomorphism follows from the uniqueness of the map φ by the usual category theory argument about the solutions to universal problems. $\square$

Of interest, since the general case follows from it by induction and taking limits, is the case of just two groups G_1 and G_2 in which case one writes $G = G_1 \ast_A G_2$ or $G = (G_1 \ast G_2 : A_1 = A_2)$, where A_1 and A_2 are the images of A in G_1 and G_2 respectively.

Forming $G = G_1 \ast_A G_2$ corresponds to forming a space or complex as the union of two subspaces or subcomplexes which have an intersection. Next we come to the group-theoretic analogue of adding a handle, see 1.2.19.

2.2.3. Definition. Let G_0 be a group and A_1 and A_2 subgroups of G_0 which are isomorphic via some specified isomorphism. Given a presentation $\langle \mathcal{X}_0 \mid \mathcal{R}_0 \rangle$ of G_0 the group G defined by the presentation $\langle G_0, t \mid t^{-1}A_1t = A_2\rangle = \langle \mathcal{X}_0, t \mid \mathcal{R}_0,\ t^{-1}U_it = V_i\ ,\ i \in I \rangle$ is an HNN-*extension* where the *stable letter* t is a new generating symbol, and $(U_i)_{i\in I}$ and $(V_i)_{i\in I}$ represent generating systems of A_1 and A_2 which correspond under the specified isomorphism. From its definition it is clear that the construction is a solution to a universal problem and so is unique up to isomorphism. The group G_0 is called the *base group* of G. The construction has an obvious generalisation to the case of two families $A_{\lambda,1}$ and $A_{\lambda,2}$, $\lambda \in \Lambda$, of subgroups where $A_{\lambda,1}$ and $A_{\lambda,2}$ are isomorphic. In this case a family t_λ of new generators is adjoined and conjugation by t_λ induces the isomorphism between $A_{\lambda,1}$ and $A_{\lambda,2}$.

The two constructions of free product with amalgamation and HNN-extension are closely interwoven (see [Lyndon-Schupp 1977, Chap. IV]). These connections mean that results about amalgamated free products can usually be translated into results about HNN-extensions and vice-versa. One such result is the *normal form theorem.*

2.2.4. Theorem. *Let $G = G_1 *_A G_2$. Then every element g of G can be uniquely expressed in the form $g = ac_1c_2 \cdots c_n$, $n \geq 0$, where $a \in A$, and $c_1, \ldots, c_n$ come alternately from transversals for the right cosets of A in G_1 and G_2 (and are not in A so that the case $n = 0$ arises only when $g \in A$).*

2.2.5. Theorem. *Let $G = \langle\ G_0, t \mid t^{-1}A_1t = A_2\ \rangle$ and let transversals for the right cosets of A_1 and A_2 in G_0 be chosen. Then every element g of G can be written uniquely in the form $bt^{\varepsilon_1}c_1 \cdots t^{\varepsilon_n}c_n$, where*
(1) b is an arbitrary element of G_0,
(2) c_i lies in a transversal for A_1 or A_2 according as $\varepsilon_i = -1$ or $\varepsilon_i = +1$,
(3) if $c_i = 1$ then $\varepsilon_i = \varepsilon_{i+1}$.

Proof of 2.2.4. The existence of such a representation of g is easy. Any non-trivial $g \in G$ can be written as a product $g = z_1 \cdots z_m$ where the factors z_i come alternately from G_1 and G_2. Moreover except when $g \in A$, one can assume that no z_i lies in A. Then one writes $z_m = a_mc_m$ with $a_m \in A$ and c_m in the appropriate transversal and so on.

Uniqueness is established by a generalisation of the argument for Theorem 2.1.1. Let $\mathcal{W}$ be the set of all finite sequences $(a, c_1, \ldots, c_n)$, $n \geq 0$ where $ac_1 \cdots c_n$ satisfies the conditions in the statement of the theorem. To define an action of G on $\mathcal{W}$ it suffices, by the universal property, to define actions of G_1 and G_2 which agree on the common subgroup A. This is done in the obvious way. For example if $z \in G_1$ then

$$z.(a, c_1, \ldots, c_n) = \begin{cases} (a', c_0, c_1, \ldots c_n) & \text{if } c_1 \in G_2 \text{ and } za = a'c_0\ , \\ (za, c_1, \ldots, c_n) & \text{if } z \in A, \\ (a', c_1', \ldots c_n) & \text{if } c_1 \in G_1 \text{ and } zac_1 = a'c_1' \notin A, \\ (a', c_2, \ldots, c_n) & \text{if } c_1 \in G_1 \text{ and } zac_1 = a' \in A. \end{cases}$$

If $g = ac_1 \cdots c_n$ as in the statement of the theorem, then g applied to the trivial sequence (1) yields $(a, c_1, \ldots, c_n)$ and the result follows. □

Theorem 2.2.5 can be derived by a similar argument. In both cases the integer n is referred to as the *length* (relative to the given decomposition of G) of the element g.

2.2.6. Corollary. *The natural maps embed the factors of a free product with amalgamation and the base group of an HNN-extension.*

2.2.7. Examples. There are numerous examples of amalgamated free products and HNN-extensions some of which we have already encountered.

(a) Let $G_1 = \langle a_1, b_1 \mid - \rangle$ and $G_2 = \langle a_2, b_2 \mid - \rangle$. If $A_1 = \langle a_1b_1a_1^{-1}b_1^{-1} \rangle$ and $A_2 = \langle b_2a_2b_2^{-1}a_2^{-1} \rangle$ then A_1 and A_2 are infinite cyclic subgroups of free

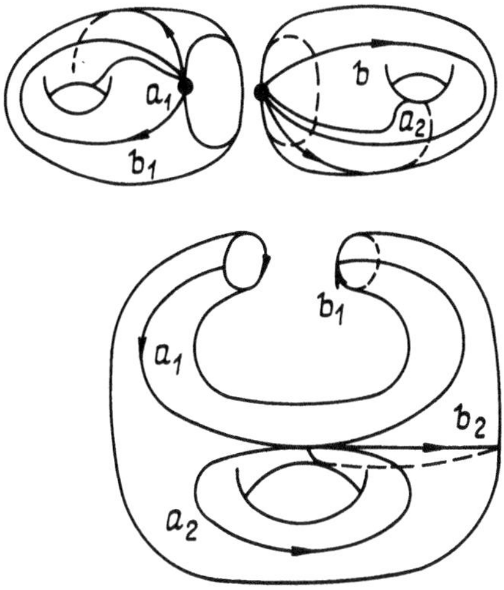

Fig. 2.2.1

groups and $G = (G_1 * G_2 \ : A_1 = A_2)$ is the fundamental group of a closed orientable surface of genus two (see 1.2.14 (c)).

If, on the other hand we put $G_0 = \langle \, b_1, a_2, b_2 \mid - \, \rangle$, $A_1 = \langle b_2 a_2 b_2^{-1} a_2^{-1} b_1 \rangle$ and $A_2 = \langle b_1 \rangle$ then A_1 and A_2 are infinite cyclic subgroups of the same group G_0, and, if we write t for a_1, then $G = \langle \, G_0, t \mid t^{-1} A_1 t = A_2 \, \rangle$ is also this fundamental group. The two presentations correspond to the two constructions of the space illustrated in Fig. 2.2.1.

(b) The free abelian group of rank two with presentation $G = \langle \, a, b \mid ab = ba \, \rangle$ is an HNN-extension with, say, base group $G_0 = \langle b \rangle = A_1 = A_2$ and $t = a$.

(c) The group with presentation $\langle \, a, b \mid a^{-1} b^2 a = b^3 \, \rangle$ is obviously an HNN-extension with $G_0 = \langle b \rangle$, $t = a$, $A_1 = \langle b^2 \rangle$ and $A_2 = \langle b^3 \rangle$ This group has an interesting property, namely it is *non-hopfian*, that is the group possesses a surjective endomorphism which is not an automorphism. To see this consider the map $a \overset{\varphi}{\mapsto} a, b \overset{\varphi}{\mapsto} b^2$. An application of φ squares both sides of the defining relation and so φ is well-defined and is surjective since the relation shows that $b = b^{-2} a^{-1} b^2 a$. If $g = a^{-1} bab a^{-1} b^{-1} ab^{-1}$, then $\varphi(g) = a^{-1} b^2 ab^2 a^{-1} b^{-2} ab^{-2}$ which is the identity since $a^{-1} b^2 a = b^3$. On the other hand it follows from the normal form theorem that g is not the identity element. By way of contrast, as we shall point out in 2.3, it is a simple consequence of the method of Nielsen transformations that a free group of finite rank is hopfian.

(d) Consider the group with presentation

$$G = \langle \, a_1, a_2, a_3, a_4 \mid a_i^{-1} a_{i-1} a_i = a_{i-1}^2, \ 1 \le i \le 4 \text{ with } i \bmod 4 \rangle.$$

This is not an HNN-extension, since upon abelianising the result is the trivial group and any HNN-extension has the infinite cyclic group as a homomorphic image. Nor does it obviously appear to be an amalgamated free product. However consider the groups

$$G_1 = \langle\ a_1, a_2, a_3 \mid a_2^{-1}a_1a_2 = a_1^2\ ,\ a_3^{-1}a_2a_3 = a_2^2\ \rangle$$

and

$$G_2 = \langle\ x, a_4, y \mid a_4^{-1}xa_4 = x^2\ ,\ y^{-1}a_4y = a_4^2\ \rangle.$$

Each is obtained from an infinite cyclic group by two successive HNN-extensions — the relevant subgroups are easily seen to be infinite cyclic in each case. If we define $A_1 = \langle a_1, a_3\rangle$ and $A_2 = \langle x, y\rangle$ then one can use the normal form theorem for HNN-extensions to show that these groups are both free of rank two on the displayed generators. One may therefore form the amalgamated free product $G = (G_1 * G_2\ :\ A_1 = A_2)$, equating $x = a_3$ and $y = a_1$. Tietze transformations then yield the initial presentation. The significance of this argument is that it demonstrates that this group is non-trivial and in particular infinite. Simple numerical arguments (see [Higman 1951]) show that no finite group can contain (non-trivial) elements satisfying the four relations of the presentations and so it follows that this group has no non-trivial finite homomorphic images. Its quotient by a maximal proper normal subgroup is therefore a finitely generated infinite simple group. (Demonstrating that non-trivial normal subgroups actually exist is not a wholly trivial matter — see [Schupp 1971]).

(e) Let $G = \langle\ a, b, c, d \mid a^nba^n = c^ndc^n\ ,\ n \geq 1\ \rangle$. Then G is the free product of two free groups of rank two amalgamating the subgroups $\langle a^nba^n : n \geq 1\rangle$ and $\langle c^ndc^n\ :\ n \geq 1\rangle$ of infinite rank (the displayed generating systems are Nielsen reduced). This group is not finitely presented. For if it were then the identity in $F(a, b, c, d)$ would induce an isomorphism from G to a group with presentation $G^* = \langle\ a, b, c, d \mid a^nba^n = c^ndc^n\ ,\ 1 \leq n \leq N\ \rangle$, for some N. Since the groups are isomorphic, the relation $a^{N+1}ba^{N+1} = c^{N+1}dc^{N+1}$ must hold in G^*. Since G^* is also an amalgamated free product, and $a^{N+1}ba^{N+1}$ does not lie in the subgroup of the free group on a and b generated by the elements $\{a^nba^n : 1 \leq n \leq N\}$, this is a contradiction.

We give one more application of the theory of HNN-extensions taken from the original paper [Higman-Neumann-Neumann 1949] where this concept was introduced.

2.2.8. Theorem (Higman, Neumann B., Neumann H.). *Every countable group H can be embedded in a group G generated by two elements.*

Proof. Let H be a countable group and let the elements of H be enumerated as $1 = h_0, h_1, h_2, \ldots$. Let G_0 be the free product $H * F$ where F is the free group with basis $\{a, b\}$. Let $A_1 = \langle b^{-n}ab^n\ :\ n = 0, 1, 2, \ldots\rangle$ and $A_2 = \langle h_na^{-n}ba^n : n = 0, 1, 2, \ldots\rangle$. An easy cancellation argument shows that A_1 is free on the given generators (the generators actually form a Nielsen reduced system) and projecting from G_0 to F shows that the same holds for A_2. Using the isomorphism that sends $b^{-n}ab^n \mapsto h_n a^{-n} b a^n$ to form $G = \langle G_0, t \mid t^{-1}A_1 t = A_2 \rangle$ yields the desired group. □

Establishing a unique normal form for elements of a group amounts to solving the word problem — provided that normal forms can be effectively calculated. For an amalgamated free product $G = (G_1 * G_2 : A_1 = A_2)$ the crucial problem in calculating normal forms is to determine when words of G_1 and G_2 define elements of the amalgamated subgroups A_1 and A_2. We refer to these as the *membership problems* for A_i in G_i, for $i = 1, 2$.

2.2.9. Theorem. *Let $G = (G_1 * G_2 : A_1 = A_2)$ (or $G = \langle G_0, t \mid t^{-1}A_1 t = A_2 \rangle$). Suppose that*

(a) *the factors G_1 and G_2 (or the base group G_0) have solvable word problem,*

(b) *the membership problems for A_1 and A_2 in the factors (or the base group) are solvable,*

(c) *the isomorphism between A_1 and A_2 is effectively computable.*

Then G has solvable word problem. □

Condition (c) is a technicality which is needed to ensure that when given a word representing an element of A_1, one can then obtain a word which represents the corresponding element of A_2.

2.2.10. Example. Let $G = \langle a_1, b_1, a_2, b_2 \mid [a_1, b_1][a_2, b_2] \rangle$. As indicated in 2.2.7 (a), this is a free product of two free groups amalgamating infinite cyclic subgroups and obviously the hypotheses of Proposition 2.2.9 are satisfied. The same applies to any free product of two free groups amalgamating a finitely generated subgroup since by Corollary 2.1.12 the menbership problem for any finitely generated subgroup of a free group is solvable (and, on general grounds from the theory of computability, any isomorphism of finitely generated groups is always effectively computable).

The normal form theorem also provides a complete description of conjugacy in amalgamated free products and HNN-extensions.

2.2.11. Proposition. *Let $G = (G_1 * G_2 : A_1 = A_2)$.*

(a) *Any element $g \in G$ is conjugate to a* cyclically reduced *element, that is an element of length 1 or an element of the form $z_1 z_2 \cdots z_n$, $n \geq 2$ whose terms come alternately form $G_1 \setminus A_1$ and $G_2 \setminus A_2$ (including z_1 and z_n).*

(b) *If u and v are conjugate cyclically reduced elements of G of length at least two then some cyclic rearrangements of u and v are conjugate by an element of the amalgamated subgroup.*

(c) *If u and v are conjugate elements of length one, then either u and v lie in the same factor and are conjugate there or there is a sequence of elements $a_1, a_2, \ldots a_n$ of the amalgamated subgroup such that, within the factors G_1 and G_2, u is conjugate to a_1, a_i is conjugate to a_{i+1} and a_n is conjugate to v.* □

Although Proposition 2.2.11 gives an essentially complete specification of when elements are conjugate in $G = (G_1 * G_2 \ : A_1 = A_2)$, and, in a descriptive sense, solves the conjugacy problem, in practice it is often not easy to apply the proposition to give an algorithmic solution of the conjugacy problem. A parallel result, with the same caveat, holds for HNN-extensions.

The normal form theorems yield a complete description of finite subgroups of amalgamated free products and HNN-extensions.

2.2.12. Theorem. *Let* $G = (G_1 * G_2 \ : A_1 = A_2)$ *or* $G = \langle\, G_0, t \mid t^{-1} A_1 t = A_2 \,\rangle$. *Then any finite subgroup of* G *is conjugate to a subgroup of a factor or of the base group, as is appropriate.* □

In view of the parallels with free groups indicated by the normal form theorems, it is not surprising that any significant commutativity within an amalgamated free product or an HNN-extension is more or less confined to the factors or the base group as appropriate. However a description that is both precise and concise is not easy to formulate and we content ourselves with two simple observations.

2.2.13. Example. Any free abelian group is an HNN-extension.

2.2.14. Proposition. *If* $G = G_1 *_A G_2$, *then* $Z(G) = A \cap Z(G_1) \cap Z(G_2)$ *where* $Z(G)$ *denotes the centre of* G, *and so on.* □

We have now essentially dealt with those matters that can be derived from the normal form alone. Deeper methods are required to describe subgroups in general. In the case of an ordinary free product $G = *_{i \in I} G_i$ a complete description was given as early as 1934 by Kurosh [Kurosh 1934].

2.2.15. Theorem. *Let* $G = \ast_{i \in I} G_i$ *be a free product. Then any subgroup* H *of* G *is expressible as a free product* $H = (\ast_{\lambda \in \Lambda} H_\lambda) * F$ *where*

(i) *for each* λ *there exists* $i(\lambda)$ *and* $z_\lambda \in G$ *such that* $H_\lambda = H \cap z_\lambda G_{i(\lambda)} z_\lambda^{-1}$;

(ii) F *is a free group and* $F \cap z G_i z^{-1} = 1$ *for all* $z \in G$ *and* $i \in I$;

(iii) *if* $H \cap z G_i z^{-1} \neq 1$ *then there is a unique* $\lambda \in \Lambda$ *such that* $H \cap z G_i z^{-1}$ *is conjugate to* H_λ. □

An elegant *proof* of 2.2.15, using coverings of 2-complexes as in 1.3.4 and 1.3.5, was given by Baer-Levi, see [ZVC 1980, 1988, 2.6]. A generalisation of the Kurosh theorem to amalgamated free products was not easily obtained. A first attempt was made by H. Neumann [Neumann 1948/9] but it was not until the work of Karrass-Solitar [Karrass-Solitar 1970], Bass-Serre [Serre 1977] that the necessary conceptual framework was developed. To describe this it is useful to recall the representation of free groups as fundamental groups of graphs, namely as the group of homotopy classes of closed paths at a basepoint.

2.2.16. Definition. *A graph of groups* is a pair $(\mathcal{G}, X)$ where X is a graph and $\mathcal{G}$ consists of a family $(G_v; v \in V(X))$ of *vertex groups* and a family $(G_\sigma; \sigma \in E(X))$ of *edge groups* where the following hold:

(1) for any edge σ, $G_{\sigma^{-1}} = G_\sigma$;

(2) for any edge σ there are monomorphisms $\kappa_\sigma : G_\sigma \to G_{s(\sigma)}$ and $\lambda_\sigma : G_\sigma \to G_{t(\sigma)}$ such that $\lambda_\sigma = \kappa_{\sigma^{-1}}$.

A *path* in $(\mathcal{G}, X)$ is a sequence $(g_0, \sigma_1, g_1, \ldots, \sigma_r, g_r)$, where $g_i \in G_{v_i}$ and $(v_0, \sigma_1, v_1, \ldots, \sigma_r, v_r)$ is a path in X. The relation $\simeq$ of *homotopy equivalence* of paths in $(\mathcal{G}, X)$ is the equivalence relation induced by the *elementary* homotopy equivalences $(\sigma, \lambda_\sigma(h), \sigma^{-1}, (\kappa_\sigma(h))^{-1}) \simeq (1)$, where $1 \in G_{s(\sigma)}$, and $(g, \sigma, 1, \sigma^{-1}, g') \simeq (gg')$. If v is a vertex of X then the homotopy classes of closed paths at v form a group under the operation of concatenation of representatives – with the natural rule that $\ldots, g)(g', \ldots \ = \ldots, gg', \ldots$. We denote this group by $\pi_1(\mathcal{G}, X, v)$ and call it the *fundamental group* of $(\mathcal{G}, X)$.

Obviously if all the vertex groups are trivial then $\pi_1(\mathcal{G}, X, v)$ is isomorphic to the usual fundamental group $\pi_1(X, v)$, and the map which deletes the group entries in any path in $(\mathcal{G}, X)$ is always an epimorphism from $\pi_1(\mathcal{G}, X, v)$ to $\pi_1(X, v)$.

2.2.17. Proposition. *Let $(\mathcal{G}, X)$ be a graph of groups and v a vertex of X. Let T be a maximal tree of X and let $E^+(X)$ be an orientation of X, i.e. a subset of $E(X)$ containing exactly one member of each pair of inverse edges of X. Then $\pi_1(\mathcal{G}, X, v)$ is the group obtained from the free product $P = (\ast_{v \in V(X)} G_v) * F$, where F is the free group on a set $\{t_\sigma : \sigma \in E^+(X)\}$ in one-one correspondence with $E^+(X)$, by adding the relations*

$$t_\sigma \lambda_\sigma(h) t_\sigma^{-1} = \kappa_\sigma(h), \quad \textit{for } \sigma \in E^+(X) \textit{ and } h \in G_\sigma,$$
$$t_\sigma = 1 \quad \textit{for } \sigma \in E(T) \cap E^+(X).$$

Proof. Write $\pi_1(\mathcal{G}, X, T)$ for the quotient of the free product P by the given relations. Then there is an obvious map $p : \pi_1(\mathcal{G}, X, v) \to \pi_1(\mathcal{G}, X, T)$ given by

$$(g_0, \sigma_1^{\varepsilon_1}, g_1, \ldots, \sigma_n^{\varepsilon_n}, g_n) \mapsto g_0 t_{\sigma_1}^{\varepsilon_1} g_1 \cdots t_{\sigma_n}^{\varepsilon_n} g_n,$$

where $\varepsilon_i = \pm 1$ and $\sigma_i \in E^+(X)$, $1 \leq i \leq n$. Conversely consider the map $f : P \to \pi_1(\mathcal{G}, X, v)$ given by

$$x \mapsto (1, \sigma_1, 1, \ldots, \sigma_r, x, \sigma_r^{-1}, 1, \ldots, \sigma_1^{-1}, 1),$$
$$t_\sigma \mapsto (1, \sigma_1, 1, \ldots, \sigma_r, 1, \sigma, 1, \tau_1, 1, \ldots, \tau_s, 1)$$

where in the first case $x \in G_u$ and $(\sigma_1, \ldots, \sigma_r)$ is the unique reduced path in T from the basepoint v to u and in the second case the edge sequences $(\sigma_1, \ldots \sigma_r)$ and $(\tau_1, \ldots, \tau_s)$ define the reduced paths from v to $s(\sigma)$ and from $t(\sigma)$ back to v. Since this map is compatible with the relations factored out, f induces a homomorphism $\tilde{f} : \pi_1(\mathcal{G}, X, T) \to \pi_1(\mathcal{G}, X, v)$, which is inverse to the map p. □

2.2.18. Corollary. *The natural map $\kappa_v : G_v \to \pi_1(\mathcal{G}, X, T)$ is an embedding.* □

It should be observed that Proposition 2.2.17 shows that the description in terms of factoring out generators which correspond to edges in a maximal tree T is independent of the choice of T — and that $\pi_1(\mathcal{G}, X, v)$ is, up to isomorphism, independent of the choice of the vertex v. We shall, without further comment identify $\pi_1(\mathcal{G}, X, v)$ with $\pi_1(\mathcal{G}, X, T)$, for any choice of T.

2.2.19. Examples

(a) Let X consist of the graph $v \overset{\sigma}{\text{——}} w$ and let $\mathcal{G} = \{(G_v, G_w), (G_\sigma)\}$. The only possible choice for T is $T = X$ and so $\pi_1(\mathcal{G}, X, v)$ is the free product $G_v * G_w * F(t_\sigma)$ modulo the relations $t_\sigma = 1$, $t_\sigma \kappa_\sigma(h) t_\sigma^{-1} = \lambda_\sigma(h)$ for $h \in G_\sigma$. Thus $\pi_1(\mathcal{G}, X, v)$ is (isomorphic to) the free product of the two vertex groups amalgamating the two copies of the edge group.

(b) Let X consist of a single vertex and a single loop at that vertex. Then $\pi_1(\mathcal{G}, X, v)$ is clearly just the HNN-extension with the vertex group as base group and the two copies of the edge group conjugated by the generator corresponding to the loop.

(c) Let X be any finite tree. Then for any graph $(\mathcal{G}, X)$ of groups, $\pi_1(\mathcal{G}, X, v)$ can be constructed by repeated formation of amalgamated free products. If now we assume X is infinite (but every vertex has finite degree), then it is an ascending union of finite subtrees X_n and $\pi_1(\mathcal{G}, X, v)$ is the ascending union of the corresponding fundamental groups $\pi_1(\mathcal{G}_n, X_n)$. Sometimes $\pi_1(\mathcal{G}, X, v)$ is called a *tree product* in these circumstances. If $(\mathcal{G}, X)$ is an arbitrary graph of groups, then $\pi_1(\mathcal{G}, X, v)$ can be regarded as an HNN-extension by a set of stable letters corresponding to edges not in a maximal tree T over the base group which is the tree product defined by the restriction of $(\mathcal{G}, X)$ to the tree T. From this discussion it follows, via the normal form theorems 2.2.4 and 2.2.5 that if X_0 is any connected subgraph of X and $(\mathcal{G}_0, X_0)$ is the restriction of $(\mathcal{G}, X)$ to X_0, then, for any vertex v of X_0, the natural map $\pi_1(\mathcal{G}_0, X_0, v) \to \pi_1(\mathcal{G}, X, v)$ is an embedding.

(d) If the finitely generated group G is $\pi_1(\mathcal{G}, X, v)$ for some graph of groups $(\mathcal{G}, X)$ then $\mathcal{G}$ may be assumed to be finite.

2.2.20. Theorem. *Let G be the fundamental group of a graph $(\mathcal{G}, X)$ of groups. Any subgroup H of G is also the fundamental group of a graph $(\mathcal{H}, Y)$ of groups whose vertex groups and edge groups are respectively of the form $H \cap zG_v z^{-1}$ and $H \cap zG_\sigma z^{-1}$, where G_v and G_σ are, respectively, vertex and edge groups of $(\mathcal{G}, X)$.*

The proof of Theorem 2.2.20 is achieved by the theory of groups acting on trees. Recall that G *acts* on a graph X *without inversion* if, for all $g \in G$, $\sigma \in E(X)$, $g\sigma \neq \sigma^{-1}$. The main structure theorem for groups acting on trees comes in two parts.

2.2.21. Theorem. *Let the group G act without inversion on the tree X and let $Y = X/G$ be the quotient graph. For any vertex v and edge σ of Y let $\tilde{v}$ and $\tilde{\sigma}$ denote a vertex and edge of X lying over v and σ respectively. Then G is (isomorphic to) the fundamental group of a graph $(\mathcal{G}, Y)$ of groups whose vertex groups and edge groups, respectively, are stabilisers of the form $G_v = Stab_G(\tilde{v})$ and $G_\sigma = Stab_G(\tilde{\sigma})$.* □

2.2.22. Theorem. *Let G be the fundamental group of a graph $(\mathcal{G}, Y)$ of groups and $E^+(Y)$ an orientation of Y. Assume that for each $\sigma \in E^+(Y)$, the map $\kappa_\sigma : G_\sigma \to G_{s(\sigma)}$ is inclusion. Let X be the graph with vertex set $V(X) = \{(gG_v, v) \mid g \in G,\ v \in V(Y)\}$ and edge set $E(X) = \{(gG_\sigma, \sigma) \mid g \in G,\ \sigma \in E(Y)\}$ with endpoints and inverses defined by*

$$s(gG_\sigma, \sigma) = (gG_{s(\sigma)}, s(\sigma)), \quad t(gG_\sigma, \sigma) = (gt_\sigma G_{t(\sigma)}, t(\sigma)) \quad \text{for } \sigma \in E^+(Y)$$

$(gG_\sigma, \sigma)^{-1} = (gG_\sigma, \sigma^{-1})$. Then X is a tree on which G acts without inversion via $h(gG_v, v) = ((hg)G_v, v)$, $h(gG_\sigma, \sigma) = ((hg)G_\sigma, \sigma)$. □

When the details have been made precise, one can show that the constructions in the above two theorems are inverse to one another. A particular case of this structure theorem is that of an amalganated free product of two factors.

2.2.21*–22*. Theorem. *The group G is the amalgamated free product of two factors if and only if G acts without inversion on a tree X such that the quotient graph X/G is a segment v—w consisting of two vertices and a single pair of edges.*

Proof. The normal form theorem 2.2.4 applied to the graph X given by the recipe of Theorem 2.2.21 shows that X is a tree on which G obviously acts.

Conversely suppose that G acts on a tree X with the graph $Y =$ v—w as quotient. Lifting back to X, we may regard Y as a subgraph of X. Let $G_v = \mathrm{Stab}_G(v)$, $G_w = \mathrm{Stab}_G(w)$ and $G_\sigma = G_v \cap G_w = \mathrm{Stab}_G(\sigma)$. Then the claim is that $G = G_v *_{G_\sigma} G_w$. The fact that X is a tree gives the required normal form for elements of G. □

Proof of Theorem 2.2.20. Let $G = \pi_1(\mathcal{G}, Y)$ and let H be a subgroup of G. Now G acts on the tree X of 2.2.22. Since H is a subgroup of G, it also acts on the tree X. If $Z = X/H$ is the quotient graph then by Theorem 2.2.21, $H = \pi_1(\mathcal{H}, Z)$ for some graph $(\mathcal{H}, Z)$ of groups. A vertex group H_z of $(\mathcal{H}, Z)$ is of the form $\mathrm{Stab}_H(\tilde{z})$ where $\tilde{z}$ is a vertex of X lying over z. Now $\tilde{z} = (gG_v, v)$ for some $g \in G$ and some $v \in V(X)$ and so

$$H_z = \mathrm{Stab}_H(\tilde{z}) = H \cap \mathrm{Stab}_G(\tilde{z}) = H \cap gG_v g^{-1}.$$

Similarly any edge group of $(\mathcal{H}, \mathcal{Z})$ is of the form $H \cap gG_\sigma g^{-1}$. □

While our aim in this section has been to examine generalisations of the idea of a free group, not surprisingly, the theory of groups acting on trees also yields a proof of the Nielsen-Schreier theorem. We say a group G acts *freely* on a graph X if every vertex of X has trivial stabiliser.

2.2.23. Theorem. *A group is free if and only if it acts freely without inversion on a tree.*

Proof. The Cayley graph, relative to some free basis of any free group, is a tree on which the group acts freely in the required manner by left multiplication on the vertices. Conversely if a group G acts freely on a tree X, then, taking a topological shortcut, we know that G is a group of covering transformations of X (see 1.2). Hence G is the fundamental group of the quotient graph and so is free. □

2.2.24. Corollary. *A subgroup of a free group is free.*

Proof. If a group acts freely on a tree then so does any subgroup. □

We conclude this section by returning to the primitive kind of cancellation argument used in Nielsen's proof of the subgroup theorem for free groups. If G is a group and $\mathcal{X}$ is a system of elements of G then a *Nielsen transformation* on $\mathcal{X}$ is just an application of one of the operations (T1) — (T3) of 2.1.8 on $\mathcal{X}$. The method of Nielsen transformations has significant applications in a wider context than just free groups. One such is the consideration of generating systems for subgroups of free products. If $G = \ast_{i\in I} G_i$ then one can again define a relation $\prec$, with appropriate properties, on systems $\{W_i\}_{i\in I}$ of (inverse pairs of) elements of G. The main results are as follows.

2.2.25. Theorem. *Let $G = \ast_{i\in I} G_i$ be a free product.*

(a) *Any finite system of elements of G can be transformed by Nielsen transformations into a system minimal relative to $\prec$.*

(b) *Let $\{W_i\}_{i\in I}$ be a system minimal relative to $\prec$ of elements of G and let H be the subgroup they generate. Then every element W of H has a representation $W = W_{i_1}^{\varepsilon_1} \cdots W_{i_r}^{\varepsilon_r}$ such that $|W| \geq |W_{i_k}|$, $1 \leq k \leq r$.* □

2.2.26. Proposition. *Let $G = \ast_{i\in I} G_i$ be a free product and suppose that G is finitely generated. Then:*

(a) *any finite system of generators of G can be transformed by Nielsen transformations into a system consisting entirely of elements of the factors of G.*

(b) $d(G) = \sum_{i=1}^{n} d(G_i)$, *where $d(G)$ denotes the minimum number of elements needed to generate G.* □

Proposition 2.2.26 is a particular form of Grushko's theorem:

2.2.27. Theorem [Grushko 1940]. *Let $G = \ast_{i\in I} G_i$ and let $\eta : F \to G$ be a epimorphism from a free group F. Then F decomposes as a free product $F = \ast_{i\in I} F_i$ with $\eta(F_i) < G_i$.* □

Given that amalgamated free products and HNN-extensions also have normal forms and lengths, one might expect to extend the Nielsen method to such groups. However strong conclusions such as those of Theorem 2.2.25 are no longer possible without some sort of restrictions on the factors (or the base group). Nonetheless applications of this approach have been made, notably in [Zieschang 1970] and [Peczynski- Rosenberger-Zieschang 1975] in determining minimal generating systems for Fuchsian groups (see Chapter 3, in particular Remark 3.2.19 for definitions).

2.2.28. Definition and Examples

(a) Let G be a group and X, Y two systems of elements of G. Then X and Y are *Nielsen equivalent* if one can be transformed into the other by a finite sequence of applications of the operations (cf. 2.1.8)

(T0) permute the elements of the system;

(T1) replace an element by its inverse;

(T2) replace an element $x \in X$ by the product xx', where $x' \in X$, $x' \neq x$.

(b) Let $G = \langle s, t \mid [s,t] = 1\rangle$. Then any system $\{x, y\}$ of generators of G is Nielsen equivalent to $\{s, t\}$. For the map $\varphi : s \mapsto x,\ t \mapsto y$ is an automorphism. Moreover every automorphism of G is induced by an automorphism of the free group $F(s,t)$ and the automorphism group of $F(s,t)$ is generated by permutation and Nielsen automorphisms (see 2.3.1). Similarly, if G is the fundamental group of a closed orientable surface of genus g, then G has only one Nielsen equivalence class of system of generators of cardinality $2g$ (see 2.4.14).

(c) Let $G(p,q) = \langle s, t \mid s^p = t^q\rangle$, $p, q \geq 2,\ p+q > 4$; $G(p,q)$ is the group of a torus knot. Then G has infinitely many distinct Nielsen equivalence classes of generating pairs [Zieschang 1977]. These classes are represented by pairs (s^a, t^b) where $\gcd(a,p) = \gcd(b,q) = \gcd(a,b) = 1$ and $0 < 2a \leq pb,\ 0 < 2b \leq qa$. By contrast the quotient of G by its centre $\langle s^p\rangle$, which is just the free product $\langle s \mid s^p\rangle * \langle t \mid t^q\rangle$ has only finitely many distinct Nielsen equivalence classes of generating pairs. Representatives for the different classes are pairs (s^a, t^b), $0 < 2a \leq p,\ 0 < 2b \leq q$ such that $\gcd(a,p) = \gcd(b,q) = \gcd(a,b) = 1$. The group $G(p,q)$ has one defining relation in terms of the generators (s,t). In terms of a generating pair (s^a, t^b), two relations are required to define $G(p,q)$ except when $a = 1$ or $b = 1$ and only one relator is necessary, see [Collins 1978].

(d) [Brunner 1976] Let $G = \langle x, y \mid y^{-1}x^{-1}yxy^{-1}xy = x^2\rangle$. Then, for every r, the presentation $G(r) = \langle x, y \mid y^{-1}x^{-2^r}yxy^{-1}x^{2^r}y = x^2\rangle$ also defines G but there is no automorphism of the free group $F(x,y)$ carrying the relator $y^{-1}x^{-2^s}yxy^{-1}x^{2^s}x^{-2}$ of $G(s)$ into the relator $y^{-1}x^{-2^r}yxy^{-1}x^{2^r}yx^{-2}$ of $G(r)$.

This means, of course, that there are infinitely many Nielsen inequivalent generating pairs of G and that G is given by a single defining relation for each such pair. If (a,b) is the generating pair associated to the original presentation of G, then $(b^{-1}a^r bab^{-1}a^{-r}b, b)$ are the generators associated to $G(r)$. (Comp. 2.4.11-12.)

An axiomatisation of the Nielsen method in [Lyndon 1963] led to the concept of length function. First we stick closely to Nielsen's idea.

2.2.29. Definition. An *integer-valued length function* on the group G is a function $L : G \to \mathbb{N}$ satisfying, in Lyndon's enumeration:

(A1') $L(g) \geq 0$ for all $g \in G$ and $L(1) = 0$;

(A2) $L(g^{-1}) = L(g)$, for all $g \in G$;

(A4) if d is defined by $d(g,h) = \frac{1}{2}(L(g) + L(h) - L(g^{-1}h))$, then $d(g,h) \geq d(g,k)$ implies $d(h,k) = d(g,k)$.

In the case when G is a free group and L the usual length function, then $d(g,h)$ just denotes the length of the maximum common initial segment of g and h. It is easy to check that the usual length functions on amalgamated free products and HNN-extensions satisfy (A1'),(A2) and (A4). The next result appears in [Chiswell 1976].

2.2.30. Theorem. *A group G has an integer-valued length function if and only if G acts on a tree.*

If G acts on a tree than it is not hard to define a length function. Fix a base vertex v_0 of the tree and let $L(g)$ be the length of the reduced path from v_0 to gv_0. □

A more general notion of length function occurs if one allows the values to lie in an arbitrary ordered abelian group Λ. We confine ourselves to the case $\Lambda = \mathbb{R}$, first studied in [Harrison 1972]. A good account appears in [Shalen 1987].

An $\mathbb{R}$-*tree* is a non-empty metric space X in which any two points v and w are joined by a unique arc, denoted by $[v,w]$, which is isometric to a real interval of length $d(v,w)$, where d is the metric on X. The connection with real-valued length functions is given in [Chiswell 1976], [Imrich 1979], [Alperin-Moss 1985] by

2.2.31. Theorem. *A group acts as a group of isometries of an $\mathbb{R}$-tree if and only if it has a real-valued length function satisfying* (A1'),(A2), *and* (A4). □

An obvious example of an $\mathbb{R}$-tree is the geometric realisation of an ordinary (simplicial) tree with each edge regarded as a segment of length 1. This example shows that a free group acts freely on an $\mathbb{R}$-tree.

The study of group actions on $\mathbb{R}$-trees is not motivated solely by the formal generalisation from the theory of groups acting on simplicial trees via the notion of length function but also by the fact that such actions appear in the theory of hyperbolic manifolds and there are strong formal parallels between the theory of group actions on $\mathbb{R}$-trees and the theory of group actions on hyperbolic space which are first noted in [Tits]. Also the theory of group actions on $\mathbb{R}$-trees has been used to give a proof of a theorem of Thurston on geodesic laminations and Teichmüller space.

§ 2.3. Automorphisms of Free Groups

Let F be a free group with basis $\mathcal{X} = (x_1, \ldots, x_n)$. To define an endomorphism $\alpha : F \to F$ it suffices to choose a system $(u_1, \ldots u_n)$ of elements of F and set $\alpha(x_i) = u_i$. The following endomorphisms are all clearly automorphisms since, in each case, it is obvious how to define a second endomorphism whose composite with the original is just the identity.

2.3.1. Definition. (a) For any permutation π of $1, 2, \ldots, n$ and $\varepsilon_i = \pm 1, i = 1, 2, \ldots, n$ let $\varphi : (x_1, \ldots, x_n) \mapsto (x_{\pi(1)}^{\varepsilon_1}, \ldots, x_{\pi(n)}^{\varepsilon_n})$. This is a *permutation automorphism.*

(b) If $\rho : (x_1, \ldots, x_n) \mapsto (x_1x_2, x_2, \ldots, x_n)$ then $\rho' : (x_1, \ldots, x_n) \mapsto (x_1x_2^{-1}, x_2, \ldots, x_n)$ is the inverse of ρ.

The automorphisms in (b) are examples of *Nielsen automorphisms* which are those automorphisms such that for some i, j, with $j \neq i$, and $\varepsilon = \pm 1$

$$x_i \mapsto x_i x_j^{\varepsilon}, \quad x_k \mapsto x_k \ , \ (k \neq i).$$

We shall denote this automorphism by $(x_i + x_j^{\varepsilon}, x_j^{\varepsilon})$ — it should be observed that $(x_i + x_j^{\varepsilon}, x_j^{\varepsilon})$ can be obtained from $(x_1 + x_2, x_2)$ and a permutation automorphism. Often $x_j \mapsto x_i^{\varepsilon} x_i$, $x_k \mapsto x_k$ is also called a Nielsen automorphism.

2.3.2. Theorem. *The automorphism group* Aut F *of a free group* F *of finite rank is generated by the permutation and Nielsen automorphisms.*

Proof. Let $\alpha \in \mathrm{Aut}F$ with $\alpha(x_i) = v_i$. Then $\mathcal{V} = (v_1, \ldots, v_n)$ is a basis of F. By Proposition 2.1.9, some finite sequence of Nielsen transformations will send $\mathcal{V}$ into a Nielsen reduced system $\mathcal{U} = (u_1, \ldots, u_n)$, which is also a basis of F. Now it follows from Proposition 2.1.7 (a) that there is a permutation automorphism ψ such that $\psi : (x_1, \ldots, x_n) \mapsto (u_1, \ldots, u_n)$ so we may suppose that some sequence of Nielsen transformations sends $(v_1, \ldots, v_n)$ to $(x_1, \ldots, x_n)$. Now each application of a Nielsen transformation can be mirrored by application of a suitable automorphism and the result follows by induction. For example if we have

$$(v_1, v_2, \ldots, v_n) \quad \overset{(T2)}{\longrightarrow} \quad (v_1v_2, v_2, \ldots, v_n) \to \ldots \to (x_1, \ldots, x_n)$$

then

$$(x_1, \ldots, x_n) \quad \overset{(x_1+x_2,x_2)}{\longmapsto} \quad (x_1x_2, x_2, \ldots, x_n) \overset{\alpha}{\longmapsto} (v_1v_2, v_2, \ldots, v_n)$$

and the induction hypothesis applies to $\alpha \circ (x_1 + x_2, x_2)$. $\square$

Much more difficult to establish is [Nielsen 1924a].

2.3.3. Theorem. *The automorphism group of a free group of finite rank is finitely presented.*

We shall sketch a proof of this theorem after 2.3.13 below.

If F is free of rank n then there is a natural epimorphism $\eta : F \to F/[F,F]$ and since $[F,F]$ is characteristic there is induced a natural map

$$\tilde{\eta} : \text{Aut } F \to \text{Aut } F/[F,F] = \text{GL}(n, \mathbf{Z}).$$

2.3.4. Theorem [Nielsen 1924b], [Magnus 1934]. *Let* $F = \langle x_1, ..., x_n \mid - \rangle$. *The natural map* $\tilde{\eta} : \text{Aut } F \to \text{GL}(n, \mathbf{Z})$ *is surjective and its kernel* $IA(F)$ *is finitely generated, namely by the automophisms*

$$(x_i|x_j) : \begin{cases} x_i \mapsto x_j^{-1} x_i x_j & \\ & \qquad (\textit{where } j \neq i), \textit{ and} \\ x_k \mapsto x_k & \textit{if } k \neq i, \end{cases}$$

$$(x_i \mid [x_j, x_k]) : \begin{cases} x_i \mapsto x_i [x_j, x_k] & \\ & \qquad (\textit{where } j, k \neq i)\ . \\ x_l \mapsto x_l & \textit{if } l \neq i \end{cases} \qquad \square$$

Defining relations for $IA(F)$ are not known except in the trivial case when F has rank two and thus $IA(F) = \langle\ (x_1 \mid x_2)\ ,\ (x_2 \mid x_1)\ \rangle = \text{Inn } F$. Using this with the presentation of $GL(2, \mathbb{Z})$ in 1.1.4 (h) gives a presentation of Aut F_2. It is conjectured that if F has rank more than two, then $IA(F)$ is not finitely presented.

2.3.5. Example. The braid group B_n defined in Example 1.1.4 (k) has an interesting representation in terms of automorphisms. Recall that the generators of B_n are σ_i, $1 \leq i \leq n-1$ and σ_i exchanges the i-th and $(i+1)$-th strings, see 5.3.13. An easy argument shows that an action of B_n on a free group F with basis $(x_1, \ldots, x_n)$ is given by

$$\begin{aligned} \sigma_i(x_i) &= x_i x_{i+1} x_i^{-1}, \\ \sigma_i(x_{i+1}) &= x_i, \\ \sigma_i(x_k) &= x_k \quad \text{for } k \neq i, i+1. \end{aligned}$$

The image in Aut F of B_n is exactly the subgroup

$$\text{Stab}(x_1 x_2 \cdots x_n) = \{\alpha \in \text{Aut } F\colon\ \alpha(x_1 x_2 \cdots x_n) = x_1 x_2 \cdots x_n\}$$

and an elegant geometric argument (see [Burde 1963]) shows that the representation is faithful.

A significant method in the theory of automorphisms of free groups was introduced by Whitehead in [Whitehead 1936a,1936b].

2.3.6. Theorem. *There is an algorithm to determine of any two* m*-tuples* $(u_1, \ldots, u_m)$ *and* $(v_1, \ldots, v_m)$ *of elements of a free group* F *of finite rank whether or not there exists an automorphism of* F *which carries one* m*-tuple to the other.*

2.3.7. Corollary. *There is an algorithm to determine of an element u of a free group F of finite rank whether or not u is* primitive, *i.e. a member of a basis of F.* □

Whitehead's argument uses a representation of a free group as the fundamental group of a certain 3-manifold; see also [Goldstein-Turner 1984]. We shall follow an algebraic account taken from [Higgins-Lyndon 1974]. We shall, without further explanation, concentrate on the case $m = 1$ — the general case is reduced to this case by a simple trick. It is convenient to work modulo the application of inner automorphisms and thus with conjugacy classes rather than elements of F. We use the following terminology.

2.3.8. Definition

(a) An automorphism α of F is called a *Whitehead automorphism* of F when there is a (fixed) $x \in \mathcal{X} \cup \mathcal{X}^{-1}$ such that $\alpha(s) \in \{s,\ sx,\ x^{-1}s, x^{-1}sx\}$ for any $s \in \mathcal{X}$; here $\alpha(x) = x$. If we define $A = \{s \in \mathcal{X} \cup \mathcal{X}^{-1} \mid \alpha(s) = sx \text{ or } \alpha(s) = x^{-1}sx\ \} \cup \{x\}$ then α is uniquely determined by the pair (A, x) and we shall sometimes use this as a notation for α. It should be observed that there are only finitely many Whitehead automorphisms.

(b) A *cyclic word* $w \in F$ is a cyclically ordered, reduced string of elements of $\mathcal{X} \cup \mathcal{X}^{-1}$. It represents a conjugacy class of elements of F. Clearly, automorphisms of F act on the set of cyclic words. The *length* $|w|$ of the cyclic word w is the number of letters from $\mathcal{X} \cup \mathcal{X}^{-1}$ in w.

(c) Two cyclic words u, v are called *equivalent* if there is an automorphism $\alpha \in \text{Aut } F$ such that $v = \alpha(u)$.

(d) A cyclic word w is called *minimal* if it has minimal length among all cyclic words which are equivalent to w. (Notice that the definition of minimality makes extensive use of the specific set of generators; hence, to be minimal is not a group theoretic property of an element.)

The main result is the following.

2.3.9. Whitehead Theorem. *Let u and v be cyclic words of F and $\alpha \in$ Aut F be such that $\alpha(u) = v$. Then α can be expressed as a product $\alpha = \rho_1\rho_2\cdots\rho_n$ of permutation and Whitehead automorphisms such that for some p, q, $1 \le p \le q \le n$,*

(a) $$|\rho_{i-1}\cdots\rho_1(u)| > |\rho_i\cdots\rho_1(u)|\ , \qquad 1 \le i \le p\ ,$$

(b) $$|\rho_{j-1}\cdots\rho_1(u)| = |\rho_j\cdots\rho_1(u)|\ , \qquad p+1 \le j \le q\ ,$$

(c) $$|\rho_{k-1}\cdots\rho_1(u)| < |\rho_k\cdots\rho_1(u)|\ , \qquad q+1 \le k \le n\ . \quad \square$$

The inequalities (a), (b) and (c) are illustrated by the following diagram in which length is plotted upwards.

The algorithm of 2.3.6 is derived as follows. If the elements u and v are not minimal, then, by 2.3.9, a bounded sequence of applications of the finitely

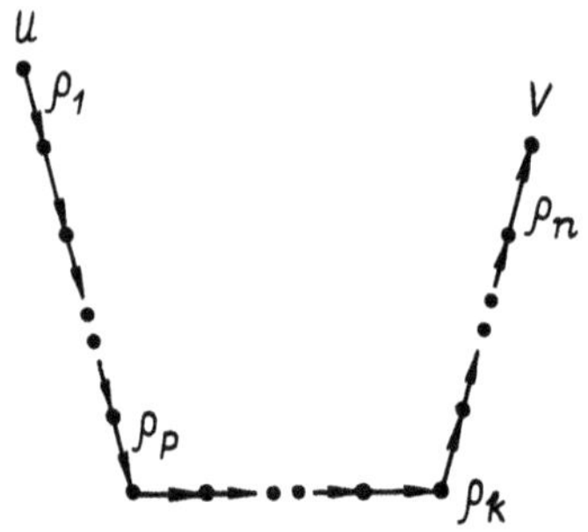

Fig. 2.3.1

many Whitehead automorphisms will replace u and v by minimal words u_0 and v_0. If u_0 and v_0 are equivalent under Aut F, then they must have the same length and, by 2.3.9, u_0 can be transformed into v_0 by a sequence of applications of permutation and Whitehead automorphisms in such a way that the lengths of the intermediate words obtained are the same as the length of u_0 and v_0. This means that the number of automorphisms in the sequence can be bounded in terms of the length of u_0 and the result follows.

We illustrate the theorem with examples.

2.3.10. Example. Let $\mathcal{X} = \{s,t\}$. Then a non-trivial Whitehead automorphism must fix one of the generators, say s, and map t to $ts^{\pm 1}$, $s^{\mp 1}t$ or $s^{\mp 1}ts^{\pm 1}$. However, the last pair of automorphisms do not alter a cyclic word since they are inner and the first two types are equal modulo inner automorphisms. Hence we can restrict ourselves to the following Whitehead automorphisms which are in fact just Nielsen automorphisms:

$$(t+s^{\varepsilon}, s^{\varepsilon}): \; s \mapsto s, \; t \mapsto ts^{\varepsilon}; \qquad (s+t^{\varepsilon}, t^{\varepsilon}): \; s \mapsto st^{\varepsilon}, \; t \mapsto t \quad \text{where } \varepsilon = \pm 1.$$

Let $u = s^3t^5$ and $v = s^3t^3s^3t^2$. It is easily checked that both words are minimal; if we apply, e.g., $(s+t^{-1}, t^{-1})$ to u we obtain $(st^{-1})^3t^5 = st^{-1}st^{-1}st^4$ and the length has increased by 1 (this is in fact the worst case). Since the words have different lengths it follows that u and $v^{\pm 1}$ are not equivalent. A similar argument shows that any word s^at^b, $a, b \geq 2$, is minimal.

Now consider $[s,t] = sts^{-1}t^{-1}$. Simple computations show that $[s,t]$ is fixed by all four of the automorphisms displayed above. Hence all Whitehead automorphisms fix $[s,t]$, regarded as a cyclic word, and, in particular, do not decrease its length. A permutation automorphism maps $[s,t]$ either to a conjugate of itself or to a conjugate of $[s,t]^{-1}$. Thus we have proved

2.3.11. Proposition. *The commutator* $[s,t] = sts^{-1}t^{-1}$ *is a minimal word in* $F(s,t)$. *Moreover any element equivalent to* $[s,t]$ *is conjugate to* $[s,t]$ *or* $[s,t]^{-1}$. □

2.3.12. Proposition. *Given $v, w \in F(s,t)$ with $[v,w] = [s,t]$, then v, w form a free basis of $F(s,t)$.*

Proof. Replacing v or w by $v^{\pm 1} w^{\pm 1}$ preserves the set consisting of conjugacy classes of $[v,w]^{\pm 1}$. Hence we may assume that (v,w) is Nielsen reduced and then it follows from Proposition 2.1.7 (a) that $\{v,w\} \subset \{s, s^{-1}, t, t^{-1}\}$. □

This result was first obtained by Nielsen in [Nielsen 1918] where he introduced the method that now bears his name.

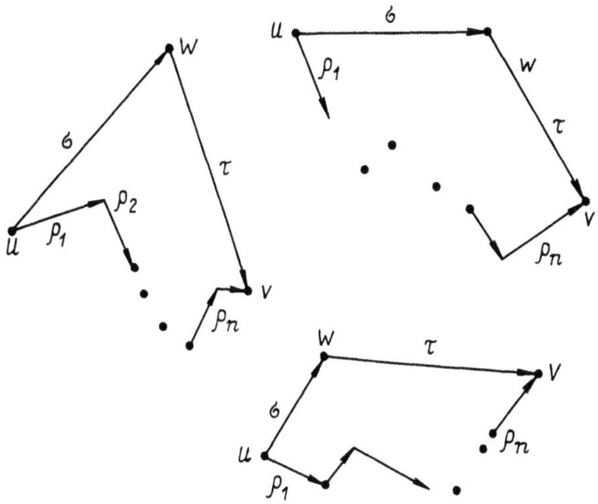

Fig. 2.3.2

We now discuss the proof of the Whitehead Theorem. The main lemma of the Higgins-Lyndon version of the proof is the following.

2.3.13. Peak-Reduction Lemma. *Let σ and τ be permutation or Whitehead automorphisms, and let u, w and v be cyclic words such that*

(1) $\sigma(u) = w \ , \quad \tau(w) = v \ ,$

(2) $|u| \leq |w| \geq |v| \ ,$

(3) $|u| < |w|$ *or* $|w| > |v| \ .$

Then $\tau\sigma$ can be expressed as a product $\tau\sigma = \rho_n \ldots \rho_2 \rho_1$ of permutation and Whitehead automorphisms so that $|\rho_i \ldots \rho_1(u)| < |w|$ for $1 \leq i \leq n-1$. □

The Peak-Reduction Lemma is illustrated by the above diagram, with length again plotted vertically. The proof is based on an efficient calculation of the number of letters which vanish or have to be added when a Whitehead automorphism is applied to a cyclic word.

Close examination of the detailed argument shows that at most four Whitehead or permutation automorphisms appear in the lower part of the Peak-Reduction diagram and this means that a system of defining relations for the Whitehead and permutation automorphisms (which generate the whole automorphism group of the free group) can be chosen among the relations of length at most six. This is the basis of the proof given in [McCool 1974] of Theorem 2.3.2 — the automorphism group of a finitely generated free group is finitely presentable. The presentation in [McCool 1974] is not the same as that found in [Nielsen 1924a] but in [McCool 1975a] the latter is derived from the presentation in [McCool 1974].

The Peak Reduction Lemma, together with the observation of McCool, also enables one to prove that other automorphism groups are finitely presentable. Let $w \in F = F(\mathcal{X})$ be some specified element (or a finite system of elements or cyclic words). We want to find the *stabiliser* $\mathrm{Stab}(w) = \{\alpha \in \mathrm{Aut}\ F : \alpha(w) = w\}$. Replacing w by its image under an automorphism simply conjugates the stabiliser and so we may assume that w is a minimal word. A presentation for the stabiliser is obtained by constructing a 2-complex as follows. Take an vertex and denote it by w. Apply a Whitehead or permutation automorphism α to w. If the cyclic word $\alpha(w)$ is minimal and different from w add a vertex with name $\alpha(w)$ and an edge running from w to $\alpha(w)$ with label α. If $\alpha(w) = w$, again an edge with label α is added. This procedure is applied for all Whitehead and permutation automorphisms at w and iterated at new vertices. Since new vertices are always minimal the procedure halts after a finite number of steps. The generators of the fundamental group of the graph thus obtained, written as words in their labels, give a system of generators for $\mathrm{Stab}(w)$. The method of [McCool 1975b] gives a system of short relations where only those Whitehead or permutation automorphisms appear which are labels of edges in the graph. If we add for each of these a disc with the corresponding boundary, we obtain a finite 2−complex whose fundamental group is $\mathrm{Stab}(w)$. Thus we obtain:

2.3.14. Theorem. *The stabilizer of an element of a finitely generated free group F (or of a finite system of elements or cyclic words) is finitely presented. Moreover, there is a procedure to determine a finite presentation of the stabilizer.* □

As an application we determine the stabilizer of the cyclic word $s^a t^b$ with $1 < a < b$. As we have seen in 2.3.10 this word is minimal, and every Whitehead automorphism which is not an inner automorphism increases the length. Hence, the only automorphisms which preserve length are permutation and inner automorphisms and the equivalent cyclic words of minimal length are $s^{\pm a} t^{\pm b}$, $s^{\pm b} t^{\pm a}$. This means that in the graph described above only permutation automorphisms and inner automorphisms appear as labels and the stabiliser is just the group Inn F of inner automorphisms of F. If, instead, we consider the "linear" word w, it follows that the stabilizer consists of those inner automorphisms with a power of $s^a t^b$ as conjugating factor [Schreier 1924].

2.3.15. Example . It can be proved (see Theorem 3.4.5) that every automorphism of

$$\pi_1(S_g) = \langle t_1, u_1, \ldots, t_g, u_g \mid \prod_{j=1}^{g} [t_j, u_j] \rangle,$$

see 1.2.6 (c), is induced by an automorphism of the free group

$$F_{2g} = \langle T_1, U_1, \ldots, T_g, U_g \mid - \rangle$$

which either preserves the conjugacy class of the cyclic word $\Pi_* = \prod_{j=1}^{g} [T_j, U_j]$ or maps it into its inverse. Hence, the outer automorphism group,

$$\mathrm{Out}(\pi_1(S_g)) = \mathrm{Aut}(\pi_1(S_g))/\mathrm{Inn}(\pi_1(S_g)),$$

is isomorphic to $\mathrm{Stab}(\Pi_*)/\mathrm{Inn}\, F$. Since the latter group is finitely presentable the group $\mathrm{Out}(\pi_1(S_g))$ is finitely presentable. This group has an important topological interpretation — see Theorem 3.4.19.

2.3.16. Equivalence of Subgroups under Automorphism. A striking generalisation of Whitehead's argument was introduced in [Gersten 1984]. Let H be a finitely generated subgroup of the free group $F = F(\mathcal{X})$ of finite rank. Let X be the coset graph of H — that is the covering of the bouquet of loops, whose fundamental group is F, corresponding to the subgroup H (see 1.3.4). The *core* X_0 of X is the smallest subgraph of X containing the basepoint whose fundamental group has the same rank as $\pi_1(X)$ — in effect X_0 is the smallest subgraph containing all the reduced generating paths of $\pi_1(X)$. Define the *complexity* of H to be the number of vertices in X_0. The complexity of H depends only on the conjugacy class of H and, if w is a cyclically reduced word of F, then the core graph of the subgroup $H = \langle\, w \,\rangle$ is just a circuit of length $|w|$.

The analysis of how complexity changes under application of Whitehead automorphisms can be generalised from the case of words and the Peak Reduction procedure carried through so that analogues of the preceding results all hold. We state one instance.

2.3.17. Theorem. *Let F be free of finite rank. Then there is an algorithm to determine of any two subgroups H and K whether or not there is an automorphism of F carrying the conjugacy class of H into the conjugacy class of K.* □

Much of the recent theory of automorphisms of free groups is inspired by attempts to simulate the theory of automorphisms of surface groups. A theorem which has roots in the theory of surfaces is that of fixed points of automorphisms of free groups.

Given F free and $\alpha \in \mathrm{Aut}\, F$, the *fixed point subgroup* of α is the group

$$\mathrm{Fix}(\alpha) = \{u \in F \mid \alpha(u) = u\}.$$

Calculation with simple examples suggests that when F has finite rank, then $\mathrm{Fix}(\alpha)$ also has finite rank and further evidence to support this came in [Jaco-Shalen 1977] where it is proved that if α is a *geometric* automorphism, i.e. is induced by a homeomorphism of a surface with boundary, then $\mathrm{rank}(\mathrm{Fix}(\alpha)) \leq \mathrm{rank}(F)$. Yet further support came in [Dyer-Scott 1975] where the characterisation of finitely generated virtually free groups, i.e. groups with a free subgroup of finite index, given in 6.2.12 was exploited to prove that if α is a periodic automorphism of F, then $\mathrm{Fix}(\alpha)$ is a free factor of F. The full result was first established in [Gersten 1983,1987].

2.3.18. Theorem. *Let F be a free group of finite rank. Then the fixed point subgroup of any automorphism of F is also of finite rank.*

The arguments in [Gersten 1987], like many which break difficult new ground, are complicated and subsequently several alternative proofs and generalisations have been given. Perhaps the simplest proof is to be found in [Goldstein-Turner 1986].

Proof of 2.3.18. Let $\alpha \in \mathrm{Aut}\ F$ and let $H = \mathrm{Fix}(\alpha)$. Let X be the coset graph of H – then by the theory of coverings $H \cong \pi_1(X)$. We assign *names* to the vertices of X as follows. Let p be a path from the basepoint corresponding to the coset H to a given vertex; then in traversing p one reads off a word w in the basis $\mathcal{X}$ of F. The name assigned to the endpoint of p is the reduced form of $\alpha(w)^{-1}w$. Since in traversing a closed path starting at the basepoint one reads off an element of H, the name is independent of the path chosen. To show that $\pi_1(X)$ is finitely generated one assigns a *direction* to each geometric edge of X. For this consider an edge $u \overset{x}{—} v$ of X with label $x \in \mathcal{X} \cup \mathcal{X}^{-1}$ where, say, $u = Hg$ and $v = Hgx$ for $g \in F$. Now if u has name z then it follows that v has name $\alpha(x)^{-1}zx$. Then the geometric edge (formally consisting of the inverse pair of edges which have u and v as endpoints) is directed from u to v if the displayed generator x is cancelled when $\alpha(x)^{-1}zx$ is reduced to normal form and otherwise from v to u. If the word z is long in comparison with $\alpha(x)$ then the edge will be directed towards u. Since there are only finitely many words of the form $\alpha(x)$, $x \in \mathcal{X}$, this shows that with finitely many exceptions, there is, at each vertex, at most one outwardly directed edge. Now a graph in which the geometric edges can be directed in such a way that at *every* vertex there is at most one outwardly directed edge contains at most one circuit. It follows that if finitely many edges are deleted from X then the result is a union of trees. Hence $\pi_1(X)$ is finitely generated. □

All the proofs so far published of Theorem 2.3.18 establish finiteness but do not give a bound on the rank of $\mathrm{Fix}(\alpha)$ which is independent of α. At the time of writing a structure theorem for automorphisms of a free group F of finite rank has been announced by M. Bestvina and M. Handel which has as a consequence the strong finiteness result that rank $\mathrm{Fix}(\alpha) \leq \mathrm{rank} F$.

The Whitehead method can be generalized to free products of groups to obtain a peak-reduction lemma as well as a result on the stabilizer of an element; see [Collins-Zieschang 1984,1987].

§ 2.4. One–Relator Groups

The classical example of a one–relator group is the fundamental group of a closed orientable surface $\pi_1(S_g) = \langle\, a_1, b_1, \ldots, a_g, b_g \mid [a_1, b_1][a_2, b_2] \ldots [a_g, b_g]\,\rangle$. The importance of this group in topology is unquestioned and Dehn's solutions of the word and conjugacy problems can be regarded as the coming of age of combinatorial group theory. It was also Dehn who suggested that significant results generalising those for surface groups could be obtained for arbitrary groups given by a single defining relator. The theory begins with two classic results of Magnus [Magnus 1930, 1931].

2.4.1. Theorem (Freiheitssatz). *Let $G = \langle \mathcal{X} \mid R \rangle$ be a group presentation with a single defining relator R, with R cyclically reduced. Let $\mathcal{Y}$ be a subset of $\mathcal{X}$ which omits a generator appearing in R. Then $\mathcal{Y}$ is a basis for a free subgroup of G.*

To illustrate the argument let $G = \langle a_1, b_1, a_2, b_2 \mid [a_1, b_1][a_2, b_2] \rangle$ and $\mathcal{Y} = \{b_1, a_2, b_2\}$. Then one can regard G as an HNN-extension $\langle \mathcal{Y}, a_1 \mid a_1 b_1 a_1^{-1} = [b_2, a_2] b_1 \rangle$ with base group $\langle \mathcal{Y} \mid - \rangle$. By the normal form theorem, the natural map embeds the free group $\langle \mathcal{Y} \mid - \rangle$ in G. The core of Magnus' argument - in the form first given in [Moldavanskij 1967] is contained in the above illustration. The second theorem of Magnus is:

2.4.2. Theorem. *Any group given by a single defining relation has solvable word problem.*

As will be apparent from our discussion, Magnus actually proves a slightly stronger result. Both theorems are proved by induction on the length of the relator in a way which we now illustrate (see [Lyndon-Schupp 1977] for a clear and detailed account). We take $G = \langle \mathcal{X} \mid R \rangle$, where $\mathcal{X} = \{a, b, c\}$ and $R \equiv b^2 a b^2 a c^{-2} b^3 a^{-3} c^2 a c^2$. This relator has zero exponent sum in the generator a and this is the crucial case in the argument. Using Tietze transformations, one can introduce new generators $b_0 = b$, $b_1 = aba^{-1}$, $b_2 = a^2 b a^{-2}$, $c_j = a^j c a^{-j}$ $(j \in \mathbb{Z})$. After further Tietze transformations, in particular, expressing the relator R in terms of these new generators, one obtains a new presentation for G:

$$\langle\, b_0, b_1, b_2, c_j, a \mid b_0^2 b_1^2 c_2^{-2} b_2^3 c_{-1}^2 c_0^2 = 1,\ a b_i a^{-1} = b_{i+1},\ a c_j a^{-1} = c_{j+1}\,\rangle$$

where $i = 0, 1$ and $j \in \mathbb{Z}$. By the induction hypothesis the subgroups $\langle b_0, b_1, c_j \ :\ j \in \mathbb{Z} \rangle$ and $\langle b_1, b_2, c_j \ :\ j \in \mathbb{Z} \rangle$ of the one–relator group

$G_0 = \langle b_0, b_1, b_2, c_j : j \in \mathbb{Z} \mid b_0^2 b_1^2 c_2^{-2} b_2^3 c_{-1}^2 c_0^2 \rangle$ are free on the displayed generators and so G is given as an HNN-extension with G_0 as base group and a as stable letter. It follows immediately that $\{b, c\}$ $(= \{b_0, c_0\})$ is a basis for a free subgroup of G. To see that $\{a, c\}$ is also a basis for a free subgroup, note firstly that any reduced word W giving a relation over $\{a, c\}$ must have zero exponent sum in a since it is a consequence of the original relator R. This means that W can be expressed as a reduced word in terms of the generators $\{c_j : j \in \mathbb{Z}\}$. Again by the induction hypothesis, no such relation W can occur. Finally to show that $\{a, b\}$ is a free basis one must exchange the roles of b and c in the process of Tietze transformations.

The case when no generator has zero exponent sum in the relator is reduced to the previous case by a trick. Suppose that $G = \langle a, b, c \mid R(a, b, c) \rangle$ and we want to show that $\{a, c\}$ is a free basis. Let α be the exponent sum of a in R and let β be the exponent sum of b in R. Let $G^* = \langle x, y, c \mid R(x^\beta, yx^{-\alpha}, c) \rangle$. If R^* is the result of cyclically reducing $R(x^\beta, yx^{-\alpha}, c)$ then $G^* = \langle x, y, c, \mid R^* \rangle$ and R^* has zero exponent sum in x so that the method of the first case can be used.

The same inductive method also yields a solution to the word problem for a one-relator group. Recall from 2.2.9 that in giving a method to solve the word problem for an HNN–extension $G = \langle G_0, t \mid tAt^{-1} = B \rangle$ two main conditions were assumed, namely

(a) G_0 has solvable word problem,

(b) the problem of membership in the subgroups A and B is solvable.

For one-relator groups this means that the inductive asssumption must be that for a subset $\mathcal{Y}$ which omits an element of the generating set $\mathcal{X}$ that occurs in the relator R, the problem of membership in the subgroup $\langle\ \mathcal{Y}\ \rangle$ must be solvable. Notice that with $\mathcal{Y}$ empty this is asking that the word problem be solvable. In our example

$$G = \langle a, b, c \mid b^2 ab^2 ac^{-2} b^3 a^{-3} c^2 ac^2 \rangle$$

we obtained

$$G = \langle\ G_0, a \mid ab_i a^{-1} = b_{i+1},\ ac_j a^{-1} = c_{j+1} : i = 0, 1,\ j \in \mathbb{Z}\ \rangle$$

with $G_0 = \langle\ b_0, b_1, b_2, c_j : j \in \mathbb{Z} \mid b_0^2 b_1^2 c_2^{-2} b_2^3 c_{-1}^2 c_0^2\ \rangle$. By the inductive hypothesis the problems of membership in the subgroups $\langle b_0, b_1, c_j\ :\ j \in \mathbb{Z} \rangle$ and $\langle b_1, b_2, c_j\ :\ j \in \mathbb{Z} \rangle$ are solvable and hence the word problem for G is solvable. Of course more must be squeezed out to continue the induction but the normal form for HNN-extensions is sufficiently powerful to obtain the desired information. □

The same general method and the torsion theorem for HNN-extensions provide a satisfactory account of torsion in one–relator groups.

2.4.3. Theorem. *Let $G = \langle \mathcal{X} \mid R \rangle$ where R is cyclically reduced.*

(a) *If R is not a proper power in the free group $F(\mathcal{X})$, then G is torsion-free.*

(b) *If* $R \equiv S^m$, $m > 1$ *and* S *is not a proper power in* $F(\mathcal{X})$, *then* (i) S *is (a representative of) an element of order* m *in* G ; (ii) *any element of finite order in* G *is conjugate to a power of* S. *Moreover any finite subgroup of* G *is conjugate to a subgroup of the cyclic group* $\langle S \rangle$. □

In our earlier discussion we have ignored the initial steps in the induction, and in particular the situation when R involves only one generator. This case, in practice, always requires a separate, but easy argument. If for instance $G = \langle a, b, c, | \ a^m = 1 \rangle$, where $m \geq 1$ then Theorem 2.4.3 follows from the classification of torsion in free products. The general case is reduced, inductively, to examples such as this via the observation that in the case when R has exponent sum zero so that G is expressible as an HNN-extension over a one–relator base group G_0, then the relator R of G is an m-th power if and only if the relator R_0 of G_0 is an m-th power.

One-relator groups with torsion are in general somewhat easier to deal with than those without torsion. The next theorem, due to Newman [Newman 1968] and Gurevic [Gurevic 1973] indicates why this is so.

2.4.4. Theorem. *Let* $G = \langle \mathcal{X} \mid R \rangle$ *where* $R \equiv S^m$, *with* $m > 1$, *and* S *is not a proper power. Let* U *and* V *be reduced words over* $\mathcal{X}$ *which represent the same element of* G *and suppose that* V *omits a letter that occurs in* U. *Then some cyclic permutation of* U *contains two disjoint subwords, each of length at least* $\frac{(m-1)}{m} \cdot |R|$, *which are also subwords of* R *or* R^{-1}. □

2.4.5. Corollary. *The word problem for a one-relator group with torsion is solvable by Dehn's algorithm, see* 4.1.1. □

The powerful Theorem 2.4.4 leads to (see [Newman 1968]):

2.4.6. Theorem. *The conjugacy problem for a one-relator group with torsion is solvable.* □

Until recently little was known in general for the conjugacy problem for one–relator groups without torsion. However it now appears that geometric methods (see 4.1.11) may lead to a solution. The idea of applying geometric methods to one–relator groups was introduced by Lyndon who used cancellation diagrams to give an alternative proof of the Freiheitssatz.

Already above we have dealt with finite subgroups of one–relator groups. We turn now to other questions about subgroups. Firstly we look at commutativity. Since they are essentially built from free groups by forming HNN-extensions, it is not surprising that one-relator groups exhibit little commutativity (see 2.1.5). The exact limits are given by the next two results ([Moldavanskij 1967], [Karrass-Solitar 1971], [Murasugi 1967]) and are proved by applying the subgroup theorem for HNN-extensions.

2.4.7. Theorem. *Any soluble subgroup of a one-relator group* G *is either locally cyclic or metabelian of the form* $\langle a, b \mid aba^{-1} = b^m \rangle$. *Furthermore, if* G

has torsion, then any soluble subgroup is either cyclic or the infinite dihedral group $\langle a, b \mid aba^{-1} = b^{-1} \rangle$. □

(A group is *soluble* if it has a finite series of normal subgroups, descending from the whole group to the trivial subgroup, such that the successive quotients are abelian. A group is *locally cyclic* if every finitely generated subgroup is cyclic and is *metabelian* if its commutator subgroup is abelian.)

2.4.8. Theorem. *Let* $G = \langle \mathcal{X} \mid R \rangle$.
(a) *If* $|\mathcal{X}| \geq 3$, *then* G *has trivial centre.*
(b) *If* $|\mathcal{X}| = 2$, *and* G *is not abelian then the centre of* G *is trivial or infinite cyclic.* □

We indicate below an elementary argument that specifies in terms of the relator when $G = \langle \mathcal{X} \mid R \rangle$ is abelian. An algorithm is also known (see [Baumslag-Taylor 1968]) for determining when a one-relator group has non-trivial centre but in this case no simple characterisation in terms of the relator has been obtained. The difficulty seems to arise from examples such as $G = \langle a, b \mid a^2b^{-3}(a^4b^{-3})^2 = 1 \rangle$. The relation implies that both a^2 and b^3 are powers of a^4b^{-3} and hence that $a^2b^3 = b^3a^2$ whence the relation yields $a^{10} = b^9$ and thus a^{10} lies in the centre of G.

A theorem about subgroups whose implications are explored in 4.2 is the following result of [Brodskij 1980, 1984].

2.4.9. Theorem. *Any finitely generated subgroup of a torsion-free one-relator group has the infinite cyclic group as a homomorphic image.* □

An interesting open question is whether every finitely generated subgroup of a one–relator group must be finitely presented.

We now move to a discussion of the isomorphism problem for one-relator groups — or, more precisely, one-relator presentations. No solution to the general problem of deciding when two one-relator presentations define the same group is known. However the more restricted problem of deciding when an arbitrary one–relator presentation defines the group given by some specific one–relator presentation has been solved in a number of cases.

We note first the following simple observation.

2.4.10. Lemma. *Let* $G = \langle \mathcal{X} \mid R \rangle$, *let* $\alpha \in \operatorname{Aut} F(\mathcal{X})$ *and let* $R^* = \alpha(R)$. *Then the group* $G^* = \langle \mathcal{X} \mid R^* \rangle$ *is isomorphic to* G. □

For a while it seemed plausible that the converse to this might hold, namely that if two one-relator presentations defined isomorphic groups, then there would be an automorphism of the free group carrying one defining relator to the other, see [Magnus-Karrass-Solitar 1966, p. 401]. Some evidence for this was provided by yet another result of Magnus, the proof of which also relies on the standard inductive procedure.

2.4.11. Theorem. *If the elements R and R^* of the free group F have the same normal closure then R^* is a conjugate of $R^{\pm 1}$.* □

However counterexamples to the converse of Lemma 2.4.10 were given in [Zieschang 1970], [McCool-Pietrowski 1971].

2.4.12. Example. Let $F = F(s,t)$ be free with basis (s,t) and let $u = s^3t^5$ and $v = s^3t^3s^3t^2$. By 2.3.10, u and $v^{\pm 1}$ are not equivalent under Aut F. Now consider the one-relator group $G = \langle s,t \mid s^3t^5\rangle$. We introduce a new generator $z = t^2$ and perform the following Tietze transformations which give presentations of the same group G:

$$\begin{aligned} G &= \langle s,t,z \mid zt^{-2},\ s^3t^5\rangle = \langle s,t,z \mid t^2z^{-1},\ s^3t^6t^{-1}\rangle \\ &= \langle s,t,z \mid t^2z^{-1},\ s^3z^3t^{-1}\rangle = \langle s,z \mid (s^3z^3)^2z^{-1}\rangle\ . \end{aligned}$$

This proves that the groups $\langle s,t \mid s^3t^5\rangle$ and $\langle s,t \mid s^3t^3s^3t^2\rangle$ are isomorphic. As noted above, the relator in the first presentation is not equivalent to the relator in the second presentation or its inverse. The above counterexample can easily be generalized to the groups $\langle s,t \mid s^p = t^q\rangle$, $p,q \geq 2$ of 2.2.28 (c) and all non-equivalent one-relator presentations of these groups can be determined, see [Collins 1978]. The example 2.2.28 (d) of [Brunner 1976] gives a group with infinitely many non-Nielsen-equivalent one-relator presentations. (See also 5.3.10.)

Nonetheless Theorem 2.4.11 does have useful applications.

2.4.13. Theorem. *Let $G = \langle \mathcal{X} \mid R\rangle$. Then G is a free group if and only if R is a primitive element in the free group $F(\mathcal{X})$ (or is the empty word). Hence, by 2.3.7, there is an algorithm to decide if G is free.* □

The fundamental groups of surfaces can be picked out among one-relator presentations by the following result of [Zieschang 1966].

2.4.14. Proposition. *Let $G_0 = \langle a_1, b_1, \ldots, a_g, b_g \mid [a_1,b_1]\ldots[a_g,b_g]\rangle$. Then $G = \langle \mathcal{X} \mid R\rangle$ is isomorphic to G_0 if and only if $|\mathcal{X}| = 2g$ and there is an isomorphism of the two free groups involved carrying $[a_1,b_1]\ldots[a_g,b_g]$ to $R^{\pm 1}$. The analogous result holds for the non-orientable surface group $G_0 = \langle\ a_1,\ldots,a_g \mid a_1^2\cdots a_g^2\ \rangle$.* □

One further positive result is due to Pride [Pride 1977].

2.4.15. Proposition. *Let $G = \langle\ a,b \mid R\ \rangle$ where $R \equiv S^m$, with $m \geq 2$, and S is not a primitive. Then for any pair (x,y) of generators of G there exist words U and V of $F(a,b)$ representing x and y and an automorphism of $F(a,b)$ carrying (U,V) to (a,b).* □

2.4.16. Corollary. *The isomorphism problem is solvable for the class consisting of all presentations of the form $\langle\ a,b \mid R\ \rangle$ with $R \equiv S^m$ where $m \geq 2$.* □

A by-product of 2.4.15 is the fact that two–generator one–relator groups with torsion are hopfian, i.e. every surjective endomorphism is an automorphism. This confirms a special case of the conjecture of G. Baumslag that all one-relator groups with torsion are hopfian — or even residually finite, i.e. the intersection of all subgroups of finite index is the trivial subgroup. Further confirmation of how torsion leads to significant results is the following application in [Fine-Howie-Rosenberger 1988] of a powerful result of [Culler-Morgan 1987].

2.4.17. Theorem. *Any one relator group with torsion which has at least three generators can be decomposed, in a non–trivial way, as an amalgamated free product.* □

2.4.18. Example [Rosenberger 1977]. An interesting decomposition of a two-generator one-relator group with torsion is obtained as follows. Let

$$H = \langle s_1, s_2, s_3, s_4 \mid s_1^2 = s_2^2 = s_3^2 = s_4^3 = s_1s_2s_3s_4 = 1 \rangle.$$

Now H is a non-trivial amalgamated free product $(\langle s_1, s_2\rangle * \langle s_3, s_4\rangle : \langle s_1s_2\rangle = \langle (s_3s_4)^{-1}\rangle)$. On the other hand, see 3.2.18, H is generated by $\{x_1, x_2\}$ where $x_1 = s_1s_2$ and $x_2 = s_3s_1$. Further the commutator $[x_1, x_2] = (s_1s_2s_3)^2$ and so $[x_1, x_2]^3 = 1$. Hence there is an epimorphism from the group $G = \langle a, b \mid [a, b]^3 = 1\rangle$ to H given by $a \mapsto x_1,\ b \mapsto x_2$. However one can lift back the decomposition of H to give a non-trivial decomposition of G as an amalgamated free product. Similar arguments apply for $G = \langle a, b \mid [a, b]^{2k+1} = 1\rangle$ whence it follows that if m is not a power of two, then $G = \langle a, b \mid [a, b]^m = 1\rangle$ also has a non-trivial decomposition. It is an open question whether or not $G = \langle a, b \mid [a, b]^{2^n} = 1\rangle$ has such a decomposition — although since the relator has exponent sum 0 on both a and b, G is an HNN-extension in various ways.

Our final topic in the theory of one–relator groups is cohomology (see 6.1 for definitions). Following the introduction of cohomology for groups in the 1940's and the elementary calculation of the cohomology groups of free groups and finite cyclic groups, the successful determination by Lyndon of cohomology groups for one–relator groups was among the earliest results in the area. The principal conclusions are consequences of the following.

Let $G = \langle \mathcal{X} \mid R\rangle$ and let N be the normal closure of R in $F(\mathcal{X})$. Then G acts on the abelianisation $\bar{N} = N/[N, N]$ by *lifting and conjugating*, i.e. if $g \in G$ is represented by $U \in F(\mathcal{X})$ then, for any $P \in N$, $g.P[N, N] = UPU^{-1}[N, N]$ is a well-defined action of G on $\bar{N}$. We write $\mathbb{Z}G$ for the integral group ring of G (see 4.2).

2.4.19. Proposition. *Let* $G = \langle \mathcal{X} \mid R\rangle$.

(a) *If G is torsion-free, then* $\bar{N} \cong \mathbb{Z}G$.

(b) *If G is not torsion-free and $R \equiv S^m$ with $m \geq 2$ and S not a proper power, then there is a short exact sequence of $\mathbb{Z}G$-modules* $0 \to \mathbb{Z}G(s-1) \to \mathbb{Z}G \to \bar{N} \to 0$, *where s denotes the element of G represented by S.* □

2.4.20. Corollary. *Let* $G = \langle \mathcal{X} \mid R \rangle$.

(a) *If* G *is torsion-free then* $\operatorname{cd}(G) \leq 2$ *and by Theorem* 6.2.16 $\operatorname{cd}(G) = 2$ *unless* G *is free. (For the definition of the cohomological dimension see* 6.1.18.*)*

(b) *If* $R \equiv S^m$ *with* $m \geq 2$ *and* S *not a proper power, then for* $n \geq 3$ *and any* $\mathbb{Z}G$*-module* M, $H^n(G, M) \cong H^n(W, M)$ *where* W *is the cyclic subgroup of* G *generated by the element* s *which is represented by* S. □

Lyndon's argument is still based on the Magnus method for analysing one–relator groups, although the details do become much more intricate. A somewhat simpler account may be found in [Chiswell-Collins-Huebschmann 1981]. A strong form of the above proposition was found by Cohen and Lyndon [Cohen-Lyndon 1963]:

2.4.21. Theorem. *Let* $G = \langle \mathcal{X} \mid \mathcal{R} \rangle$ *with* $R \equiv S^m$ *where* S *is not a proper power and let* $\mathcal{T}$ *be a transversal for the subgroup* $\langle S \rangle N$ *where* N *is the normal closure of* R *in* $F(\mathcal{X})$. *Then the set* $\{URU^{-1} : U \in \mathcal{T}\}$ *is a free basis for* N. □

Chapter 3
Surfaces and Planar Discontinuous Groups

§3.1. Surfaces

Surfaces appear in different fields of mathematics: in differential geometry, in complex analysis as Riemann surfaces, in algebraic geometry and topology. Most topological problems on surfaces can be treated within combinatorial group theory thanks to some basic results on surfaces. In the theory of Riemann surfaces discontinuous groups appear; again they have a rich combinatorial structure which we describe here.

Topologically, a surface is defined to be a 2-dimensional manifold [Novikov 1986, pp. 37-39]. Now, by [Rado 1924], every surface can be triangulated and thus a 2-dimensional cell complex can be realised on the surface. The kind of 2-complex realisable on a surface can be characterised by the properties given in Definition 3.1.1 below and initially we shall work exclusively in a combinatorial framework with such *surface complexes*. Later, however, we shall also discuss such notions as curves and isotopy on surfaces and for this it is convenient to assume that surface complexes have actually been realised on a topological surface.

3.1.1. Let C be a connected 2-dimensional complex with the following properties:

(i) every vertex (0-cell) is in the boundary of at least one edge;

(ii) every oriented edge (1-cell) appears in the boundaries of the oriented faces (2-cells) at least once and at most twice.

A *boundary edge* appears only once in the boundaries of the 2-cells and the set of all boundary edges, with their endpoints, is called *the boundary of* C, denoted by ∂C.

The oriented edge σ' is called a *neighbour* of the oriented edge σ if the path $\sigma^{-1}\sigma'$ is a subpath of the boundary of a face. A boundary edge has only one neighbour and this characterizes boundary edges, except in the case when the edges σ, σ' have the same initial vertex which is of degree 2.

A sequence $\sigma_1, \ldots, \sigma_k$ of different oriented edges with a common initial vertex is called a *star* if $\sigma_j, 1 < j < k$ has the edges σ_{j-1} and σ_{j+1} as neighbours and $\sigma_{j-1} \neq \sigma_{j+1}$. Thus at most σ_1 and σ_k can be boundary edges. The star is called *closed* if, for $k > 2$, σ_1 and σ_k are neighbours and, for $k = 2$ or 1, $\sigma_1^{-1}\sigma_2$ or $\sigma_1^{-1}\sigma_1$, respectively, appears twice in the boundaries of faces. Now we can formulate our last assumption:

(iii) Any two edges σ, τ with common initial vertex v can be connected by a star $\sigma = \sigma_1, \sigma_2, \ldots, \sigma_k = \tau$ around v.

Clearly, if a complex C fulfils the conditions (i), (ii), (iii) then every complex obtained by cutting and pasting from C also fulfils these conditions. Having a closed star or being a boundary edge is preserved when passing to a homeomorphic 2-complex.

3.1.2. Definition

(a) A *surface complex* is a 2-complex C which satisfies (i), (ii) and (iii). We will say that two surface complexes define the *same surface type* or, briefly, *surface* S if they are homeomorphic, see 1.2.15.

(b) A surface complex is called *compact* if it consists of finitely many cells. A surface complex C having no boundary edges is called *closed* if it is compact and is called *open* if it has infinitely many cells.

(c) A surface is called *orientable* if, for every pair $\{\varphi, \varphi^{-1}\}$ of faces, one of the pair can be chosen to be "positive" so that each directed edge which is not a boundary edge appears exactly once in the positive boundary path of a positive face. The system of positive faces is called an *orientation*. The star $\sigma_1, \ldots, \sigma_k$ arround v is *positive* if $\sigma_i^{-1}\sigma_{i+1}$, $1 \leq i < k$ is part of the positive boundary of a positive face. At each vertex v there is a maximal positive star and its inverse is a maximal negative star.

It is easily seen that cutting and pasting, see 1.2.15, preserve the properties of being "compact", "closed" or "orientable". The boundary edges, with their endpoints, form a graph the connected components of which are called *boundary components of the surface.* Each such component is homeomorphic to a circle or a line.

3.1.3. Examples: Closed Surfaces. *The complexes* S_g *and* N_g *in* 1.2.6 (c), (d) *are closed surfaces and every closed surface is homeomorphic to some* S_g *or some* N_g, see 3.1.5.

In S_g the star at v is $\tau_1, \mu_1^{-1}, \tau_1^{-1}, \mu_1, \tau_2, \mu_2^{-1}, \tau_2^{-1}, \mu_2, \ldots, \tau_g, \mu_g^{-1}, \tau_g^{-1}, \mu_g$ and is closed. The Euler characteristic is $\chi(S_g) = 2 - 2g$. Hence, S_g and S_h with $g \neq h$ are not homeomorphic. The number g is called the *genus of* S_g. The surface S_g is orientable.

In N_g the star around v is $\nu_1^{-1}, \nu_1, \nu_2^{-1}, \nu_2, \ldots, \nu_g^{-1}, \nu_g$ and is closed. Morover $\chi(N_g) = 2 - g$; hence, N_g and N_h are not homeomorphic if $g \neq h$. The number g is called the *genus of* N_g. The surface N_g is not orientable. Hence, *N_g and S_h are not homeomorphic.*

3.1.4. Example: Compact Surfaces. Let the complex $S_{g,r}$ (or $N_{g,r}$, $r > 0$) have $r+1$ vertices $v, v_1, \ldots, v_r$, and $2r + 2g$ (or $2r + g$, respectively) pairs of edges

$$\{\rho_j^{\pm 1}, \sigma_j^{\pm 1} : 1 \leq j \leq r\} \cup \{\tau_j^{\pm 1}, \mu_j^{\pm 1} : 1 \leq j \leq g\}$$
$$(\text{or } \{\rho_j^{\pm 1}, \sigma_j^{\pm 1} : 1 \leq j \leq r\} \cup \{\nu_j^{\pm 1} : 1 \leq j \leq g\}, \quad \text{respectively,})$$

and one pair of faces $\varphi^{\pm 1}$ with the following boundary conditions: the τ_j, μ_j, ν_j start and end at v, σ_j runs from v to v_j, ρ_j from v_j to v_j, and

$$\partial\varphi = \prod_{i=1}^{r} \sigma_j \rho_j \sigma_j^{-1} \cdot \prod_{j=1}^{g} [\tau_j, \mu_j] \quad (\text{or } \prod_{i=1}^{r} \sigma_j \rho_j \sigma_j^{-1} \cdot \prod_{j=1}^{g} \nu_j^2, \text{ respectively}).$$

This defines an orientable (or non-orientable, respectively) compact surface $S_{g,r}$ (or $N_{g,r}$, respectively) with r boundary components $\rho_1, \ldots, \rho_r$; the number g is again called the *genus*. Write also $S_{g,0}$ for S_g and $N_{g,0}$ for N_g.

Homeomorphic surface complexes obviously have the same number r of boundary components. The Euler characteristic is

$$\chi(S_{g,r}) = (r+1) - (2g + 2r) + 1 = 2 - 2g - r$$

and is an invariant of the surface. Hence two surfaces $S_{g,r}, S_{g',r'}$ are homeomorphic only if $g = g', r = r'$ and thus g is also an invariant of the surfaces. A similar argument can be applied to the non-orientable surfaces $N_{g,r}$.

Starting with an arbitrary finite surface complex one can cut and paste as in 1.2.15 until there is only one vertex in each boundary component and one additional vertex not on the boundary. Further cutting and pasting will yield a complex having only one pair of faces except in the case of a closed orientable surface of genus 0 (when one obtains two pairs of faces and one pair of edges). A more delicate argument then gives one of the canonical forms $S_{g,r}$ or $N_{g,r}$, see [ZVC 1980,1988, 3.2]. This gives part (a) of the following theorem; part (b) is a direct consequence of 1.2.10 and 1.2.16.

3.1.5. Classification Theorem of Compact Surfaces

(a) *Any finite surface complex is homeomorphic to exactly one of the complexes $S_{g,r}, N_{g,r}$.*

(b) *If an orientable or non-orientable compact surface has genus g and r boundary components then its fundamental group is isomorphic to*

$$\pi_1(S_{g,r}) = \langle s_1, \dots, s_r, t_1, u_1, \dots, t_g, u_g \mid \prod_{i=1}^{r} s_i \prod_{j=1}^{g} [t_j, u_j] \rangle \quad \textit{or}$$

$$\pi_1(N_{g,r}) = \langle s_1, \dots, s_r, v_1, \dots, v_g \mid \prod_{i=1}^{r} s_i \prod_{j=1}^{g} v_j^2 \rangle, \ \textit{respectively.}$$

(c)
$$H_1(S_{g,r}) = \begin{cases} \mathbb{Z}^{2g} & \textit{if } r = 0, \\ \mathbb{Z}^{2g+r-1} & \textit{if } r > 0; \end{cases}$$

$$H_1(N_{g,r}) = \begin{cases} \mathbb{Z}_2 \oplus \mathbb{Z}^{g-1} & \textit{if } r = 0, \\ \mathbb{Z}^{g+r-1} & \textit{if } r > 0. \quad \square \end{cases}$$

If $r > 0$ then by Tietze transformations one of the generators and the single defining relation can be omitted and hence the fundamental group is free of rank $2g+r-1$ or $g+r-1$, respectively. This gives an algebraic argument for the invariance of the genus. For $r = 0$ one obtains, by abelianizing, the group $H_1(S_g) \cong \mathbb{Z}^{2g}$ and $H_1(N_g) \cong \mathbb{Z}^{g-1} \oplus \mathbb{Z}_2$. This shows that the group $\pi_1(N_g)$ is not a free group. The fact that the group $\pi_1(S_g)$ is not free can be shown in different ways. On the one hand it follows from the theory of one-relator groups, see 2.4.13. An elementary direct proof is as follows: Since $\pi_1(S_g)$ is generated by $2g$ elements and the factor group $H_1(S_g)$ has rank $2g$ the group $\pi_1(S_g)$ has rank $2g$ and could only be isomorphic to the free group of rank $2g$. By 1.3.5, any subgroup of index 2 in the free group is a free group of rank $2 \cdot (2g-1) + 1 = 4g - 1$. However, using the Reidemeister-Schreier method one can prove that $\pi_1(S_g)$ contains a subgroup of index 2 of rank $4g-2$; in fact, this is true for all subgroups of index 2, see 3.1.9.

3.1.6. Proposition. *Closed surfaces are homeomorphic if and only if their fundamental groups are isomorphic.* $\square$

3.1.7. Notation. Certain surfaces and any complex defining them have standard names. These are: $S_{0,1}$ — *disc;* $S_{0,2}$ — *annulus;* $S_{0,0}$ — *2-sphere* S^2*;* $S_{1,0}$ — *torus* $S^1 \times S^1$*;* $N_{1,0}$ — *projective plane* P^2*;* $N_{1,1}$ — *Möbius band;* $N_{2,0}$ — *Klein bottle.*

In the following by an (*orientable* or *non-orientable*) *surface group* is meant a group isomorphic to the fundamental group of a closed (orientable or non-orientable, respectively) surface. An important tool in the topological theory of surfaces is given by the theory of coverings, and this can again be applied to their fundamental groups. From the definitions we immediately obtain the following theorem and its consequences which are the analogues of 1.3.2 (b) and 1.3.5.

3.1.8. Theorem. *Let $p : \hat{S} \to S$ be a covering. Then $\hat{S}$ is a surface if and only if S is a surface.* □

3.1.9. Corollary. *Let G be the fundamental group of a closed surface S and U a subgroup of finite index. Then U is isomorphic to a surface group. If S is an orientable surface of genus g then U is isomorphic to the fundamental group of an orientable surface of genus $[G : U] \cdot (g-1) + 1$. If S is a non-orientable surface of genus g then U is isomorphic to the fundamental group of a non-orientable surface of genus $[G : U] \cdot (g-2) + 2$ or to the fundamental group of an orientable surface of genus $\frac{1}{2}[G : U] \cdot (g-2) + 1$. If $G = \pi_1(N_g)$ then G contains a characteristic subgroup of index 2 which is isomorphic to $\pi_1(S_{g-1})$.* □

The calculations prior to 3.1.6 parallel the above geometric theorem. Indeed, the basis of the Reidemeister-Schreier method in 1.3.7 was the use of coverings. Theorem 3.1.8 and Corollary 3.1.9 can also been proved purely algebraically using a modified Reidemeister-Schreier method, but that proof is much less transparent, see [ZVC 1980, 1988, 4.14].

For later use, let us introduce the notion of intersection number. Fix an orientation of the closed (topological) surface S_g. For two curves α, β on S_g in general position, i.e. they intersect one another transversely in simple points of intersection (one can make this more precise using dual complexes on S_g), one defines the *algebraic intersection number* $\alpha \cdot \beta$ by counting $+1$ at every intersection point where β crosses α from right to left, and -1 where β crosses α from left to right and adding all these numbers. The sum obtained is invariant with respect to homotopic deformations and so yields a mapping $\nu : \pi_1(S_g) \times \pi_1(S_g) \to \mathbb{Z}$. Clearly

$$\alpha \cdot \beta = -\beta \cdot \alpha \text{ and } \alpha\alpha' \cdot \beta = \alpha \cdot \beta + \alpha' \cdot \beta.$$

A consequence is that ν can be factored through the homology group $H_1(S_g) = \pi_1(S_g)^{ab}$; hence:

3.1.10. Proposition. *There is a bilinear form $\nu : H_1(S_g) \times H_1(S_g) \to \mathbb{Z}$, called the* intersection form, *with the following properties:*

(a) $\nu(a, b) = -\nu(b, a)$ *for* $a, b \in H_1(S_g)$.

(b) *If* $\pi_1(S_g) = \langle t_1, u_1, \dots t_g, u_g \mid \prod_{j=1}^{g}[t_j, u_j]\rangle$ *then*

$$\nu(t_i, t_j) = \nu(u_i, u_j) = \nu(t_i, u_k) = 0, \quad \nu(t_i, u_i) = 1 \text{ for } 1 \leq i, j, k \leq g,\ i \neq k$$

where t_i, u_i also mean t_i^{ab}, u_i^{ab}. □

§ 3.2. Planar Discontinuous Groups

The universal covering of a surface is a simply connected surface. If we consider only surfaces without boundary it turns out that there are only two types of simply connected surfaces: the orientable closed surface of genus 0 and the plane — as long as we are only concerned with topological, (real) differentiable or combinatorial properties of the surface. This result is a consequence of the Schönflies Theorem the combinatorial form of which is easily proved, see [ZVC 1980, 1988, 7.4.1].

3.2.1. Schönflies Theorem. *Let S be a simply connected surface without boundary and $\gamma \subset S$ a simple closed path. If S is not compact then S is the union of a disc and an infinite surface whose intersection γ is the boundary of each, and if S is compact then γ bounds two discs.* □

An immediate consequence of 3.2.1 is, see [ZVC 1980,1988, 4.1.7]:

3.2.2. Theorem. *Let $\mathbb{E}$, $\mathbb{E}_1$ be simply connected surface complexes without boundary. If both are finite or both are infinite then they are homeomorphic. In other words, there are only two simply connected surfaces without boundary: the* 2*-sphere and the* plane. □

If we consider a surface without boundary then its universal cover must be the sphere, i.e. a finite complex, or the plane. Since the fundamental group acts freely on the universal cover it follows that only the projective plane and the sphere itself have the sphere as universal cover. All other surfaces can be obtained from complexes on the plane by factoring out by the action of a group of automorphisms, the group of covering transformations. More generally, let us consider arbitrary automorphism groups of planar complexes (or *planar nets*) and try to classify them.

First we will describe the types of automorphisms of a planar net $\mathbb{E}$. Fix an orientation of $\mathbb{E}$. An automorphism $\alpha : \mathbb{E} \to \mathbb{E}$ which maps positive faces of $\mathbb{E}$ to positive faces is called *orientation preserving.* If it maps some positive face to a negative face then it maps every positive face to a negative face and is called *orientation reversing.* The property of being orientation preserving does not depend on the choice of the orientation. It follows directly from the definition that a transformation preserving orientation and fixing a (directed) edge is the identity. If it fixes a vertex or a 2-cell it behaves, in geometric language, like a *rotation* around the vertex or the "centre" of the 2-cell, respectively. A transformation of finite order preserving the orientation is of one of these types or else is "a rotation of order 2 around the centre of an edge". After a suitable subdivision of $\mathbb{E}$ we may assume that orientation preserving automorphisms of finite order are rotations around a vertex. In a similar way one can see that, perhaps after a suitable subdivision of the 2-cells of $\mathbb{E}$, an orientation reversing automorphism of finite order leaves fixed the edges and vertices of a line and has order 2. This line divides the plane into two parts and the automorphism interchanges them like a *reflection in this line.* Moreover, automorphisms of infinite order do not fix vertices and do not fix or invert edges or 2-cells.

Two (oriented) cells of $\mathbb{E}$ are called *(G-)equivalent* if there is an element of G mapping one to the other. An important notion for group actions is given next.

3.2.3. Definition. Let $\mathbb{E}$ be a connected surface complex and let G be a group of automorphisms of it. A connected subcomplex D of $\mathbb{E}$ is called a *fundamental domain of* G if it contains exactly one 2-cell from each equivalence class together with their boundaries. We say that G *has compact fundamental domain* if the number of G-equivalence classes of 2-cells is finite, that is, D consists of a finite number of 2-cells together with their boundaries.

The following statements are easily proved:

3.2.4. Definition and Proposition. *A group G of automorphisms of a planar net $\mathbb{E}$ is called a* planar discontinuous group.

(a) *Every planar discontinuous group G has a fundamental domain. Any fundamental domain of G is simply connected.*

(b) *$\mathbb{E}/G$ is a surface. The group G has compact fundamental domain if and only if the surface $\mathbb{E}/G$ is compact. The projection $p : \mathbb{E} \to \mathbb{E}/G$ is a homomorphism of complexes which is a covering if and only if G is torsion-free. In this case G coincides with the group of covering transformations of the covering. The surface $\mathbb{E}/G$ has non-empty boundary if and only if G contains reflections, i.e. G contains orientation reversing automorphims of finite order. If G has torsion but does not contain reflections then p: $\mathbb{E} \to \mathbb{E}/G$ is a "branched covering".* □

We shall see below that planar discontinous groups all occur as groups of motions of the euclidean or the Bolyai-Lobachevskii (hyperbolic) plane. Then "rotations" and "reflections" will indeed be rotations and reflections in the corresponding geometry.

The usual analytic notion of a *branched covering*, namely, a mapping which has almost the same properties as a covering but may, at some points, behave like the mapping $z \mapsto z^n$, $n \in \{2, 3, ...\}$ of the unit disc to itself, also has a combinatorial analogue, see [Reidemeister 1932]; [ZVC 1980,1988, 3.3.1], but we omit details. In this case G also coincides with the group of covering transformations but there is no analogue of Theorem 1.3.4.

3.2.5. Hypothesis. *In the following we will restrict ourselves to the case where all transformations are orientation preserving, in particular, where the quotient surface $\mathbb{E}/G$ is orientable.*

Just as for the classification of surfaces, one can modify a complex $\mathbb{E}$ in the plane in a G-invariant manner. Since $\mathbb{E}/G$ is a surface, there is a sequence of cut and paste operations which reduces the number of 2-cells to one and

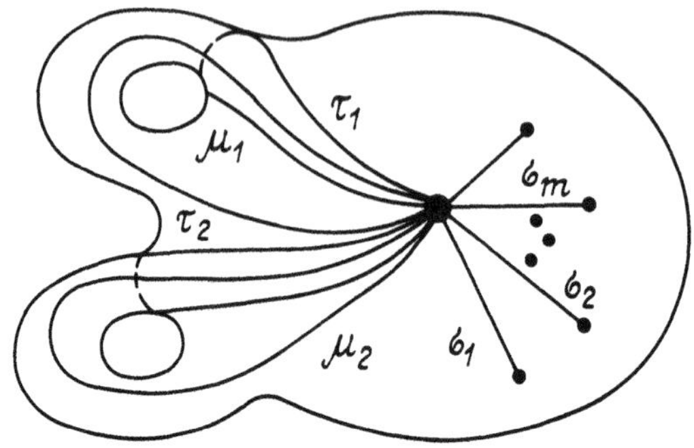

Fig. 3.2.1

minimalises the number of vertices, and then transforms the resulting surface into canonical form. Each operation can be lifted back to (infinitely many elementary) transformations in $\mathbb{E}$ at all the cells of $\mathbb{E}$ lying over the cells of $\mathbb{E}/G$ involved in the operation. Care must be taken not to delete the image of the fixed point of a rotation of order 2. The final quotient complex looks like Fig. 3.2.1 and we obtain the following canonical forms for the fundamental domains.

3.2.6. Theorem. *A planar discontinuous group with compact fundamental domain may be realized by a pair* $(\mathbb{E}, G)$ *in which any two positive faces are* G*−equivalent, only the identity element of* G *leaves a face fixed, and the boundary path of a face has the form:*

$$(*) \qquad \prod_{i=1}^{m} \sigma_i' \sigma_i^{-1} \prod_{j=1}^{g} \tau_j' \mu_j^{-1} \tau_j^{-1} \mu_j'$$

Here edges denoted by the same greek letter and index (e.g. σ_i' *and* σ_i*) are* G*-equivalent.* □

Suppose that $g(D) \cap D$ is 1-dimensional. Then there are two possibilities:

(a) $g(D) \cap D$ consists of one edge $\sigma \in \mathbb{E}$. Then $g^{-1}(D) \cap D = g^{-1}(\sigma) \neq \sigma^{\pm 1}$ and $g^{-1}(\sigma)$ is the only other edge in D which is G-equivalent to σ.

(b) $g(D) \cap D$ consists of two edges $\sigma, \tau \in \mathbb{E}$ and the two edges have a common vertex v. Then $g^{\varepsilon}(\sigma) = \tau$, with $\varepsilon = \pm 1$, and g is a rotation with rotation centre v. In particular, g has finite order.

In both cases we obtain a pair of G-equivalent edges in the boundary of D. It follows easily that no third edge in the boundary of D can be equivalent to those of the pair. If for every such pair we take an automorphism mapping one member to the other, i.e. g as above, then we obtain a system of generators for G. For, applying them and their inverses to D we obtain all translates of D which have a common edge with D. If now $h(D)$ is such a neighbouring domain and $D' \cap h(D)$ contains an edge then $h^{-1}(D')$ is also a neighbour of D: $h^{-1}(D') = g^{\pm 1}(D)$ for a generator g. Hence, $D' = hg^{\pm 1}(D)$. Iterating this procedure, we get the whole tesselation $\mathbb{E}$ of the plane. In particular this proves that we obtain a system of generators of G by looking at those transformations moving D to a neighbour. By simple arguments of a similar kind one sees that a system of defining relations arises from the stars of the inequivalent vertices of D. (This method of determining a presentation can be formalized using the notion of a dual complex yielding a (modified) Cayley diagram for the presentation in question.)

The final result is given in the following theorem.

3.2.7. Theorem. *A planar discontinuous group G of orientation preserving transformations has the following structure:*

3.2.8 $$G = \langle s_1, \ldots, s_m, t_1, u_1, \ldots, t_g, u_g \mid s_1^{h_1}, \ldots, s_m^{h_m}, \prod_{i=1}^{m} s_i \cdot \prod_{j=1}^{g} [t_j, u_j] \rangle.$$

No proper subword of a defining relator is a relation. □

The assertion about the subwords of relators is claimed only for planar discontinuous groups, not for all groups given by a presentation of the above form. The generators denoted by s_i represent mappings of finite order which fix some vertex, thus, behave like a rotation. For a planar discontinuous group G neither the form of its fundamental domain D nor its presentation are invariants and there arises the question of classifying planar groups. Another problem is whether a given presentation (or formal form of a fundamental domain) can be realized by a planar discontinuous group. Moreover, the relation between algebraic und geometric properties has to be considered.

For planar discontinuous groups, two types of equivalence present themselves:

3.2.9. (a) There is an *(algebraic) isomorphism* between the groups.

(b) There is a *geometric isomorphism* between the groups, that is there are realizations $(\mathbb{E}, G)$ and $(\mathbb{E}', G')$ and an isomorphism h: $\mathbb{E}' \to \mathbb{E}$ such that $x \mapsto h^{-1}xh$ defines an isomorphism from G to G'.

Clearly, a geometric isomorphism is algebraic. Classification with respect to geometric isomorphism is easily carried out. If G and G' are geometrically isomorphic, then the surfaces $\mathbb{E}'/G'$, $\mathbb{E}/G$ have the same genus ($g' = g$) and the "branching properties" are the "same". For the algebraic classification of planar discontinuous groups we have the following theorem:

3.2.10. Theorem. *If two planar discontinuous groups are isomorphic then they are also geometrically isomorphic.*

Proof. A nice geometric argument shows that the elements of finite order are conjugate to the powers of the s_i, that the subgroups $\langle s_i \rangle$ are maximal finite subgroups and no two are conjugate. So the numbers $h_1, \dots, h_m$ are algebraic invariants. Furthermore, the normal closure U of $s_1, \dots, s_m$ is a characteristic subgroup and therefore its abelianized factor group $(G/U)/[G/U, G/U] \cong \mathbb{Z}^{2g}$ is an algebraic invariant of G, and so $2g$ is an invariant too. The geometric arguments on elements of finite order can be avoided, see 3.2.15. □

The existence problem can be attacked in two ways, either by using analytic/geometric constructions or by applying arguments of combinatorial group theory and topology. Let us now first formulate the results, then sketch the analytic proof and finally give more details of the combinatorial approach.

3.2.11. Theorem. *A group G defined abstractly by a presentation as in* Theorem 3.2.8 *occurs as a planar discontinuous group with compact fundamental domain if it is of infinite order. This is the case if and only if $\mu(G) \geq 0$, where the* measure $\mu(G)$ of G *is defined by* $\mu(G) = 4g - 4 + 2\sum_{i=1}^{m}(1 - \frac{1}{h_i})$.

If $\mu(G) < 0$ then either $g = 0$, $m \leq 2$ or $g = 0$, $m = 3$ and $\frac{1}{h_1} + \frac{1}{h_2} + \frac{1}{h_3} > 1$. In these cases the groups obtained are either trivial or can be realized by groups of motions of the sphere S^2. The groups of the last form with rotation orders (h_1, h_2, h_3) are the dihedral groups D_n of order $2n$ (case $(2,2,n)$) and the platonic groups: tetrahedral group $(2,3,3)$ of order 12, octahedral group $(2,3,4)$ of order 24, dodecahedral group $(2,3,5)$ of order 60.

3.2.12. Geometric Approach. We will illustrate the geometric proof of existence with the simplest examples, namely the triangle groups where $g = 0$, $m = 3$: Construct a triangle ABC with sides a, b, c and angles $\frac{\pi}{p}, \frac{\pi}{q}, \frac{\pi}{r}$; this can be done in the euclidean plane if $\frac{\pi}{p} + \frac{\pi}{q} + \frac{\pi}{r} = \pi$, on the sphere if the sum is bigger than π and in the Bolyai-Lobachevskij (hyperbolic) plane if the sum is smaller than π. Take the three reflections (in the correponding geometries) in the lines determined by the faces of the triangles and also denote them by a, b, c. Then $a^2 = b^2 = c^2 = 1$, $(ab)^r = (bc)^p = (ca)^q = 1$. Applying the group $H = \langle a, b, c \rangle$ generated by a, b, c to the triangle ABC gives a tesselation of the euclidean plane (or sphere or hyperbolic plane, respectively) into triangles which are drawn for the euclidean cases $(2,3,6), (2,4,4), (3,3,3)$ in Fig. 3.2.2-.4 and the group H acts on this tesselation. The triangle is a fundamental domain. By the general procedure in the proof of 3.2.7 one obtains the presentation

$$\begin{aligned} H &= \langle a, b, c \mid a^2, b^2, c^2,\ (ab)^r, (bc)^p, (ca)^q \rangle \\ &= \langle a, b, c, a', e \mid a^2, b^2, c^2, a'^2, (ab)^r, (bc)^p, (ca')^q, aea'^{-1}e^{-1}, e \rangle \ . \end{aligned}$$

Let us now consider the subgroup $D \triangleleft H$ of transformations preserving the orientation; of course, it has index 2 and $a, b, c \notin D$. Using geometric arguments or the Reidemeister-Schreier method one sees that $s_1 = bc$, $s_2 = ca$, $s_3 = ab$ generate D and satisfy the relations $s_1^p = s_2^q = s_3^r = 1$ and $s_1 s_2 s_3 = (bc)(ca)(ab) = b \cdot c^2 \cdot a^2 \cdot b = 1$; this gives the presentation $D = \langle s_1, s_2, s_3 \mid s_1^p, s_2^q, s_3^r, s_1 s_2 s_3 \rangle$ showing that the groups with $g = 0$, $m = 3$ can

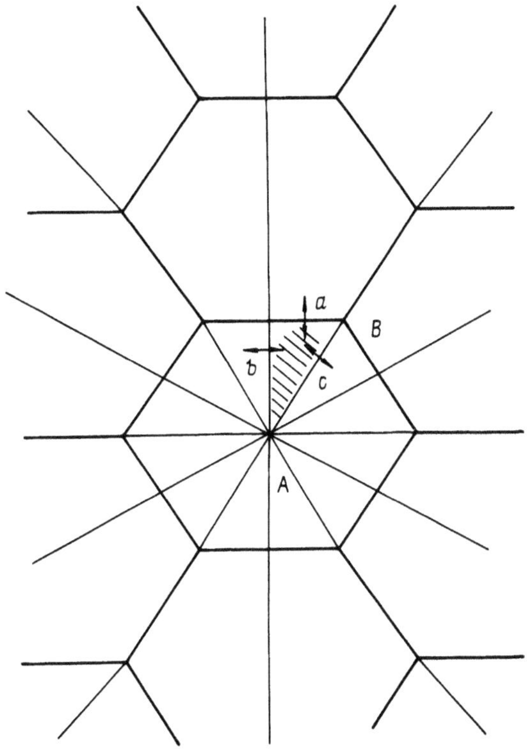

Fig. 3.2.2

be realized by discontinuous groups of the euclidean plane if $\frac{1}{h_1} + \frac{1}{h_2} + \frac{1}{h_3} = 1$ and of the hyperbolic plane if $\frac{1}{h_1} + \frac{1}{h_2} + \frac{1}{h_3} < 1$.

The same construction can be applied to all groups with a presentation 3.2.8 by first constructing some convex polygon on the sphere, in the euclidean or Bolyai-Lobachevskij plane with the properties that 1) the sides are denoted by the symbols from 3.2.6 (*), where sides with equivalent names (e.g. σ_i, σ_i') have the same length, and 2) the angle at the corner where the sides σ_i', σ_i meet is $2\pi/h_i$, and 3) the sum of all remaining angles is 2π. The sum of all the angles is $\alpha = \sum_{i=1}^{m} \frac{2\pi}{h_i}$. The number of vertices of the polygon is $N = 2m+4g$. Such a polygon exists on the sphere if $\pi(N-2) < \alpha$, in the euclidean plane if $\pi(N-2) = \alpha$, and in the Bolyai-Lobachevskij plane if $\pi(N-2) > \alpha$. (Of course, the notions convex and length used depend on the geometry.) For details see [ZVC 1980,1988, 6.4].

3.2.13. Combinatorial Approach. Let G be a group with a presentation 3.2.8. For G we construct a *modified Cayley diagram* $\mathbb{E}^*$, obtained from the proper Cayley diagram by identifying all faces with the same boundary. We will show that $\mathbb{E}^*$ is a surface complex. To see this we need the following assertion the proof of which is postponed:

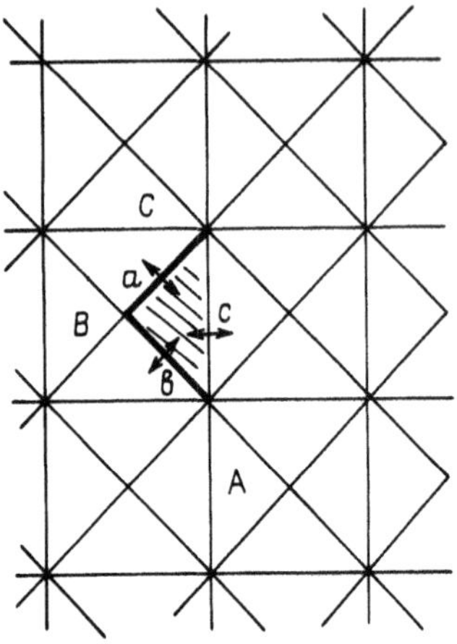

Fig. 3.2.3

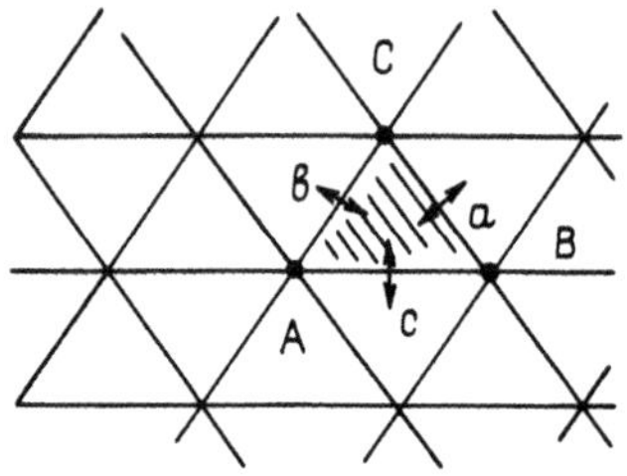

Fig. 3.2.4

3.2.14. Lemma. *If G is infinite then no proper subword of a defining relator is a relation of G. The groups not mentioned in* 3.2.11 *are infinite.*

Now we can check the conditions 3.1.1 (i - iii) that $\mathbb{E}^*$ is a surface complex. *Each directed edge appears in the boundaries of exactly two 2-cells and is traversed once in each boundary.* This is clear for generators which only appear twice (including inverses) in the relations. For the power relations $s_i^{h_i}$ this follows from 3.2.14, because then the same edge cannot be multiply traversed by a power relation. Thus a face corresponding to one of these relations has a given edge appearing at most once in its boundary.

The edges emanating from some vertex constitute a star. The neighbours of a symbol X are defined to be the symbols which follow X^{-1} in the relations, or the inverses of the symbols which precede X. A symbol then has two neighbours. Writing the symbols in succession, with each symbol flanked by its neighbours, one obtains the cycle

$$S_1, S_1^{-1}, \dots, S_m, S_m^{-1}, T_1, U_1^{-1}, T_1^{-1}, U_1, \dots, T_g, U_g^{-1}, T_g^{-1}, U_g.$$

This proves that the complex $\mathbb{E}$ is a surface. Since each closed path is a relation and thus a product of conjugates of the defining relations, the fundamental group $\pi_1(\mathbb{E})$ is trivial and it follows that $\mathbb{E}$ must be a complex on the plane or on the sphere. The group G acts on this net in the obvious way. This finishes the sketch of a combinatorial proof of Theorem 3.2.11. □

3.2.15. Some Algebraic Properties of Planar Groups. To show 3.2.14 we first consider the case $g = q = 0,\ m \geq 4$. Then

$$G = \langle s_1, \dots, s_m \mid s_1^{h_1}, \dots, s_m^{h_m}, \prod_{i=1}^{m} s_i \rangle$$
$$= \langle s_1, s_2 \mid s_1^{h_1}, s_2^{h_2} \rangle *_{\langle s_1 s_2 \rangle = \langle (s_3 \dots s_m)^{-1} \rangle} \langle s_3, \dots, s_m \mid s_3^{h_3}, \dots, s_m^{h_m} \rangle.$$

Both factors are free products of finite cyclic groups and from the solution of the word problem 2.2.4 it follows that $s_1 s_2$ as well as $s_3 \dots s_m$ has infinite order; hence the decomposition of G above is in fact a decomposition into a free product with amalgamated infinite cyclic subgroups. It follows from 2.2.4 that proper subwords of the defining relations are not relations and this gives 3.2.14. Similar arguments apply except for the cases $g = 0$, $m \le 3$; $g = 1$, $m \le 1$. For $g = 1 = m$ we pass to a quotient group by introducing the relations $u_1^2, s_1 t_1^2$ and obtain the presentation $\langle t_1, u_1 \mid t_1^{2h_1}, u_1^2, t_1^{-1} u_1 t_1^{-1} u_1^{-1} \rangle$ of the dihedral group D_{2h_1} of order $2h_1$ where it is trivial to check that no proper subword of a defining relation is a relation. For $g = 0$, $m \le 3$ the groups with $m \le 2$ are excluded (the groups are finite cyclic groups). For the remaining case we use the triangles ABC from 3.2.12 to obtain $s_i^k \ne 1$ if $k \not\equiv 0 \bmod h_i$.

The decomposition of a planar discontinous group into a free product with amalgamated subgroups evidently opens up the possibility of proving algebraic properties of G by algebraic methods and solving the word and conjugacy problem, applying 2.2.9 and 2.2.11. We will not do this here, but will use geometric arguments instead. As we have seen at the beginning of the section, the elements of finite order are characterized as those which have a fixed vertex (rotations). If a vertex v is fixed under the transformation $b \in G$ then, for an arbitrary $x \in G$, $x(v)$ is fixed under $a = xbx^{-1}$. Conversely, if v is fixed under b and $x(v)$ under a, then xbx^{-1} and a have the same fixed vertex and, hence, they are powers of the same element. Moreover, xbx^{-1} has the same rotation centre as b only when x itself is a rotation about this vertex, and so x and b are powers of the same element and hence commute. The different powers of b are therefore not conjugate to each other. This gives the following theorem.

3.2.16. Theorem. *Let G be an infinite group with a presentation* 3.2.8, *i.e. $\mu(G) \ge 0$ for the measure $\mu(G) = 4g - 4 + 2 \cdot \sum_{i=1}^{m}(1 - \frac{1}{h_i})$ of G.*

(a) *An element of finite order in G is conjugate to a power of s_i; s_i^a and s_j^b are conjugate if and only if $i = j$ and $a \equiv b \bmod h_i$.*

(b) *A torsionfree planar discontinuous group is the fundamental group of a surface.*

(c) *The measure $\mu(G)$ of G is an algebraic invariant.* □

Next we will deal with the rank of Fuchsian groups. Consider, e.g., the group $H = \langle s_1, \dots, s_5 \mid s_1^2,\ s_2^3,\ s_3^5,\ s_4^7,\ s_5^{11}, s_1 \cdot \ldots \cdot s_5 \rangle$. Clearly, $d(H) \le 4$. This group is neither trivial nor cyclic, but has trivial abelianisation. Is the rank 2, 3 or 4? It seems to be quite difficult to decide this algebraically.

Let us first examine a geometric analogue. A side σ of a fundamental domain B of a planar group G defines an element $x \in G$ such that

$\partial B \cap \partial x(B) = \sigma$ and the pairs of sides $\sigma, x(\sigma)$ of B determine a system of generators of G as we have seen in the arguments before 3.2.7. Let us call such a system a *system of geometric generators* and let r_B denote the number of pairs in the system. The minimum of the r_B over all fundamental domains B is call the *geometric rank of* G. If G has torsion we can take one of the rotation centres as base point and find a fundamental domain with $2m+4g-2$ sides which will define $m+2g-1$ geometric generators and give the following presentation:

$$G = \langle s_1, \dots, s_{m-1}, t_1, u_1, \dots, t_g, u_g \mid s_1^{h_1}, \dots, s_{m-1}^{h_{m-1}}, \ (\prod_{j=1}^{g} [t_j, u_j] \prod_{i=1}^{m-1} s_i)^{h_m} \rangle.$$

This gives an upper bound for the geometric rank, and it turns out that it is also a lower bound.

3.2.17. Theorem. *The planar group above has geometric rank* $2g+m-1$ *if* $m > 0$ *and* $2g$ *if* $m = 0$.

Proof. The case $m = 0$ is trivial. Consider the Euler characteristic of $\mathbb{E}/G$, an orientable closed surface of genus g. If F is a fundamental domain then there is one face, r_B edges and at least m vertices since there are m different conjugacy classes of maximal finite cyclic groups. Hence $2-2g = \chi(\mathbb{E}/G) \geq m - r_B + 1 \Longrightarrow r_B \geq 2g+m-1$. □

These arguments can also be used to determine the geometric rank of all planar discontinuous groups which do not contain reflections. A problem posed originally by Nielsen is to decide whether the geometric and the algebraic rank of a Fuchsian group (see Remark 3.2.19 below) coincide. Let us give the final result.

3.2.18. Theorem [Peczynski-Rosenberger-Zieschang 1975]. *Let*

$$G = \langle s_1, \dots, s_m, t_1, u_1, \dots, t_g, u_g \mid s_1^{h_1}, \dots, s_m^{h_m}, \ \prod_{i=1}^{m} s_i \prod_{j=1}^{g} [t_j, u_j] \rangle,$$

where $2 \leq h_1 \leq h_2 \leq \dots \leq h_m$. *Then the rank*

$$d(G) = \begin{cases} 2g & \text{if } m = 0, \\ m-2 & \text{if } g = 0, \ m \text{ is even}, \ h_1 = \dots = h_{m-1} = 2, \ h_m \text{ odd}, \\ 2g+m-1 & \text{for the other cases of infinite } G. \end{cases}$$

Remarks on the proof. Let us consider the group:

$G = \langle s_1, \dots, s_{2l} \mid s_1^2 = \dots s_{2l-1}^2 = s_{2l}^{2k+1} = s_1 \dots s_{2l} = 1 \rangle$, where $m = 2l, k > 0$.

Define $x_{2i-1} = s_1 s_{2i}$, $x_{2i} = s_{2i+1} s_1$ for $1 \leq i \leq l-1$. Then

$$\prod_{i=1}^{m-2} x_i = s_1 s_2 s_3 \dots s_{m-1} s_1, \quad \prod_{i=1}^{m-2} x_i^{-1} = s_2 s_3 \dots s_{m-1}; \text{ hence}$$

$$\prod_{i=1}^{m-2} x_i \cdot \prod_{i=1}^{m-2} x_i^{-1} = (s_1 \dots s_{m-1})^2 = s_m^{-2}.$$

Since the order of s_m is odd it follows that s_m lies in $\langle x_1, \dots, x_{m-2} \rangle$ and thus so does $s_1 \dots s_{m-1} = s_m^{-1}$. The first equation above implies that s_1 does as well and so, by the definitions of the x_i, this holds for the other s_i, too. Now the claim follows since $G^{ab} \cong \mathbb{Z}_2^{m-2}$.

That the given numbers are upper bounds for the other cases is clear. To prove that they are also lower bounds needs an unpleasant argument using the Nielsen method for amalgamated free products. □

3.2.19. Remarks. (a) If $\mu(G) > 0$ then G is a *Fuchsian group*, that is, it can be realized as a group of motions of the hyperbolic plane $\mathbb{H}^2$. If $\mu(G) = 0$ it is a crystallographic group, that is, it can be realized by a group of motions of the euclidean plane.

(b) The theory developed above can be extended to groups which contain orientation reversing elements (NEC-groups), see [Wilkie 1966], [ZVC 1980,1988, Chap. 4]. Extensions to the case of finitely generated groups with non-compact fundamental domain can be found in [Macbeath-Hoare 1976], [ZVC 1980,1988, 4.11].

§ 3.3. Subgroups of Planar Groups

Let $\mathbb{E}$ be a planar net, G a group acting on $\mathbb{E}$ and H a subgroup of G. Then H also acts on $\mathbb{E}$. If all elements of G preserve orientation then this is true for the elements of H. If the index is finite and if G has a compact fundamental region then so does H. This gives the first statement of the following theorem.

3.3.1. Theorem. *Let G have a presentation* 3.2.8 *and let $H < G$, $[G : H] < \infty$. Then: H also has a presentation* 3.2.8. *Moreover, $\mu(H) = [G : H] \cdot \mu(G)$ (Riemann-Hurwitz formula).*

If G is torsion free, that is G is the fundamental group of a closed surface $C = \mathbb{E}/G$ by 3.2.16 (b), then H is also the fundamental group of a closed surface $C' = \mathbb{E}/H$ and we obtain a covering $C' \to C$ of order $[G : H]$ and the equation for μ is the Euler characteristic formula of 1.3.2 (c) since in this case $\mu(G) = 4g - 4 = -2 \cdot \chi(C)$, $\mu(H) = -2 \cdot \chi(C')$. This argument can be generalized to the other groups G using branched coverings. The formula can also be proved purely algebraically using a modified Reidemeister-Schreier method, see [ZVC 1980,1988, 4.14.22]. □

Consider a closed surface C and a group Γ of *symmetries* of the complex C, that is automorphisms of C. Then we obtain a mapping $p'\colon C \to C/\Gamma$. The mappings of Γ can be lifted to the universal covering $\tilde{p}\colon \mathbb{E} \to C$ and we obtain a discontinuous group G acting on $\mathbb{E}$. It contains the group of covering transformations, isomorphic to and identified with $\pi_1(C)$, and $[G : \pi_1(C)] = |\Gamma|$. In this context there arise the following problems:

3.3.2. Problems

(a) Given a closed surface S does it have symmetries except the identity? More precisely: Does there exist a complex C on S with non-trivial symmetries?

(b) Describe and classify the symmetry groups on a surface. In particular, find bounds for the order of the group of symmetries. Is this number always finite?

(c) Which discontinuous groups of the plane can be obtained in the way described above. In other words, which discontinuous groups of the plane contain a surface group? Given a planar group, find all subgroups of finite index isomorphic to the fundamental group of a surface!

3.3.3. Examples

(a) For any $n > 1$, a rotation about the polar axis through $2\pi/n$ and reflection in the equator generate a dihedral group of symmetries of the sphere. Moreover the sphere has the groups of symmetries of the regular polyhedra given in 3.2.11. These are the only possibilities for groups of orientation preserving symmetries.

(b) Constructing a torus by rotation of a circle around an axis and dividing this figure into n congruent annuli we see that the torus has the cyclic and dihedral groups as groups of symmetries; as in (a) there is no upper bound for the order of the symmetry groups of complexes on the torus. For the full determination of the possible symmetry groups on a torus see [Zieschang 1981, Chap. 2], [ZVC 1988, Chap. 8].

The answer to 3.3.2 (c) was given by Bungaard-Nielsen and R.H. Fox:

3.3.4. Theorem. *Every discontinuous group of the plane has the fundamental group of a surface as a subgroup of finite index.* □

In the *proof* one can restrict oneself to the case of groups containing only orientation preserving transformations, i.e. of the form $G = \langle s_1, \ldots, s_m, t_1, \ldots, u_g \mid s_1^{h_1}, \ldots, s_m^{h_m}, \prod_{i=1}^m s_i \cdot \prod_{j=1}^g [t_j, u_j] \rangle$. The elements of finite order are conjugate to the powers of the s_i. So it suffices to find a homomorphism φ of G to a finite group E such that $\varphi(s_i)$ has order h_i, $1 \leq i \leq m$. Suitable finite groups are found, for instance, among subgroups of $\mathrm{GL}(2, k)$, k a finite field (Macbeath, see [ZVC 1980,1988, 4.10]). □

The theorem is a special case of the Selberg Lemma, see [Selberg 1960]:

3.3.5. Selberg Lemma. *Any finitely generated subgroup of $GL(m,\mathbb{C})$ contains a torsion free subgroup of finite index.* □

Let us now consider problem 3.3.2 (b) for the case of closed orientable surfaces S_γ, $\gamma \geq 2$ and a group A of symmetries which preserve the orientation of S_γ. This can be lifted to the universal cover of S_γ to get a planar group D. Then $[D : \pi_1(S_\gamma)] = |A|$. From Theorem 3.3.1 we obtain, in the notation of 3.2.8,

$$0 < 2\gamma - 2 = |A| \cdot \left[\sum_{i=1}^{m}(1 - \frac{1}{h_i}) + 2g - 2\right] = \frac{1}{2}|A| \cdot \mu(D) .$$

Therefore $\mu(D) > 0$ and this implies that if $g > 0$ then $\mu(D) > \frac{1}{2}$ and if $g = 0, m \geq 5$ then $\mu(D) \geq \frac{1}{2}$. By checking the cases $g = 0, m = 3, 4$ one obtains $\mu(D) \geq \frac{1}{42}$ and the lower bound $\frac{1}{42}$ is realised for and only for the triangle group $D(2,3,7) = \langle s_1, s_2, s_3 \mid s_1^2,\ s_2^3,\ s_3^7,\ s_1 s_2 s_3\rangle$. This shows that the situation is different to the case of surfaces of genus ≤ 1 and we have

3.3.6. Corollary

(a) *Let S be a closed orientable surface of genus $\gamma \geq 2$ and A a group of orientation preserving symmetries of a complex on S. Then $|A| \leq 84 \cdot (\gamma - 1)$* [Hurwitz 1893].

(b) *If G is a torsionfree normal subgroup of the triangle group $D(2,3,7)$ of finite index, then G is isomorphic to the fundamental group of a closed orientable surface of genus γ where $[D(2,3,7) : G] = 84 \cdot (\gamma - 1)$. There are infinitely many γ where the group is obtained that way and where the group is not contained in another normal subgroup isomorphic to the fundamental group of a surface* [Macbeath 1969].

(c) *If $\pi_1(S_\gamma)$ is not isomorphic to a normal subgroup of $D(2,3,7)$ then any group of orientation preserving symmetries of a complex on S_γ has order $\leq 40 \cdot (\gamma - 1)$.* □

Next we will find lower bounds for the order of symmetry groups on a surface. Consider the triangle group $D(4,2k,2) = \langle s_1, s_2 \mid s_1^4 = s_2^{2k} = (s_1 s_2)^2 = 1\rangle$, $k \geq 2$ and $H_k = \langle a, b \mid a^4 = b^{2k} = (ab)^2 = (a^{-1}b)^2 = 1\rangle$ and the homomorphism ρ: $D(4,2k,2) \to H_k$, $s_1 \mapsto a$, $s_2 \mapsto b$. The group H_k has order $8k$ and contains an abelian subgroup A_k of order $4k$ which is uniquely determined if $k \geq 3$. Using the Reidemeister-Schreier method it follows that ker ρ has rank $2(k-1)$ and, thus, has a presentation $\langle t_1, u_1, \ldots, t_{k-1}, u_{k-1} \mid \prod_{i=1}^{k-1}[t_i, u_i]\rangle$. The kernel of the epimorphism $D(2, 2\gamma+1, 2(2\gamma+1)) \to \mathbb{Z}_{2(2\gamma+1)}$, $s_1 \mapsto 2\gamma+1$, $s_2 \mapsto 2\gamma$, $s_3 \mapsto 1$, is isomorphic to the fundamental group of a closed orientable surface of genus γ.

3.3.7. Proposition. *Let S_γ be a closed orientable surface of genus γ.*

(a) *The group $H_{\gamma+1}$ of order $8 \cdot (\gamma + 1)$, the abelian group A_k of order $4 \cdot (\gamma + 1)$, and $\mathbb{Z}_{2(2\gamma+1)}$ act effectively on S_γ, i.e. only the neutral element acts as the identity.*

(b) *Let $N(\gamma)$, $N_a(\gamma)$ and $N_c(\gamma)$ denote the maximal orders of a group, an abelian and a cyclic group, respectively, of orientation preserving symmetries of a closed orientable surface of genus γ. Then*

$$8 \cdot (\gamma + 1) \leq N(\gamma) \leq 84 \cdot (\gamma + 1), \ N_a(\gamma) = 4 \cdot (\gamma + 1), \ N_c(\gamma) = 2 \cdot (2\gamma + 1) .$$

The upper and lower bound in the inequality are realised for infinitely many γ. □

References for these results are in [ZVC 1980,1988, 4.10]. In addition, there is an extensive literature on this subject with sharper results than mentioned above. This theory can quite easily be generalized to the case when orientation reversing homeomorphisms also occur; we will just mention the expressions connected with this theory: *NEC-groups* and *Kleinian surfaces.*

On a surface there are self-homeomorphisms which induce "outer" automorphisms on the fundamental group but the identity on homology, for instance a homeomorphism of S_2 fixing the generators t_1, u_1 and conjugating t_2, u_2 by $[t_1, u_1]$. The situation is different for automorphisms of finite order.

3.3.8. Proposition. *Let G be a planar group with compact fundamental domain and let $N \triangleleft G$ be a normal subgroup isomorphic to the fundamental group of a closed orientable surface of genus $\gamma \geq 2$. Then $[G : N] < \infty$ and G/N acts effectively on $N/[N,N]$ by $y \cdot [N,N] \mapsto x^{-1}yx \cdot [N,N]$, $x \in G$, $y \in N$.*

Proof (assuming that the elements of G preserve orientation). $[G : N] < \infty$ follows from the fact that otherwise N operates on a planar net with non-compact fundamental domain and, hence, is free. If the action is not effective then there exists a non-trivial element $x \in G$ with $x^p \in N$, for some prime p, commuting with all $y \in N$. The group H generated by N and x has finite index in G and is, hence, also a planar group with compact fundamental domain. Since x operates trivially on N it follows that $H/[N,N] \cong \mathbb{Z}^{2\gamma} \oplus B$ where B is either trivial or isomorphic to $\mathbb{Z}_p$. This implies that $[N,N] = [H,H]$. The group H has a presentation 3.2.8 with $h_i = p$ $(1 \leq i \leq m)$. Abelianizing H gives $g = \gamma$, and the Riemann-Hurwitz formula 3.3.1 implies

$$4\gamma - 4 = p \cdot \left[2m(1 - \frac{1}{p}) + 4g - 4 \right] \iff 2 - 2\gamma = m$$

since $p \neq 1$, and this contradicts the assumptions $\gamma \geq 2, \ m \geq 0$. □

3.3.9. Corollary [Hurwitz 1893]. *Let A be a finite group acting on the closed orientable surface S_γ of genus $\gamma \geq 2$. Then the induced action of A on the homology group $H_1(S_\gamma)$ is effective.* □

For further results and references see [ZVC 1980,1988, 4.15].

§ 3.4. Automorphisms of Fuchsian Groups

In this section we view surfaces topologically and use terms such as homeomorphisms with their usual topological meaning. A translation from the topological standpoint to the combinatorial one is provided by, for instance, the simplicial approximation theorem which ensures that the fundamental group of a surface S can be calculated as the fundamental group of a complex realized on S. This theorem also ensures that self-homeomorphisms of S give rise to automorphisms of an appropriate surface complex.

Let S be a surface. By 1.2.9(b), a self-homeomorphism $f : S \to S$ preserving the basepoint induces automorphisms of $\pi_1(S)$ and $H_1(S)$. Now the following questions arise:

3.4.1. Problems

(a) *Which automorphisms of $\pi_1(S)$ or $H_1(S)$ are induced by homeomorphisms of S?*

(b) *How do homeomorphisms which induce the same automorphism of $\pi_1(S)$ (or $H_1(S)$) differ?*

(c) *Describe the group of self-homeomorphisms of a surface and the induced group of automorphisms of the fundamental or homology group.*

(d) *What is the connection between geometric properties of curves on S and algebraic properties of the corresponding elements in $\pi_1(S)$ (or $H_1(S)$, respectively)?*

When "lifting" these questions to the universal cover similar problems arise for planar groups. These problems can be attacked by geometrical-topological methods. Here we will describe a more combinatorial group-theoretical approach. Basic for this is the following observation: Let S be a closed orientable surface. If there is a homeomorphism $f : S \to S$ fixing the basepoint $v_\#$ then f is isotopic to a homeomorphism $f' : (S, v_\#) \to (S, v_\#)$ which maps a "small" disc D with $v_\# \in \partial D$ onto itself and, hence, induces a homeomorphism $f'' : \overline{S - D} \to \overline{S - D}$. Now $\pi_1(\overline{S - D}, v_\#) = \langle t_1, u_1, \ldots, t_g, u_g \mid - \rangle$ and $\partial D \in \prod_{i=1}^{g} [t_i, u_i]$. Since $f''(\partial D) = \partial D$, it follows that $f''_\#(\prod_{i=1}^{g} [t_i, u_i])$ is conjugate to $(\prod_{i=1}^{g} [t_i, u_i])^{\pm 1}$. Conversely, if an automorphism $f''_\#$ of the free group is given which maps the product of commutators to itself then there is a homeomorphism f'' and this extends to an self-homeomorphism f of S.

3.4.2. Binary Products. Let $G = \langle s_1, s_2, \ldots \mid \mathcal{R} \rangle$ be a group and let $x_1, x_2, \ldots, x_n$ be elements of G where $x_i = w_i s_i^{\varepsilon_i} w_i^{-1}$, $\varepsilon_i = \pm 1$ for $1 \le i \le m \le n$.

(a) Let $\mathcal{X} = (X_1, \ldots, X_n)$ be a system of symbols and $\Pi_\mathcal{X} = \Pi_\mathcal{X}\,(X_1, \ldots, X_n)$ a word where each symbol $X_1, \ldots, X_m$ appears exactly once (with exponent 1 or -1) and each $X_{m+1}, \ldots, X_n$ exactly twice. Then $\{x_1, \ldots, x_n; \Pi_\mathcal{X}\}$ is called a *binary product* with *factors* X_i (or x_i). The binary product is called *alternating* if every X_i, $m < i \le n$ appears once with exponent $+1$ and once with exponent -1.

(b) The following processes are called *bifurcations*.

(1) If $\Pi_{\mathcal{X}}$ has a place $\dots X_i X_k \dots, 1 \le i \le m$ then define $y_i = x_k^{-1} x_i x_k$, $y_j = x_j$, $j \neq i$, new symbols $Y_1, \dots, Y_n$ and a new word $\Pi_{\mathcal{Y}}$ by replacing, in $\Pi_{\mathcal{X}}$, the subword $X_i X_k$ by $Y_k Y_i$ and X_j by Y_j elsewhere. The inverse procedure.

(2) If $\Pi_{\mathcal{X}} = \dots X_i X_k \dots X_i^{\varepsilon} \dots$, $i > m, \varepsilon \in \{1, -1\}$ then let $y_i = x_i x_k, y_j = x_j, j \neq i$ and $\Pi_{\mathcal{Y}} = \dots Y_i \dots (Y_i Y_k^{-1})^{\varepsilon} \dots$. Similarly for situations $X_k X_i$.

(c) Two binary products are *related* if one can be converted into the other by finitely many renumberings of the first m and last $n-m$ factors, by replacing factors by their inverses, and by bifurcations.

From the definitions it follows immediately that the factors of related binary products generate the same subgroup of G and that alternating products remain alternating. In geometric language, a system of cuts of a surface S is a system of curves such that when the surface is "cut open" along these curves the result is a disc. Such a system of cuts of a surface S determines a binary product — read off the curves and the boundary edges of S as they occur around the boundary of the disc — which is alternating when the surface is orientable. Now a bifurcation of the binary product corresponds to a "bifurcation" of the surface, that is cutting the only face of the surface once by an edge and pasting the two pieces together by deleting one of the original edges. If a binary product stems from a system of cuts of a surface then all related binary products likewise stem from systems of cuts of the surface. If $m = n$ then bifurcations correspond to the generators σ_k of the braid group B_n, see 2.3.5: the bifurcation described in 3.4.2 (b,1) corresponds to σ_{i-1}. A Nielsen transformation can be applied to a binary product with factors in a free group if some factor cancels half or more of its neighbour and ultimately one obtains a *Nielsen reduced* binary product. (For an application see 4.2.14, 4.2.16.) Here we will apply this to a situation related to surfaces: Let $\hat{G}$ have free generators $\hat{s}_1, \dots, \hat{s}_m$ and either $\hat{t}_1, \hat{u}_1, \dots, \hat{t}_g, \hat{u}_g$ or $\hat{v}_1, \dots, \hat{v}_g$ and define $\Pi_* = S_1 \dots S_m \prod_{i=1}^{g} [T_i, U_i]$ or $\Pi_* = S_1 \dots S_m V_1^2 \dots V_g^2$, according as we are dealing with an orientable or non-orientable surface. In order to avoid continually distinguishing the two cases, we write the generators of $\hat{G}$ as $\hat{z}_1, \dots, \hat{z}_n$, and the binary product is written as $\{\hat{z}_1, \dots, \hat{z}_n; \Pi_*\}$.

3.4.3. Lemma. *If $\{\hat{x}_1, \dots, \hat{x}_{n'}; \Pi_{\mathcal{X}}\}$ is a binary product in $\hat{G}$ with $n' \le n$, $\hat{x}_i = \hat{w}_i \hat{z}_{r_i}^{\varepsilon_i} \hat{w}_i^{-1}$ where $i \le m'$, $1 \le r_i \le m$, $\varepsilon_i = \pm 1$ and $\Pi_{\mathcal{X}}(\hat{x}) = \Pi_*(\hat{z})$ in $\hat{G}$ then $m' = m, n' = n$ and $\{\hat{x}_1, \dots, \hat{x}_{n'}; \Pi_{\mathcal{X}}\}$ is related to $\{\hat{z}_1, \dots, \hat{z}_n; \Pi_*\}$.* □

3.4.4. Corollary. *If $\hat{\alpha}$: $\hat{G} \to \hat{G}$ is an endomorphism with $\hat{\alpha}(\hat{s}_i) = \hat{w}_i \hat{s}_{r_i}^{\varepsilon_i} \hat{w}_i^{-1}$, $1 \le i \le m$ and $\hat{\alpha}(\Pi_*(\hat{z})) = \hat{w} \Pi_*(\hat{z})^{\varepsilon} \hat{w}^{-1}$, $\varepsilon_i, \varepsilon \in \{1, -1\}$, then $\hat{\alpha}$ is an automorphism.* □

For $m = 0$, $g = 1$ this corollary is from [Nielsen 1918], for $g = 0$ it is known from the theory of braids. Now let $G = \langle z_1, \dots, z_n \mid \Pi_* \rangle$ be the fundamental group of a surface, that is $G = \langle s_1, \dots, s_m, t_1, u_1, \dots, t_g, u_g \mid$

$\prod_{i=1}^m s_i \cdot \prod_{i=1}^g [t_j, u_j]\rangle$ or $G = \langle s_1, \ldots, s_m, v_1, \ldots, v_g \mid \prod_{i=1}^m s_i \cdot \prod_{i=1}^g v_j^2\rangle$. Let $\hat{G} = \langle \hat{z}_1, \ldots, \hat{z}_n \mid -\rangle$ be "the free group in the generators of G" and consider the natural epimorphism $\hat{G} \to G$, $\hat{z}_i \mapsto z_i$. Let $\omega : G \to \mathbb{Z}_2 = \{1, -1\}$ send the v_i to -1 and all other generators to 1, i.e. $\omega(x) = 1$ if the curves from x are two-sided and $\omega(x) = -1$ if they are one-sided. By calculating binary products in the fundamental groups of surfaces the following theorem can be proved.

3.4.5. Theorem. *Each automorphism $\alpha : G \to G$ with $\alpha(s_i) = w_i s_{r_i}^{\varepsilon_i} w_i^{-1}$, $1 \le i \le m$ is induced by an automorphism $\hat{\alpha} : \hat{G} \to \hat{G}$ with $\hat{\alpha}(\hat{s}_i) = \hat{w}_i \cdot \hat{s}_{r_i}^{\varepsilon_i} \cdot \hat{w}_i^{-1}$ and $\hat{\alpha}(\Pi_*(\hat{z})) = \hat{w} \cdot \Pi_*(\hat{z}) \cdot \hat{w}^{-1}$, where $\hat{w}_i, \hat{w} \in \hat{G}$ and $\omega(w_i)\varepsilon_i = \omega(w)\varepsilon = \pm 1$.* □

By 3.4.3 and 3.4.5 the two binary products $\{\alpha(z_1), \ldots, \alpha(z_n); \Pi_*\}$ and $\{z_1, \ldots, z_n; \Pi_*\}$ are related. Since both result from systems of cuts of the surface S and these systems are of the same type it follows that there is a homeomorphism inducing α:

3.4.6. Dehn-Nielsen Theorem. *An automorphism α of the fundamental group G of a compact surface is induced by a homeomorphism of the surface if and only if $\alpha(s_i) = w_i s_{r_i}^{\varepsilon_i} w_i^{-1}$, $1 \le i \le m$. In this case $\begin{pmatrix} 1 & \ldots & m \\ r_1 & \ldots & r_m \end{pmatrix}$ is a permutation and $w_i \in G$, $\omega(w_i)\varepsilon_i = \varepsilon = \pm 1$. If the surface is orientable then the homeomorphism preserves the orientation if and only if $\varepsilon = +1$.* □

An easy consequence of Theorem 3.4.5 is the following generalization.

3.4.7. Theorem. *Each automorphism of the planar group $G = \langle z_1, \ldots z_n \mid \Pi_* \rangle$ is induced by an automorphism $\hat{\alpha}$ of the free group $\hat{G} = \langle \hat{z}_1, \ldots, \hat{z}_n \mid -\rangle$ with the following properties:*

(a) $\hat{\alpha}(\hat{s}_i) = \hat{w}_i \hat{s}_{r_i}^{\varepsilon_i} \hat{w}_i^{-1}$, $1 \le i \le m$, $h_{r_i} = h_i$, *with* $\begin{pmatrix} 1 & \ldots & m \\ r_1 & \ldots & r_m \end{pmatrix}$ *a permutation,*

(b) $\hat{\alpha}(\Pi_*(\hat{z})) = w \Pi_*^{\varepsilon} w^{-1}$ *with* $\omega(w_i)\varepsilon_i = \omega(w)\varepsilon = \pm 1$. □

Theorem 3.4.7 allows one to generalize the geometric Dehn-Nielsen Theorem 3.4.6 to planar groups without reflections. If there are no rotations s_i then lifting the homeomorphism of the surface $\mathbb{E}/G$ to $\mathbb{E}$ will give the necessary homeomorphism of $\mathbb{E}$. For the case with rotations delete "small equivariant" discs around the rotation centres and their images on $\mathbb{E}/G$, apply Theorem 3.4.6 to find a homeomorphism on $\mathbb{E}/G$ and lift it back to $\mathbb{E}$ minus the discs. The mapping thus obtained can be extended to $\mathbb{E}$:

3.4.8. Theorem. *Each automorphism α of a planar discontinuous group G with compact fundamental domain and without reflections is induced by a homeomorphism of the plane, i.e. there exists a homeomorphism f: $\mathbb{E} \to \mathbb{E}$*

such that $\alpha(g) = f \circ g \circ f^{-1}$ *for* $g \in G$. *Also* f *preserves the orientation of* $\mathbb{E}$ *if and only if* $\omega(w)\varepsilon = +1$. □

For references for the above results and historical remarks, see [ZVC 1980, 1988, 5.8+5.11].

Theorems 3.4.6 and 3.4.8 give satisfying answers to question 3.4.1 (a) restricted to the case of fundamental groups. For the homological problem we restrict ourselves to the case of closed orientable surfaces so as to use intersection numbers (see 3.1.12). The main result is the following.

3.4.9. Theorem. *Let* $t_1, u_1, \dots, t_g, u_g$ *be a system of canonical generators of* $\pi_1(S_g)$. *Let the same symbols be used for the induced basis of the homology groups* $H_1(S_g)$. *Let* $f : S_g \to S_g$ *be a homeomorphism.*

(a) *Let the induced mapping* f_*: $H_1(S_g) \to H_1(S_g)$ *be described relative to the above basis by the matrix* A. *Let* $\varepsilon = 1$ *if* f *preserves orientation, otherwise* $= -1$. *Then:*

$$A^t\ K\ A = \varepsilon K, \text{ where } K = \begin{pmatrix} E & 0 & \dots & 0 \\ 0 & E & \dots & 0 \\ & & \ddots & \\ 0 & 0 & \dots & E \end{pmatrix} \text{ and } E = \begin{pmatrix} 0 & 1 \\ -1 & 0 \end{pmatrix}.$$

(b) *Every matrix* A *with the above properties can be derived from a homeomorphism* $f : S_g \to S_g$. *(The group of matrices* A *satisfying* $A^tKA = K$ *is conjugate in* $SL(2g, \mathbb{Z})$ *to Siegel's modular group* [Siegel 1939].*)* □

As first applications of the theorems on the existence of homeomorphisms let us mention some results on simple closed curves on surfaces. We will restrict ourselves to the case of orientable closed surfaces, for the general case see [ZVC 1980,1988, 3.5]. Since an arbitrary simple closed curve which is not nullhomologous belongs to some canonical system of cuts the following assertion is a simple consequence of Theorem 3.4.9.

3.4.10. Corollary. *An element* $0 \neq x = \sum_{j=1}^{g}(a_j t_j + b_j u_j) \in H_1(S_g)$ *can be realized by a simple closed curve if and only if* $\gcd(a_1, \dots, a_g, b_1, \dots, b_g) = 1$. □

Describing the elements of $\pi_1(S_g)$ which contain simple closed curves is less easy. From the construction of canonical systems of cuts it follows immediately that for any two simple closed curves which are not nullhomologous there is a homeomorphism of S_g mapping one of the curves to the other. A simple closed nullhomologous curve γ disconnects S_g into two orientable surfaces of genera g_1 and g_2, $g = g_1 + g_2$ with γ as one boundary curve; if two such curves give the same genera, then one can be transformed into the other by a homeomorphism of S_g. Theorem 3.4.6 now implies the following characterization of simple closed curves.

3.4.11. Proposition. *Let* $\pi_1(S_g) = \langle t_1, u_1, \dots, t_g, u_g \mid \prod_{j=1}^{g}[t_j, u_j]\rangle$. *An element* $x \in \pi_1(S_g)$ *contains a simple closed curve not homologous to zero if and only if there is an automorphism* α *of* $\pi_1(S_g)$ *with* $\alpha(x) = t_1$. *The class* x *contains a simple closed curve separating* S_g *into two components of genera* g_1, g_2 *where* $g = g_1 + g_2$ *if and only if there is an automorphism* α *with* $\alpha(x) = \prod_{j=1}^{g_1}[t_j, u_j]$. □

Of course, this proposition does not give an effective procedure to decide whether a given class $x \in \pi_1(S_g) = \langle t_1, u_1, \dots, t_g, u_g \mid \prod_{j=1}^{g}[t_j, u_j]\rangle$ contains a simple closed curve or not. Algorithms for this can be based on Dehn's solution of the word or conjugacy problems for these groups, see 4.1.2. By geometric arguments one obtains:

3.4.12. Lemma. *Assume that* x *contains a simple closed curve* γ. *Let* Σ^* *be a system of cuts on* S_g *dual to the canonical system defining the generators* $t_1, \dots, u_g$, *that is every curve of* Σ^* *meets* Σ *in a single point. Then* γ *can be isotopically deformed into a curve* γ^* *which intersects* Σ^* *according to the representative of the conjugacy class of* x. □

Hence, to decide whether an element $x \in \pi_1(S_g)$ contains a simple closed curve one determines a representative W_x of the conjugacy class of x and checks whether there is a simple closed curve whose successive intersections with Σ^* parallel the successive letters of W_x. This can be checked geometrically by drawing arcs on a disc with $4g$ sides. An algebraic method is given by Theorem 3.4.5. Again let $\hat{G} = \langle \hat{t}_1, \hat{u}_1, \dots, \hat{t}_g, \hat{u}_g \mid - \rangle$. If there is an automorphism $\alpha : \pi_1(S_g) \to \pi_1(S_g)$ with, say, $\alpha(x) = t_1$ (or $\alpha(x) = \prod_{j=1}^{k}[t_j, u_j]$ for some $k < g$), then there is an automorphism $\hat{\alpha} : \hat{G} \to \hat{G}$ with $\hat{\alpha}(\prod_{j=1}^{g}[\hat{t}_j, \hat{u}_j]) = L \cdot \prod_{j=1}^{g}[\hat{t}_j, \hat{u}_j] \cdot L^{-1}$ and $\hat{\alpha}(W_x) = \hat{t}_1$ (or $\hat{\alpha}(W_x) = \prod_{j=1}^{k}[\hat{t}_j, \hat{u}_j]$, respectively). One can decide if such a $\hat{\alpha}$ exists by using the Whitehead algorithm 2.3.9. The arguments can be generalized to surfaces with boundary and to non-orientable ones. For the "small" surfaces there are other proofs.

3.4.13. Theorem. *It can be decided in a finite number of steps whether a given conjugacy class in the fundamental group of a compact surface contains a simple closed curve or not.* □

There are several proofs of this theorem with geometric arguments, see [ZVC 1980,1988, 5.15.8], [Birman-Series 1984]. Another problem is to decide when two simple closed curves on a surface are isotopic; this problem has a simple answer obtained by using the solution 4.1.2 of the conjugacy problem:

3.4.14. Theorem (Baer). *If two simple closed curves on the surface* S *which do not bound discs are homotopic then they are isotopic, too.* □

In 3.4.14 the homotopy and the isotopy may move the basepoint. There arises the question whether, for a basepoint preserving homotopy, the isotopy can be chosen in such a form that the basepoint remains fixed throughout. This question almost always has a positive answer, but there are exceptions:

3.4.15. Theorem. *Let γ, δ be two simple closed curves with the same initial point v which bound neither discs nor Möbius strips. If there is a homotopy $f_t : [0,1] \to S$, $t \in [0,1]$ with $f_0 = \gamma, f_1 = \delta$ and $f_t(0) = f_t(1) = v$, $0 \le t \le 1$ then γ and δ are isotopic under an isotopy which leaves v fixed.* □

Now consider a homeomorphism $f : S_g \to S_g$ which induces an inner automorphism $f_\# : \pi_1(S_g) \to \pi_1(S_g)$. Let $\Sigma = (\tau_1, \mu_1, \dots, \tau_g, \mu_g)$ be a canonical system of curves. Then we can apply Theorem 3.4.14 to $f(\tau_1)$ and it can be isotoped back to τ_1. Moreover, the constructions in the proof of 3.4.14 can be done in such a way that we find isotopies moving every curve of $f(\Sigma)$ back to Σ. After cutting S_g along Σ we obtain a homeomorphism of a disc onto itself which is the identity on the boundary; hence, it is, by the Alexander-Tietze deformation theorem, isotopic to the identity. These arguments can be generalized to all compact surfaces to prove the following assertion.

3.4.16. Baer Theorem

(a) *If a homeomorphism f of a surface S different from S^2 induces an inner automorphism of $\pi_1(S)$ then f is isotopic to the identity of S.*

(b) *If the homeomorphism h leaves the basepoint $v_\#$ of S fixed and induces the identity on $\pi_1(S, v_\#)$ then there is an isotopy of h to the identity of S which leaves the basepoint fixed throughout* □

For literature and generalisations, see [ZVC 1980,1988, 5.14].

If mappings are classified with respect to homotopy, one is led to the notion of a mapping class, and for surfaces these can be treated using the Dehn-Nielsen and the Baer theorems as essential tools. Let us first give the general definition and then collect results for surfaces.

3.4.17. Definition. Two homeomorphisms $f, g : X \to Y$ between two topological spaces *belong to the same mapping* or *homeotopy class* if they are homotopic. Denote the mapping class of f by $[f]$. The system of all mapping classes of homeomorphisms of a space X onto itself forms a group, called *homeotopy group* or *mapping class group* of X; it is denoted by $\mathcal{M}(X)$. The product is defined by $[f]\cdot[g] = [f \circ g]$. The definitions and notations are similar for pairs (X, A) of spaces.

3.4.18. On the Mapping Class Groups of Surfaces. If $\mathcal{H}(S)$ denotes the group of all homeomorphisms of the surface S onto itself and $\mathcal{I}(S)$ the subgroup of isotopies, then the Baer Theorem 3.4.16 (a) is equivalent to

(a) $$\mathcal{M}(S) \cong \mathcal{H}(S)/\mathcal{I}(S).$$

Let $G = \pi_1(S)$ and denote by $\mathrm{Aut}_*(G)$ the group of automorphisms of G which are induced by homeomorphisms of S and by $\mathrm{Inn}(G)$ the group of inner automorphisms of G. There is a homomorphism $\lambda : \mathcal{M}(S, v_\#) \to \mathrm{Aut}_*(G)$ sending every basepoint preserving homeomorphism to the induced isomorphism of the fundamental group. Mappings that can be deformed into the identity where the basepoint is not fixed during the deformation are mapped to inner automorphisms. Now we can reformulate the previous results as follows for the case when S is neither a sphere nor a disc.

(b) $\lambda : \mathcal{M}(S, v_\#) \to \mathrm{Aut}_*(G)$ is an epimorphism (Theorem 3.4.6).

(c) $\bar{\lambda} : \mathcal{M}(S) \to \mathrm{Aut}_*(G)/\mathrm{Inn}(G)$ is an isomorphism (Theorem 3.4.16).

Now the problem arises to determine the mapping class group of a surface. For "small" surfaces this can be done: the mapping class groups of the sphere and the Möbius strip are isomorphic to $\mathbb{Z}_2$, the non-trivial element containing the homeomorphisms which reverse the orientation of S^2 or of the boundary of the Möbius strip. The disc and the projective plane have trivial homeotopy group. (If only isotopic deformations are allowed in forming the classes, the disc has two classes.) For a torus the isomorphism λ maps $\mathcal{M}(S_1)$ onto $GL(2, \mathbb{Z})$, see 1.1.4 (h).

In principle, there is an algebraic method to determinate a presentation of the mapping class group of an arbitrary surface. By 3.4.18 we need only determine the group $\mathrm{Aut}_*(G)$. For a closed surface 3.4.5 gives $\mathrm{Aut}_*(G) \cong \{\hat{\alpha} \in \mathrm{Aut}(\hat{G}) : \hat{\alpha}(\Pi_*) = \hat{w}\Pi_*^{\varepsilon}\hat{w}^{-1}\}$ and a presentation of this stabilizer can be found using the Whitehead method, see 2.3.15. However, this is a cumbersome calculation and the generators and relators obtained give no insight into the homeotopy group. (The case of surfaces with boundary can also be handled that way.) A different proof is given in [Hatcher-Thurston 1980].

3.4.19. Theorem. *The mapping class group of a compact surface admits a finite presentation.* □

Generators for the mapping class group have been determined by many authors. It turns out that most of them are the so-called Dehn (or Lickorish) twists which are the identity outside a regular neighbourhood of a simple closed curve and are a twist inside the neighbourhood. Geometric methods for finding defining relations of the mapping class groups have been used by several authors, see [ZVC 1980,1988, 5.15]. For the special case of a sphere with $n+1$ holes, the mapping class group is closely related to the braid group B_n; good guides to results and literature are [Birman 1974], [Magnus 1974], [Maclachlan 1978].

Of particular interest is the conjugacy problem in the group of homeomorphisms or mapping classes of a surface because conjugate transformations are "of the same topological-geometric nature". But, so far, little information has been obtained from presentations of the mapping class group for the general case. Deep studies of J. Nielsen [Nielsen 1927,1929,1931] and Handel-Thurston [Handel-Thurston 1985] describe the main types; see also [Gilman 1981], [Miller 1982]. In the special case of the mapping class group of the sphere with n holes this problem is the same as the conjugacy problem for the braid group B_n which has been solved in [Garside 1969], [Makanin 1968].

Finite groups of mapping classes, in particular finite cyclic groups, of a surface S are quite well understood. Given such a group C, the idea is to pick a representative transformation $f_x : S \to S$ of each class $x \in C$ in such a way that the chosen transformations form a group isomorphic to C. Let us restrict ourselves here to orientation preserving mappings. In the special case of the torus the full group of mapping classes has quite a simple structure: it is $SL(2, \mathbb{Z})$, see 1.1.4 (h), and the finite subgroups are conjugate to the cyclic groups generated by A, B, B^2 and $A^2 = B^3$. All these classes can be realized by self-homeomorphisms of the torus of orders 4, 6, 3 and 2, respectively.

The problem for the more complicated surfaces is called the *Nielsen realisation problem* following partial results obtained by Nielsen [Nielsen 1942], see also [Zieschang 1981], [ZVC 1988]. Since the fundamental group G of such a surface has trivial centre the collection of all homeomorphisms within the classes of C forms a group H containing G as normal subgroup of index $|C|$. The extension H is uniquely determined by G and the action of C on G since C operates faithfully on G. By passing to the universal cover of S the group G becomes a planar discontinuous group. If C can be realized by a finite group of homeomorphisms of S then, again by lifting to the universal cover of S, H becomes a planar discontinuous group containing, obviously, G. So the Nielsen realisation problem can also be formulated in the following form: *Let G be a planar group and H a finite faithful extension of G, that is, $G \lhd H$, $[H : G] < \infty$ and for $h \in H$ the condition $hgh^{-1} = g \ \forall g \in G$ implies that $h \in G$ (in fact, then $h = 1$). Is H isomorphic to a planar group?* It turns out that the Nielsen problem has a positive solution, finally given by [Kerckhoff 1983], [Eckmann-Müller 1980], [Eckmann-Linnell 1983].

3.4.20. Theorem. *Every finite subgroup of the homeotopy group of a surface can be realized by a finite group of homeomorphisms. Every faithful finite extension of a planar group is isomorphic to a planar group.* □

For details see [Zieschang 1981], [ZVC 1988]. In dimension 3 the corresponding question does not have a positive answer in all cases, see 6.1.6 and [Zieschang 1981, 6.2.1], [ZVC 1988, 12.2.1].

§ 3.5. Relations to Other Theories of Surfaces

3.5.1. Topological and Smooth Surfaces. A topological Hausdorff space X with a countable basis is called a *topological surface* if every point $x \in X$ has a neighbourhood U such that $(U, \{x\})$ is homeomorphic to $(D^2, \{0\})$ or to $(D^2_+, \{0\})$; in the first case the point x is called *interior* and in the second case a *boundary point*; the boundary points form the *boundary* ∂X. By the domain preserving theorem [Spanier 1966, 4.8.16] it follows that homeomorphism maps interior points to interior ones. A surface X without boundary is called *closed* if it is compact, otherwise *open*.

There is another way to define topological surfaces (or, more generally, manifolds in arbitrary dimensions), namely using charts. We consider X as a set, the neighbourhoods U from above as subsets, the postulated homeomorphisms $\varphi_U : U \to D^2$ only as charts, that is only as bijective mappings, and postulate that all possible compositions

$$\varphi_{U'}\varphi_U^{-1} : \varphi_U(U' \cap U) \to \varphi_{U'}(U' \cap U) \subset \mathbb{R}^2$$

are topological. In this way we can construct a topological structure on X and it turns out that it is a topological surface as defined above.

This construction can be refined, for example by postulating that all compositions $\varphi_{U'}\varphi_U^{-1}$ are not only topological, but also (real) differentiable of class C^n, $n \in \{1, 2, ..., \infty\}$. This defines *differentiable* or *smooth surfaces.*

3.5.2. On Riemann Surfaces. Specific examples of topological or differentiable surfaces are those with a complex analytic structure, that is, different charts are related by complex analytic mappings between domains in $\mathbb{R}^2$. These surfaces are called *Riemann surfaces* and are intensively studied in complex analysis. It turns out that every topological orientable surface without boundary appears as a Riemann surface. However, on the same topological surface there are, in general, infinitely many non-holomorphic complex analytic structures; the only exception is S^2. Here there arises the *modular problem*, namely to find classifying moduli for complex analytic structures.

3.5.3. On the Hauptvermutung in Dimension 2. As we have seen, the classification of combinatorial compact surfaces is easily done using the Euler characteristic. It turns out that this is valid even for the classification of topological surfaces because of the theorem of Rado [Rado 1924-26], see [ZVC 1980,1988, 7.5.1], stating that all surfaces are triangulable and that the Hauptvermutung is true in dimension 2, that is, every topological surface admits a triangulation and any two triangulations of a surface have isomorphic subdivisions. Moreover, any two (real) differentiable structures are equivalent. Consequently, topological and real differentiable surfaces can be classified by homological invariants: orientability, Euler characteristic, boundary properties. This gives the background why central topological properties of surfaces can be obtained using just tools from combinatorial group theory.

3.5.4. On Coverings. To every topological surface there exists a universal covering which is a simply connected surface. If we restrict ourselves to the case of surfaces without boundary, it turns out that there are only two types of simply connected surfaces: the 2-sphere and the plane as long as we only take account of topological, differentiable or combinatorial properties of the surface. If the surfaces admit conformal structures then, by the Riemann mapping theorem, there are three possibilities: the sphere $\bar{\mathbb{C}}$, the euclidean plane $\mathbb{C}$ and the Bolyai-Lobatschevskij plane $\mathbb{H}^2$. If we consider coverings between Riemann surfaces the covering transformations become conformal mappings and these are well known for the surfaces above: they are rotations of S^2, similarities of $\mathbb{R}^2$ and motions of the Bolyai-Lobatschevskij plane (i.e. if this plane is realized by Poincaré model in the upper half plane these mappings are the linear fractional transformations $w = \frac{az+b}{cz+d}$, where $a, b, c, b \in \mathbb{R}$, $ad - bc = 1$).

This leads to another field of complex analysis: to the theory of discontinuous groups of the sphere or the euclidean or hyperbolic plane, namely to the symmetry groups of the Platonic regular polyhedra or to the crystallographic groups of the euclidean plane or to Fuchsian and Kleinian groups. Let us first explain the concept and related ones in a general form.

3.5.5. Discontinuous Groups. Let X be a topological space and G a group of homeomorphisms acting on X. The action is called *discontinuous at the point* $x_0 \in X$ if there is a neighbourhood U of x_0 such that $\{g \in G : g(U) \cap U \neq \emptyset\}$ is finite. The set of points $x \in X$ with the above property is called the *region of discontinuity* and the group is called *discontinuous* if this region is not empty. The action is called *free* if an equation $g(x) = x,\ g \in G,\ x \in X$ implies that $g = 1$. A group G is called a *topological group* if it admits, in addition to the group structure, a topology with the property that the group operation and taking the inverse become continuous mappings. A (topological) group is called *discrete* if the underlying topological space is discrete. A *topological group G acts on the space X* if the mapping $G \times X \to X,\ (g, x) \mapsto g(x)$ is continuous and the neutral element is mapped to the identity.

Examples of topological groups are given by matrix groups with coefficients in $\mathbb{R}$ or $\mathbb{C}$ or by Lie groups. A topological group acting discontinuously is discrete and one expects that discreteness implies discontinuity. However this is not true in general; a counterexample is the *Picard group* of linear fractional transformations f with $f(z) = \frac{\alpha z+\beta}{\gamma z+\delta}$ where $\alpha, \beta, \gamma, \delta \in \mathbb{Z} + i\mathbb{Z}$, $\alpha\delta - \beta\gamma = 1$; see [Lehner 1964, p. 96].

3.5.6. The Modular Problem for Tori. Let us illustrate these notions with groups of motions of the euclidean plane. A motion μ of the euclidean plane is expressed with respect to an orthonormal coordinate system in the form

$$\mu : \mathbb{R}^2 \to \mathbb{R}^2,\ x \mapsto Ax + b, \text{ where } A \in O(2),\ x = \begin{pmatrix} x_1 \\ x_2 \end{pmatrix},\ b = \begin{pmatrix} b_1 \\ b_2 \end{pmatrix} \in \mathbb{R}^2 .$$

Let G be a group of motions of the euclidean plane acting discontinuously and freely on $\mathbb{R}^2$. Moreover, let us assume that the transformations of G preserve orientation. Then every transformation of G is a translation and it follows that G acts discontinuously at every point of $\mathbb{R}^2$. This implies that the conformal structure of $\mathbb{R}^2$ induces a conformal structure on the quotient surface $T = \mathbb{R}^2/G$ and that T becomes a Riemann surface. Moreover,

$$c = \inf\{\|x - \mu(x)\| : x \in \mathbb{R}^2, \mu \in G \setminus \{\mathrm{id}\}\} > 0$$

and there is a translation $\mu_1 \in G$ and $x \in \mathbb{R}^2$ such that $\|x - \mu_1(x)\| = c$ (in fact, this holds for all $x \in \mathbb{R}^2$). Let $\mu_1(x) = x + \omega_1$, $\omega_1 \in \mathbb{R}^2$. For an arbitrary $\mu \in G$, $\mu(x) = x + w$, $w \in \mathbb{R}^2$ it follows easily from the minimality condition on c that either w and ω_1 are linearly independent or that $w = m \cdot \omega_1$ for some $m \in \mathbb{Z}$. If the second case never appears then $G \cong \mathbb{Z}$ and $\mathbb{R}^2/G$ is a cylinder; we will not deal with this case but assume that G is not cyclic. Next take look for a translation $\mu_2 \in G$ with a minimal shift length ω_2 among all $\mu \in G$ which are not multiples of μ_1. From the discontinuity property it follows that G is generated by μ_1, μ_2 and $G \cong \mathbb{Z}^2$. A consequence is that T is a torus. The conformal structure of this torus is determined by the two numbers ω_1, ω_2. If ν is a motion of $\mathbb{R}^2$ then $\mathbb{R}^2 \to \mathbb{R}^2$, $z \mapsto \nu(z)$ defines a biholomorphic mapping $f : \mathbb{R}^2/G \to \mathbb{R}^2/\nu G \nu^{-1}$, in other words, the two Riemann surfaces are of the same conformal type. Of course, tori defined by translations with shift parameters ω_1, ω_2 and $\omega_1, -\omega_2$ also have the same conformal structure. So we may represent the conformal type of T by translations $z \mapsto z+1$, $z \mapsto z + \omega_2/\omega_1$ where the imaginary part of $\omega = \omega_2/\omega_1$ is positive. This proves that *every Riemann surface obtained by the action of $\mathbb{Z} \oplus \mathbb{Z}$ as a group of translations of $\mathbb{C} = \mathbb{R} \oplus \mathbb{R}$ is determine by a* modulus ω *in the upper half plane.* (In fact, these are all possibilities for closed Riemann surfaces of genus 1.)

Now the modular problem is to decide which moduli represent equivalent Riemann surfaces. Clearly, we may replace the generating system μ_2, μ_1 from above by another generating pair $a\mu_2 + b\mu_1$, $c\mu_2 + d\mu_1$ where $A = \left(\begin{smallmatrix} a & b \\ c & d \end{smallmatrix}\right) \in GL(2, \mathbb{Z})$; the restriction that the imaginary part of the quotient ω is positive postulates that $A \in SL(2, \mathbb{Z})$. The effect on ω is that is transformed into $\omega' = \frac{a\omega + b}{c\omega + d}$, that is *two moduli ω, ω' represent conformal equivalent tori if one is the image of the other under a transformation of the modular group $PSL(2, \mathbb{Z})$*. This condition is also necessary ([Zieschang 1981, 21.11], [ZVC 1988, 7.4.13]). This explains the name *modular group.* It acts discontinuously on the Bolyai-Lobachevskij plane and the action can easily be described to obtain the presentation $PSL(2, \mathbb{Z}) = \langle a, b \mid a^2, b^3 \rangle$ from 1.1.4 (i), see, for instance, [Zieschang 1981, Sec. 21], [ZVC 1988, 8.3].

3.5.7. Remarks on the Modular Problem for Higher Genus. The approach to the classification of genus 1 Riemann surfaces can be generalized to arbitrary Riemann surfaces (of finite type) and this leads to the *Fricke moduli.* The space of all Fricke moduli, which corresponds to the upper half plane from above, is homeomorphic to some $\mathbb{R}^m$ (e.g., $m = 6g - 6$ for closed Riemann surfaces of genus g), see [Zieschang 1981, Chap. 3], [ZVC 1988, Chap. 9], but now the action of the generalization of the modular group is much more complicated. For an approach of a topological nature see [ZVC 1980, 1988, 6.6-7].

Using methods of analysis one constructs a conformal structure on the space of moduli, as was done first by Teichmüller, and one calls the space obtained *Teichmüller space.* Now one can apply theorems from analysis to get results on surfaces and their groups. Here we mention just the result 3.4.20 of Kerckhoff about the realization of finite groups of mapping classes. This approach can also be applied in investigations on the types of mappings on surfaces and in 3-topology.

Chapter 4 Cancellation Diagrams and Equations Over Groups

§ 4.1. Cancellation Diagrams

We begin by discussing Dehn's solution to the word problem for surface groups

$$\pi_1(S_g) = \langle a_1, b_1, \dots, a_g, b_g \mid [a_1, b_1] \cdots [a_g, b_g] \rangle$$

of genus at least two, following the account in [Stillwell 1980]. Now $\pi_1(S_g)$ can be represented as the fundamental group of a 2-complex with one vertex, $2g$ geometric edges and a single face attached by the path defined by the relator and this complex can be realised as an identification space of a polygon (see 1.2.6(d)). This means (cf. 3.2) that the Cayley diagram can be obtained by (irregularly) tesselating the plane with polygons with $4g$ sides, and attaching generators as labels of directed edges in such a way that the boundary curve of each polygon carries the above relator as its positive boundary label. Then any word in the generators determines, starting at a basepoint, a path in the 1-skeleton which is closed if and only if the word is a consequence of the defining relator.

For simplicity we take the case $g = 2$. The tesselation of the plane is constructed by using an infinite sequence $C_0, C_1, C_2, \dots$ of concentric circles of increasing radius. The first circle is subdivided into eight arcs and each arc labelled in such a way that in traversing the circle in a positive direction one reads the relator $[a_1, b_1][a_2, b_2]$. Now assume that the construction has been carried out to the point where C_n has been subdivided and labelled. Some of the vertices of the subdivision of C_n are endpoints of edges radiating out from C_{n-1}. From each vertex v of C_n either five or six edges are drawn to C_{n+1}, according as v is or is not the endpoint of an edge emanating from C_{n-1}. Each arc of C_{n+1} running between adjacent edges which radiate from a given vertex of C_n is divided into six subarcs and each arc of C_n running between adjacent edges which radiate out from distinct adjacent vertices of C_n is divided into five subarcs. In this way we extend the tesselation of the plane by octagons. Since each of the signed generators $a_1, b_1, \ldots, b_2^{-1}$ occurs exactly once in the defining relator of G it is not hard to see that the initial labelling given to C_0 can be uniquely extended to the required labelling of the 1-skeleton.

Now let W be any reduced word in the generators of G representing a relator and choose as basepoint a vertex of C_0. The closed path defined by W will reach some outermost C_k and to do so will travel outwards along a radial edge. At that point the path must then turn either right or left and follow a sequence of at least five edges, all of which lie on the boundary of a single polygon, before it returns to C_{k-1}. Thus the path label, namely W, contains at least six consecutive letters from some cyclic rearrangement of the defining relator or its inverse. Then the path is homotopic to the path in which the six edges are replaced by the remaining two edges of the octagon. This means that the following procedure is an algorithm which solves the word problem for G.

4.1.1. Dehn's Algorithm. *Search for a subword of the given word W which consists of more than half of (some cyclic rearrangement of) a defining relator or its inverse. If no such subword exists halt; otherwise replace the subword by the inverse of the remainder of the defining relator and freely reduce. Iterate this procedure. Then W is a relator if and only if the procedure halts in the empty word.* □

Similar considerations yield the following solution of the conjugacy problem for $\pi_1(S_g)$.

4.1.2. Solution of the Conjugacy Problem

(a) *An arbitrary element $x \in \pi_1(S_g)$ admits as* representive of its conjugacy class *a word W_x with the following properties:*

(i) *W_x is cyclically reduced.*

(ii) *W_x does not contain a subword which comprises more than half of a cyclic permutation of the defining relation or its inverse.*

(iii) *If W_x contains a subword which comprises half of $(\prod_{i=1}^{g}[a_i, b_i])^{\varepsilon}$, this subword contains a_1^{ε}.*

(b) *Let $g \geq 2$. If two words with the above properties represent the same conjugacy class then they coincide up to a cyclic permutation.* □

The idea underlying these algorithms lends itself to wide generalisation and to a geometric method for constructing examples of groups with particular kinds of properties (see below). The method, which is described below, is due originally to van Kampen [van Kampen 1933b] and was rediscovered and developed in [Lyndon 1966].

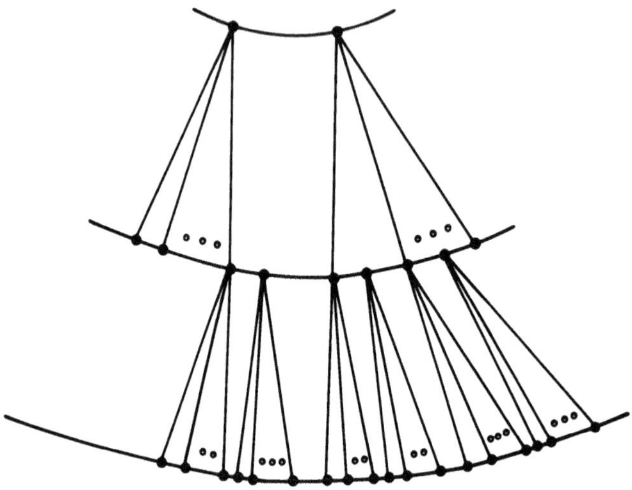

Fig. 4.1.1

4.1.3. Construction. Let $\langle \mathcal{X} \mid \mathcal{R}\rangle$ be a group presentation and let the reduced word W be a consequence of the set $\mathcal{R}$ of relators. Then, of course, W is freely equal to a product $\prod_{i=1}^{n} U_i R_i^{\varepsilon_i} U_i^{-1}$ with $R_i \in \mathcal{R}$, $i = 1, 2, \dots, n$.

A *tailed disc* over $\langle\ \mathcal{X} \mid \mathcal{R}\rangle$ is a pointed 2-complex consisting of a disc, with subdivided boundary, which has a simple path attached at a vertex. The basepoint is the unattached end of the simple path (or a vertex of the disc if the path is trivial). Moreover each oriented edge σ carries a label $f(\sigma)$ from $\mathcal{X} \cup \mathcal{X}^{-1}$, such that $f(\sigma^{-1}) = f(\sigma)^{-1}$ and the boundary circle of the disc carries the label $R^{\pm 1}$ for some $R \in \mathcal{R}$; see Fig. 4.1.2.

For each term of the product, form a tailed disc with positive boundary label $U_i R_i^{\varepsilon_i} U_i^{-1}$. Identifying the basepoints of these tailed discs gives a labelled 2-complex, which can be realised in the plane, in such a way that its positive boundary label is the product $\prod_{i=1}^{n} U_i R_i^{\varepsilon_i} U_i^{-1}$. If boundary edges with the same label and the same terminal vertex are then successively identified, the result is a pointed 2-complex K, called a *cancellation diagram* for W which, under easily satisfied hypotheses (such as the minimality of n) on the representation of the relator W as a consequence of the defining relations, satisfies the following conditions :

(1) K is connected and simply connected;
(2) K can be realised in the plane;
(3) the positive boundary label of K, beginning at the basepoint, is the reduced word W;
(4) the positive boundary label of each face is, starting at a suitable vertex, of the form $R_i^{\varepsilon_i}$.

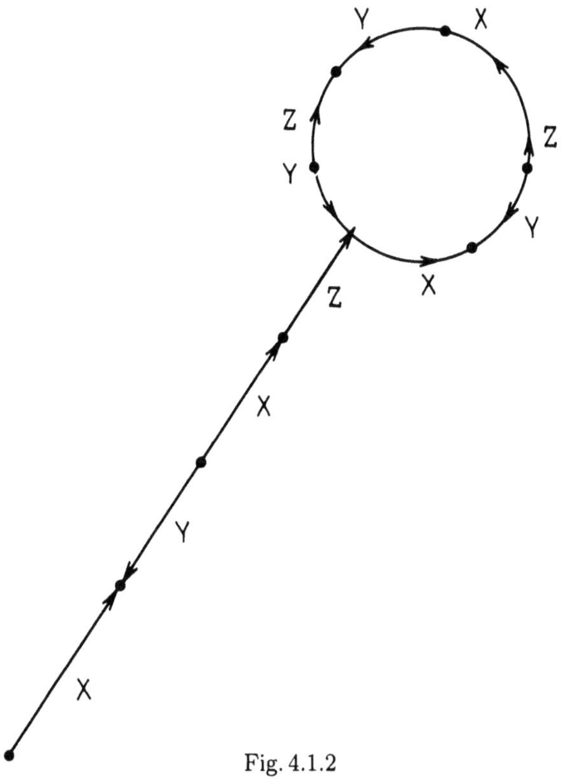

Fig. 4.1.2

Conversely given any pointed 2-complex satisfying conditions (1)-(4), its positive boundary label is a consequence of the elements of $\mathcal{R}$ as may be seen, intuitively, by unstitching the diagram to form a bouquet of tailed discs.

4.1.4. Example. Recall from 1.1.2 (c) that $A^2B^2A^{-2}B^{-2}$ is a consequence of the relation $ABA^{-1}B^{-1}$ via the equation

$$A^2B^2A^{-2}B^{-2} =$$
$$[A,B]\cdot BAB^{-1}[A,B]BA^{-1}B^{-1}\cdot B[A,B]B^{-1}\cdot B^2AB^{-1}[A,B]BA^{-1}B^{-2}.$$

The initial bouquet of tailed discs is given in Fig 4.1.3 and the final complex K is that in Fig 4.1.4. Here K is in fact embedded in the universal cover $\tilde{C}$ of the complex C but in general K is only a singular subcomplex of the universal cover $\tilde{C}$, that is there is a morphism $K \to \tilde{C}$ of complexes given by the edge labels and a random choice of base vertex in $\tilde{C}$. The advantage of working with K is that the planarity can be exploited in the same sort of way that it was exploited for $\tilde{C}$ in the case of the surface group. The key to the argument in that case was that the boundary paths of distinct regions have at most one edge in common. By an Euler characteristic argument, this forces any finite connected and simply connected subcomplex to have a region whose boundary cycle overlaps with the boundary cycle of the whole subcomplex in a relatively large number of edges. Provided that suitable *small cancellation* conditions on the set $\mathcal{R}$ of defining relators are assumed, an exactly parallel argument can be applied to any cancellation diagram K. Typical conditions guaranteeing this are as follows.

Let $\mathcal{R}$ be a set of words over $\mathcal{X}$ and let $0 < \lambda < 1$. For the sake of convenience, assume that $\mathcal{R}$ is *symmetrised*, that is $\mathcal{R}$ consists of cyclically reduced words and if $R \in \mathcal{R}$ then so does every cyclic permutation of R and R^{-1}. (Clearly any set $\mathcal{R}$ can be extended to a symmetrised set. Also it is not essential that $\mathcal{R}$ be finite.) Then $\mathcal{R}$ satisfies the *metric cancellation hypothesis* $C'(\lambda)$ if, whenever a word U is a common initial segment of two distinct elements R and R' of $\mathcal{R}$, then $|U| < \lambda|R|,\ \lambda|R'|$. The commonest applications of this condition are with $\lambda = \frac{1}{6}$. It should be observed that the symmetrised extension of the single relation defining an orientable surface group of genus g satisfies the condition $C'(\frac{1}{4g-1})$. To obtain results with $\lambda > \frac{1}{6}$ additional hypotheses are necessary. The symmetrised set $\mathcal{R}$ is said to satisfy the *triangle condition* T if, given any three elements R_1, R_2, R_3 of $\mathcal{R}$, no two of which are inverse to one another, no cancellation is possible in at least one of the words R_1R_2, R_2R_3, R_3R_1.

4.1.5. Theorem. *Let $\mathcal{R}$ satisfy the metric cancellation condition $C'(\frac{1}{6})$, or the metric cancellation condition $C'(\frac{1}{4})$ and the triangle condition T. If the cyclically reduced word W is a consequence of $\mathcal{R}$, then there exists $R \in \mathcal{R}$ such that W and R have a common subword V with $|V| > \frac{1}{2}|R|$.* □

4.1.6. Corollary. *The word problem for $\langle\ \mathcal{X} \mid \mathcal{R}\ \rangle$ is solvable by Dehn's Algorithm if $\mathcal{R}$ satisfies $C'(\frac{1}{6})$, or $C'(\frac{1}{4})$ and T.* □

Theorem 4.1.5, for the case $C'(\frac{1}{6})$ is due originally to [Tartakovskij 1949] and was strengthened in [Greendlinger 1960] who showed that except when W just consists of a single relator then some cyclic rearrangement of W has at least two non-overlapping subwords each of which constitutes more than half an element of $\mathcal{R}$ — indeed a still stronger statement actually holds. Neither Tartakovskij nor Greendlinger actually used cancellation diagrams and their methods are correspondingly more complicated. The case $C'(\frac{1}{4})$ and T was first examined by [Schiek 1956].

In his analysis of surface groups, Dehn was also able to solve the conjugacy problem, see 4.1.2. Although it is not possible to obtain an exact analogue of Dehn's solution to the conjugacy problem for presentations whose set of relators satisfy $C'(\frac{1}{6})$, or $C'(\frac{1}{4})$ and T, nonetheless a similar and extremely efficient algorithm can easily be established by the method of cancellation diagrams.

Suppose then that we have a presentation $G = \langle\ \mathcal{X} \mid \mathcal{R}\ \rangle$ with $\mathcal{R}$ symmetrised and satisfying $C'(\frac{1}{6})$, or $C'(\frac{1}{4})$ and T. Let U and V be arbitrary words, which are not conjugate in the free group $F(\mathcal{X})$ and suppose that we are trying to determine if they are conjugate in G. By Theorem 4.1.5 it suffices

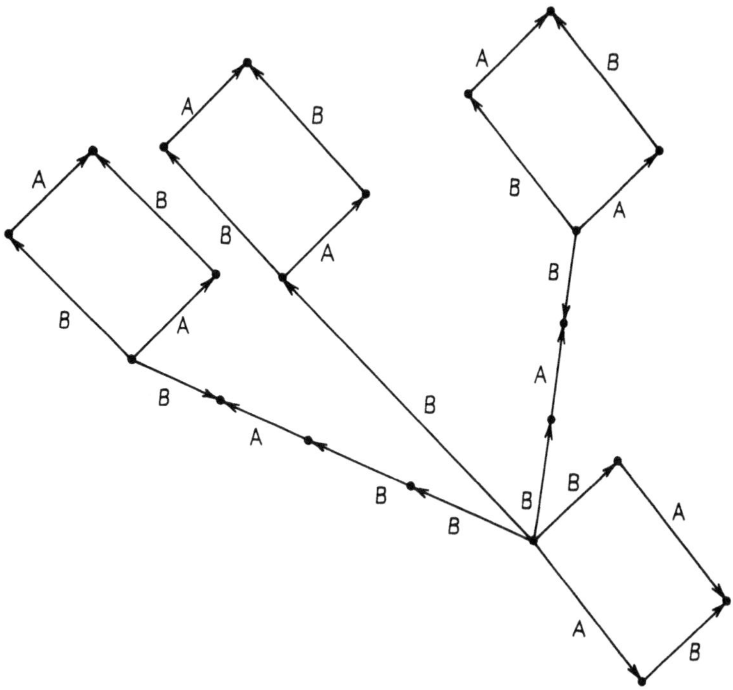

Fig. 4.1.3

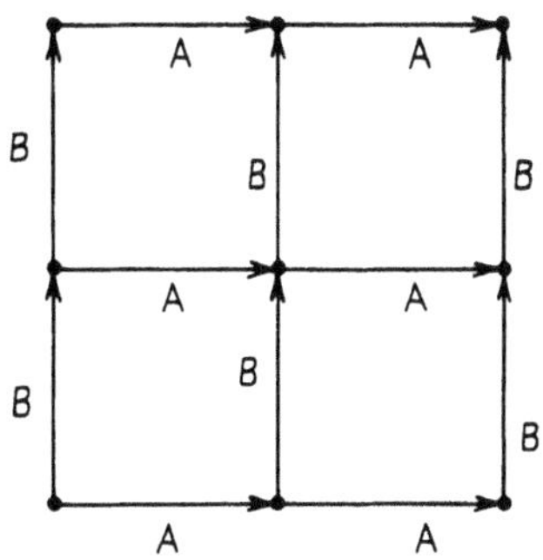

Fig. 4.1.4

to deal with the case when U and V are cyclically reduced non-trivial words, which are also $\mathcal{R}$-*reduced*, that is, no subword constitutes more than half a relator. If in fact U and V are conjugate in G then there is an equality

$$V = WUW^{-1}\prod_{i=1}^{n} W_i R_i^{\varepsilon_i} W_i^{-1}$$

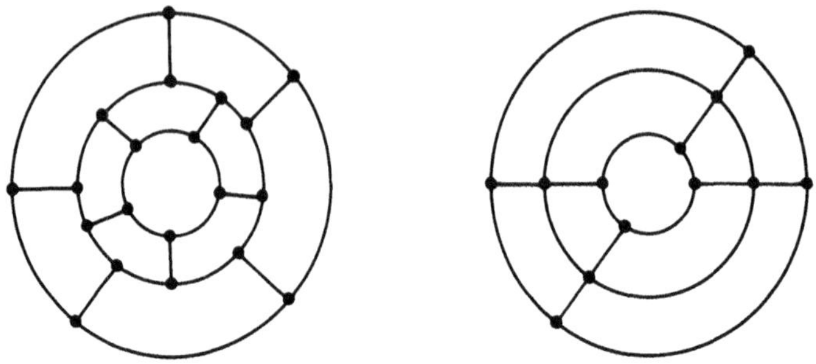

Fig. 4.1.5

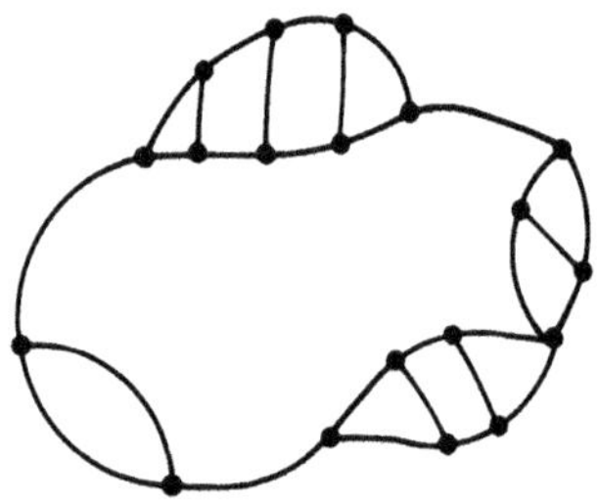

Fig. 4.1.6

in the free group $F(\mathcal{X})$. Applying Construction 4.1.3 to the right-hand side will yield a planar cancellation diagram with positive boundary label V. Furthermore one of the regions has boundary label U while the other regions have boundary labels in $\mathcal{R}$. Deleting the region with boundary label U produces an annular diagram with two boundary components (which may partly overlap — see Fig. 4.1.5,6 below) carrying the labels V and U respectively. (The condition of $\mathcal{R}$-reducedness guarantees that the boundary paths involved will be simple closed paths.) We call this a *conjugacy diagram* for U and V. An Euler characteristic argument, in the case $C'(\frac{1}{4})$ and T, then yields the following.

4.1.7. Proposition. *Let $\mathcal{R}$ satisfy $C'(\frac{1}{4})$ and T and let K be a conjugacy diagram for the $\mathcal{R}$-reduced words U and V. Then*

(a) (i) *the boundary cycle of every region of K contains an edge that is part of the boundary of K;*

(ii) *every interior vertex of K has degree two or four;*

(iii) *every region of K which has interior vertices of degree four has two such vertices.*

Furthermore

(b) *if the boundary cycle of some region of K has edges in both boundary cycles of K then*

(i) *every region has edges on both boundary cycles of* K*;*
(ii) *every interior vertex has degree two.* □

An essentially similar conclusion is obtained in the case $C'(\frac{1}{6})$. The effect of Proposition 4.1.7 is that any conjugacy diagram has one of the forms of Fig. 4.1.5, 4.1.6.

Proposition 4.1.7 also yields the following solution to the conjugacy problem for finite presentations whose set of relators satisfy $C'(\frac{1}{6})$, or $C'(\frac{1}{4})$ and T.

4.1.8. Corollary. *Let* $G = \langle \mathcal{X} \mid \mathcal{R} \rangle$ *where* $\mathcal{R}$ *satisfies* $C'(\frac{1}{6})$ *or* $C'(\frac{1}{4})$ *and* T*. If* U *and* V *are* $\mathcal{R}$*-reduced words which are conjugate in* G *(but not in* $F(\mathcal{X})$*), then some cyclic rearrangements of* U *and* V *are conjugate in* G *by an element which can be represented by a word which is a product (in* $F(\mathcal{X})$*) of at most two subwords of elements of* $\mathcal{R}$*.* □

There is also good control of torsion under small cancellation hypotheses.

4.1.9. Theorem. *Let* $G = \langle \mathcal{X} \mid \mathcal{R} \rangle$ *where* $\mathcal{R}$ *satisfies* $C'(\frac{1}{6})$*. If* W *is an element of finite order, then there exists* $R \in \mathcal{R}$ *such that* $R \equiv S^m$*, with* $m > 1$ *and* W *is conjugate to a power of* S*. In particular if no relator is a proper power then* G *is torsion-free.* □

4.1.10. Example. Proposition 4.1.7 has a nice application in [Boileau-Collins-Zieschang] to the classification of Heegaard splittings of genus 2 of certain Seifert manifolds (see 5.2.9). The problem in question is reduced to the analysis of Nielsen equivalence of pairs of generators for the triangle group $G = \langle s_1, s_2 \mid s_1^{\alpha_1} = s_2^{\alpha_2} = (s_1 s_2)^{\alpha_3} = 1 \rangle$. The issue to be resolved is when pairs $\{s_1^{p_1}, s_2^{p_2}\}$ and $\{s_1^{q_1}, s_2^{q_2}\}$, where $0 < p_i, q_i < \frac{\alpha_i}{2}$ are Nielsen equivalent. By a theorem of Nielsen, see 2.3.12, two such pairs are Nielsen equivalent only if the commutator $[s_1^{p_1}, s_2^{p_2}]$ is conjugate to the commutator $[s_1^{q_1}, s_2^{q_2}]^{\pm 1}$. Except for small values of $\alpha_1, \alpha_2, \alpha_3$ the (symmetrisation of the) above presentation satisfies $C'(\frac{1}{4})$ and T and so 4.1.7 may be applied. Suppose, for instance, that $[s_1^{p_1}, s_2^{p_2}]$ is conjugate to $[s_1^{q_1}, s_2^{q_2}]$. We want to show that the only conjugacy diagram possible is the trivial one with no regions and hence that $p_1 = q_1$ and $p_2 = q_2$. Suppose for instance that a conjugacy diagram K occurs which actually has interior vertices of degree 4. Then, say, the region D must have label $(s_1 s_2)^{\alpha_3}$, with label s_1 on the edge e, and then the region D' must have label $s_1^{-\alpha_1}$. But then the exterior boundary of K must carry $s_2(s_1 s_2)^{\alpha_3 - 2}$ as part of its label $[s_1^{p_1}, s_2^{p_2}]$ and this is imposible provided that $\alpha_3 \geq 3$.

4.1.11. Remark. The metric cancellation conditions $C'(\lambda)$ discussed above are not the most general which enable the Euler characteristic argument, which underlies the whole method, to be applied. There is also the (non-metric) condition $C(6)$ which is precisely the condition needed on $\mathcal{R}$ to ensure

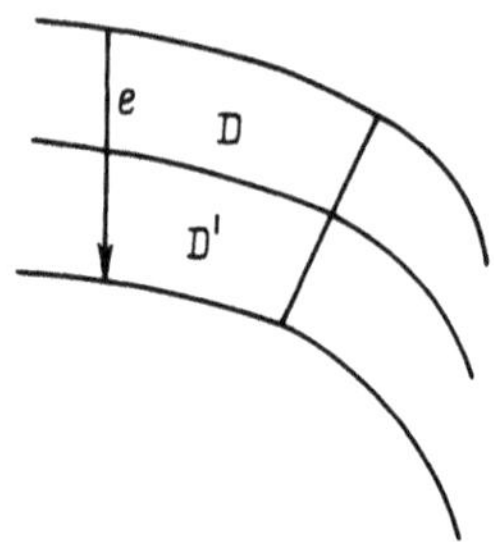

Fig. 4.1.7

that in any cancellation diagram K over $\mathcal{R}$, every region whose boundary cycle contains no boundary edges of K has degree at least 6. (Here *degree* means the number of vertices in a boundary cycle after vertices of degree 2 have been deleted.) The word and conjugacy problems are also solvable for a group given by a presentation satisfying $C(6)$ (see Remark 4.1.16).

The method of cancellation diagrams has had very significant applications in constructing examples of groups with special or unusual properties. The history of these particular questions goes back to the work of Burnside. Say that a group *has finite exponent* n if every element has finite order dividing n. Burnside proved that a finitely generated group of matrices which was of finite exponent must be finite and asked whether this held for groups in general. Positive answers to this question are presently known only for the values $n = 2, 3, 4$ and 6 [Burnside 1906], [Sanov 1940], [Hall 1957], [Adyan 1975], [Olshanskij 1979,1982]. In view of the paucity of positive results it was conjectured that for sufficiently large n, there would exist infinite finitely generated groups of exponent n. Examples of such groups were first constructed in [Novikov-Adyan 1968] by an argument of monumental complexity. A refinement of the original argument appears in [Adyan 1975] to give:

4.1.12. Theorem. *If* $n \geq 665$ *and is odd, then there exists a finitely generated infinite group of exponent* n. □

The method of Adyan and Novikov was not geometric in character but can nonetheless be described as a very general kind of small cancellation argument. It is therefore not surprising that it is possible to prove the existence of finitely generated, infinite groups of finite exponent by using cancellation diagrams. This was done by Olshanskij [Olshanskij 1982] who has used this method to construct some truly remarkable groups.

4.1.13. Theorem. *There exists a two generator infinite simple group all of whose proper subgroups are infinite cyclic.* □

4.1.14. Theorem. *There exists a two generator infinite group all of whose proper subgroups are cyclic of a fixed prime order.* □

Groups with a presentation satisfying a small cancellation condition have been put in a more general setting by the introduction of the notion of a *word hyperbolic group* described in [Gromov 1987]. This concept is defined as follows.

Let G be generated by the finite set $\mathcal{X}$. For any $g \in G$ let $|g|$ denote the length of the shortest word in $\mathcal{X}$ that represents g. Define, for any two elements g and h of G, $d(g,h) = \frac{1}{2}(|g| + |h| - |g^{-1}h|)$. Then G is called *word hyperbolic* if there exists $\delta \geq 0$ such that for any three elements g, h and k of G,

$$d(g,h) \geq \min\{d(g,k), d(h,k)\} - \delta.$$

This definition can be proved to be independent of the choice of the generating set $\mathcal{X}$.

From the point of view of cancellation diagrams, word hyperbolic groups can be characterised in the following manner.

4.1.15. Theorem [Gromov 1987, 2.3]

(a) *Let G be word hyperbolic and finitely presentable. Then for every finite presentation $G = \langle\, \mathcal{X} \mid \mathcal{R}\,\rangle$ there is a constant C such that for every consequence W of $\mathcal{R}$ there is a cancellation diagram for W whose area (see below) is bounded above by $C \cdot |W|$. In particular G has solvable word problem.*

(b) *Conversely if $G = \langle\, \mathcal{X} \mid \mathcal{R}\,\rangle$ has the property that there is a constant C such that any consequence W of $\mathcal{R}$ has a cancellation diagram whose area is bounded above by $C \cdot |W|$, then G is word hyperbolic (relative to $\mathcal{X}$).* □

The *area* of a diagram is now usually taken to be the number of regions it contains, although Gromov's original notion is a little more complex — the significant point is that if the area of a cancellation diagram for W is bounded above by $C \cdot |W|$, then there is an expression

$$W = U_1 R_1^{\varepsilon_1} U^{-1} \cdots U_n R_n^{\varepsilon_n} U_n^{-1}$$

representing W as a product of conjugates of elements of $\mathcal{R}^{\pm 1}$ determined by the cancellation diagram, with bounds for n and the lengths of the conjugating elements U_i in terms of the constant C and $|W|$.

4.1.16. Remark. It is clear that a free group is word hyperbolic — indeed with $\delta = 0$ — and it can be shown that a group with a presentation satisfying the metric cancellation condition $C'(\frac{1}{6})$ is word hyperbolic. In particular every consequence W of a set $\mathcal{R}$ relations satisfying $C'(\frac{1}{6})$ is represented by a cancellation diagram whose area is linearly bounded in terms of $|W|$. By contrast, for a presentation involving a set $\mathcal{R}$ of relations which satisfies $C(6)$ (but not $C'(\frac{1}{6})$), the word problem is actually solved by obtaining a bound for the area of cancellation diagrams which is quadratic in terms of the length of the word labelling the boundary.

Word hyperbolic groups have a number of interesting properties of which we mention only the following two.

4.1.17. Proposition [Gromov 1987, 7.4.B]. *Any finitely presented word hyperbolic group has solvable conjugacy problem.* □

4.1.18. Theorem [Gromov 1987, 5.3.C']. *Let G be a torsion-free word hyperbolic group and G_0 a finitely generated non-abelian group which is not a non-trivial free product. Then G contains at most finitely many conjugacy classes of subgroups isomorphic to G_0.* □

§ 4.2. Locally Indicable Groups and Equations Over Groups

A group is *locally indicable* if every finitely generated subgroup has the infinite cyclic group as a homomorphic image. Locally indicable groups first occur in a purely algebraic setting. If k is an integral domain and G is a group then the *group ring* kG of G is the set of all finite formal sums $\sum_{g \in G} a_g g$, $a_g \in k$ where addition is just addition of coefficients and multiplication is derived from the rule $(ag)(bh) = (ab)(gh)$ where $a, b \in k$ and $g, h \in G$. A natural question about group rings is that of determining all units — clearly, for any $g \in G$ and unit u of k, ug is a unit of kG. The question is whether there are any other units in kG. In a similar way, if $g \in G$ is a non-trivial element of finite order n, then $g-1$ and $g^{n-1}+\cdots+g+1$ are zero-divisors in kG. So again there is a natural question to ask — are such elements the only zero-divisors. In particular, if G is torsion-free, does it follow that kG has no zero-divisors. This is still an open question.

These problems about zero-divisors and units are easily settled in the affirmative when G is infinite cyclic and the justification for introducing locally indicable groups was that it provided a technical condition for carrying over the argument for the infinite cyclic case to a larger class of groups.

4.2.1. Theorem [Higman 1940]. *Let k be an integral domain and G a locally indicable group. Then the group ring kG has no zero-divisors, and no units except those of the form ug where u is a unit of k and $g \in G$.* □

Although this result was generalised to various other classes of groups, very little further work was done on locally indicable groups until the appearance of [Brodskij 1980, 1984].

4.2.2. Theorem. *Every torsion-free one-relator group is locally indicable.* □

This result, which was previously stated as Theorem 2.4.9, meant, of course, that the question of units and zero-divisors for group rings of torsion-free one-relator groups was settled. Ironically, however, this question had already been dealt with in [Lewin-Lewin] where it was proved that such a group ring was embeddable in a skew field. Thus, at first sight, Theorem 4.2.2 came too late for what was then thought of as the principal application of local indicability.

However locally indicable groups have found a major role in the theory of equations over groups.

4.2.3. Definition. If G is a group, then an *equation over* G is an equation $w = 1$ where w is an element of the free product $G * F$, where F is a free group. If $\mathcal{X}$ is a basis for F and the terms of w that lie in F are expressed as words in $\mathcal{X}$, we refer to the elements of $\mathcal{X}$ as the *variables* in the equation. An *equation* $w = 1$ *has a solution* if there exist a group H and a homomorphism $\varphi : G * F \to H$ such that $\varphi(w) = 1$ and φ restricts to a monomorphism on G.

Clearly $w = 1$ has a solution if and only if the natural map from G to the group $(G * F)/N$, where N is the normal closure of w, is an embedding. A simple example of an equation with a solution is $x^n g^{-1} = 1$ (or $x^n = g$), whenever g is not of finite order in G, since then $(G*F)/N$ is an amalgamated free product. Similarly a whole system of equations $x^{-1}g_i x = h_i \quad , \quad i = 1, 2, \dots, n$ has a solution whenever the map $g_i \mapsto h_i$ induces an isomorphism of the subgroups generated by $\{g_i : 1 \le i \le n\}$ and $\{h_i : 1 \le i \le n\}$. Here of course $(G * F)/N$ is just an HNN–extension. If the subgroups generated by $\{g_i : 1 \le i \le n\}$ and $\{h_i : 1 \le i \le n\}$ are not isomorphic, then the system has no solution. There are a number of unsolved problems in the area of equations over groups that have close connections with topology.

The most famous (or notorious) is the Kervaire-Laudenbach problem.

4.2.4. Conjecture. *For any group G and any $w \in G * C$, with C infinite cyclic, the quotient $(G * C)/N$, where N is the normal closure of w, is non-trivial.*

It should be observed that the conjecture is trivially valid for those elements w having zero exponent sum in the generator of C. A generalisation of this conjecture is therefore given by considering systems of independent equations; a system $(w_i = 1 : 1 \le i \le m)$ of equations in the variables $\{x_1, x_2, \dots, x_n\}$ is *independent* if the matrix (σ_{ij}) of exponent sums, where σ_{ij} denotes the exponent sum of x_j in w_i, has rank m.

4.2.5. Conjecture. *Any independent system of equations over a group G has a solution.*

The concept of independence can be extended to infinite systems by requiring that every finite subsystem be independent. There are currently two major results known in the direction of this conjecture.

4.2.6. Theorem [Gerstenhaber-Rothaus 1962]. *Any independent system of equations over a compact connected Lie group has a solution.* □

In purely group-theoretic terms the strongest condition under which this theorem can be applied is that every finitely generated subgroup is residually finite. This includes many types of group but equally excludes many familiar groups. The second major result brings us back to our theme.

4.2.7. Theorem [Brodskij 1980,1984], [Howie 1981]. *Any independent system of equations over a locally indicable group has a solution.*

The *proofs* by the different authors proceed along superficially dissimilar lines. The argument in [Brodskij 1980,1984] is group theoretic in character, that in [Howie 1981] topological. Since we seek to stress the links between topology and group theory, we outline the latter.

Asking whether a system $\mathcal{W} = (w_i = 1 : i \in I)$ of equations over G has a solution amounts to asking whether the natural map from G to the quotient $(G * F)/N$, where F is the free group on the variables and N is the normal closure of $\{w_i : i \in I\}$, is an embedding. One can therefore think of taking a presentation of G and constructing a presentation of $(G * F)/N$ by adjoining new generators $x_1, x_2, \ldots, x_n$ and relations $w_i = 1$, $i \in I$. Looked at from this point of view, there is a natural way to associate with the system $\mathcal{W}$ a pair (K, L) of 2–complexes, with $L \subset K$, by taking K to be the standard 2–complex realising the given presentation of $(G * F)/N$ and L to be the subcomplex corresponding to the presentation of G. The problem and its solution have an easy translation into the language of 2–complexes which turns out as follows. The system of equations is independent if and only if the second homology group $H_2(K, L; \mathbb{Z})$ of the pair (K, L) is zero and the system has a solution if and only if the natural homomorphism from the fundamental group $\pi_1(L)$ to the fundamental group $\pi_1(K)$ is an embedding. With this conceptual framework available there is a technique derived from the proof of the Sphere Theorem 5.1.6 in [Papakyriakopoulos 1957b] that lends itself to the given situation.

Formally a *tower* over a 2-complex K is a 2-complex K' and a map $g : K' \to K$ which is an alternating composite of inclusions and coverings. For the present purpose only coverings with infinite cyclic covering group will be considered in which case one refers to a C_∞-tower. A *tower lifting* of a map $f : X \to K$ is a commutative triangle

$$\begin{array}{ccc} & & K' \\ & {}^{f'}\nearrow & \downarrow g \\ X & \xrightarrow{f} & K \end{array}$$

where g is a tower. A tower lifting as above is *maximal* if the only tower lifting of f' is that in which the tower is the identity on K'. The principal, but easy, technical step is then:

4.2.8. Proposition. *Let X be a finite 2–complex and $f : X \to K$ a combinatorial map of 2-complexes, that is, each cell of X is mapped homeomorphically to a cell of K. Then there is a maximal C_∞-tower lifting of f.* □

The classical Freiheitssatz of Magnus, Theorem 2.4.1, can be interpreted as a theorem about equations over groups. The theorem asserts that *if* $G =$

$\langle \mathcal{X} \mid r = 1 \rangle$, *then any subset of* $\mathcal{X}$ *which omits a generator occurring in* r *is a free basis of the subgroup it generates.* This says precisely that any equation in a single variable (corresponding to the omitted generator) over a free group has a solution. The method of proof is closely connected with the theory of towers developed above. As explained in 2.4, the proof proceeds by expressing the one–relator group G as an HNN–extension over a base group with shorter relator — or as a subgroup of such a group. From the standpoint of towers these two possibilities correspond to infinite cyclic covers and inclusions and the whole structure of the proof corresponds to building a C_∞–tower. With hindsight it is therefore not surprising that Theorem 4.2.7 leads to a substantial generalisation of the classical Freiheitssatz.

Let A and B be two groups. A *one-relator product* of A and B is a quotient $G = (A * B)/N$ of the free product $A * B$ by the normal closure N of a single cyclically reduced element r of length at least 2.

4.2.9. Theorem. *Let* G *be a one–relator product of* A *and* B. *If* A *and* B *are locally indicable, then* A *and* B *are naturally embedded in* G.

Proof. An easy reduction shows that it suffices to consider the case when A and B are finitely generated and hence have infinite cyclic images. Passing to an infinite cyclic image of B, say, then reduces the problem to that when B is actually infinite cyclic on, say x. Just as in the classical proof there are two cases according as x does or does not have exponent sum zero in the relator r. In the latter case Theorem 4.2.7 shows that A is embedded in G. The former case is dealt with exactly as in the classical proof. □

Simple examples show that some form of hypothesis is necessary on the groups A and B if the conclusion of the theorem is to hold – the most elementary is given by taking A and B to be cyclic of orders 2 and 3 respectively and $r = ab$. The trick in this example is the use of torsion and this leads to:

4.2.10. Conjecture. *Let* A *and* B *be torsion-free and* G *a one-relator product of* A *and* B. *Then* A *and* B *are naturally embedded in* G.

Following on from the generalisation of the Freiheitssatz, a whole theory of one-relator products has been developed, see [Howie 1987]. Some typical theorems are the following.

4.2.11. Theorem. *Let* G *be a one-relator product of the locally indicable groups* A *and* B. *If the relator is not a proper power in* $A * B$ *then* G *is torsion-free and locally indicable.* □

4.2.12. Theorem. *Let* G *be a one-relator product of the groups* A *and* B *by a relator of the form* $r = s^m$ *where* $m \geq 4$. *Then* A *and* B *are naturally embedded in* G. □

This theorem reflects the fact that, as in the classical case, the theory is somewhat easier when the relator is a proper power.

In the above discussion we have examined equation *over* groups, that is, given a group G and an equation $w = 1$, with coefficients in G, we have sought to embed G in a group H in which the equation has a solution. A similar but contrasting situation occurs if one seeks to find solutions already in G (a more detailed account appears in [Lyndon-Schupp 1977]).

4.2.13. Definition. As before we consider an equation $w = 1$ where $w \in G * F$ and F is free with basis $\mathcal{X}$. We say $w = 1$ has a *solution in* G if there exists a homomorphism $\varphi : G * F \to G$ such that $\varphi(w) = 1$ and φ restricts to the identity on G.

Of particular interest is the case when G itself is a free group and w is a *quadratic* word of F, i.e. w contains each variable exactly twice. A typical result is

4.2.14. Proposition. *Let E be free on $a_1, \dots, a_{2g}$.*
(a) *The equation*

$$[x_1, x_2] \cdot \ldots \cdot [x_{2m-1}, x_{2m}] = [a_1, a_2] \ldots [a_{2g-1}, a_{2g}]$$

has a solution in E if and only if $m \geq g$.
(b) *The equation*

$$x_1^2 \cdot \ldots \cdot x_m^2 = [a_1, a_2] \cdot \ldots \cdot [a_{2g-1}, a_{2g}]$$

has a solution in E if and only if $m \geq 2g + 1$. □

A convenient concept for equations with solutions in the group of coefficients is the following:

4.2.15. Definition. Let G be a group. Then the *inner rank of* G is

$$\mathrm{Ir}(G) = \max\{d(E) : E \text{ is a free epimorphic image of } G\}.$$

If F is free with basis $\mathcal{X}$ and $w \in \mathcal{X}$ then $\mathrm{Ir}(w) = \mathrm{Ir}(G)$ where $G = \langle \mathcal{X} \mid w = 1 \rangle$. If E is free and $\varphi : E * F \to E$ defines a solution of $w = 1$ in E then rank $\varphi(F) \leq \mathrm{Ir}(w)$ and clearly $\mathrm{Ir}(w)$ is the maximal possible rank of $\varphi(F)$, where φ defines a solution in E.

4.2.16. Theorem. *Let $w \in F(\mathcal{X})$ be strictly quadratic in $\mathcal{X} = \{x_1, \dots, x_n\}$. Then $\mathrm{Ir}(w) = \left[\frac{n}{2}\right]$, the integer part of $\frac{n}{2}$.* □

Both 4.2.14 and 4.2.16 are established by applying the Nielsen method to binary products, see 3.4.2. Similar methods also yield 4.2.17 below.

Let S_g be a closed orientable surface of genus g and let E be a free group. Call two epimorphisms $\varphi_i : \pi_1(S_g) \to E$, $i = 1, 2$ *T-equivalent* if they differ by an automorphism of $\pi_1(S_g)$ and an automorphism of E, i.e. $\varphi_2 = \zeta\varphi_1\alpha$, $\alpha \in \mathrm{Aut}\ \pi_1(S_g)$, $\zeta \in \mathrm{Aut}\ E$. Note that by 4.2.16, the rank of E is at most g.

4.2.17. Proposition [Zieschang 1964]. *For any closed orientable surface* S_g *and free group* E *of rank* $d \leq g$ *there is one* T*-equivalence class of epimorphisms* $\varphi : \pi_1(S_g) \to E$. *If* $\pi_1(S_g) = \langle a_1, b_1, \ldots, a_g, b_g \mid [a_1, b_1] \ldots [a_g, b_g] \rangle$ *and* $E = \langle c_1, \ldots, c_d \mid - \rangle$ *then*

$$\begin{aligned} a_i &\mapsto c_i, \quad 1 \leq i \leq d, \\ a_i &\mapsto 1, \quad d+1 \leq i \leq g, \\ b_i &\mapsto 1, \quad 1 \leq i \leq g \end{aligned}$$

is a representative epimorphism. □

Chapter 5
3-Manifolds and Knots

§ 5.1. Fundamental Groups of 3-Manifolds

Among the usual invariants of algebraic topology the fundamental groups carry most information for 3-manifolds. Consider, e.g., a closed orientable 3-manifold M^3. Then the 0-th and 3-rd homology groups are isomorphic to $\mathbb{Z}$, and the first is the abelianized $\pi_1(M^3)$. By Poincaré duality, see [Novikov 1986, p. 52], $H^1(M) \cong H_2(M)$ and by the universal coefficient theorem $H^1(M) \cong \mathrm{Hom}(H_1(M), \mathbb{Z}) \oplus Ext(H_0(M^3), \mathbb{Z}) \cong \mathbb{Z}^{p_1}$ where p_1 is the first Betti number. All other homology and cohomology groups are trivial. So homology and cohomology is to a great extent determined by the first homology group (here we do not consider the ring structure of $H^*(M^3)$). Now the following questions arise:

Which groups can be fundamental groups of 3*-manifolds?*

To what extent does the fundamental group characterize the manifold?

Let us first deal with the first question. Only a few groups are fundamental groups of 2-manifolds. The situation is entirely different for dimension 4 because all finitely presentable groups appear as fundamental groups of 4-manifolds (see below). Hence it is not surprising that there is no obvious answer to the question of what happens in dimension 3.

5.1.1. Theorem. *For* $n \geq 4$, *every finitely presentable group* $G = \langle s_1, \ldots, s_m \mid R_1, \ldots, R_q \rangle$ *is isomorphic to the fundamental group of a closed orientable* n*-manifold.*

Proof. Clearly $\pi_1(S^1 \times S^{n-1}) \cong \mathbb{Z}$, with generator the homotopy class of the path $t \mapsto (e^{2\pi i t}, 0, \ldots, 0, 1)$. For the connected sum $M_0 = S^1 \times S^{n-1} \# \ldots \# S^1 \times S^{n-1}$, we have $\pi_1(M_0) = \langle s_1, \ldots, s_m \mid - \rangle$. For $r \in \pi_1(M_0)$ there is an embedding $f : S^1 \times D^{n-1} \to M_0$ such that $t \mapsto f(e^{2\pi i t}, 0)$ represents the homotopy class r. (Thus far the argument applies for $n \geq 3$, but not for $n = 2$; however the next step is only possible for $n > 3$.) The fundamental groups of the spaces M_0 and $M_1 = M_0 \setminus f(S^1 \times D^{n-1})$ are isomorphic. Since $S^1 \times \partial D^{n-1}$ is homeomorphic to $\partial D^2 \times S^{n-2}$ and $S^1 \times S^{n-2}$ one may glue $D^2 \times S^{n-2}$ to M_1 by $(x, y) \mapsto f(x, y)$, $x \in S^1$, $y \in S^{n-2}$ and to obtain a manifold M_2 with $\pi_1(M_2) \cong \langle s_1, \ldots, s_m \mid r \rangle$. By iterating this construction one realizes the given presentation by an n-manifold. □

Let us now consider 3-manifolds. A *handlebody H_g of genus g* is homeomorphic to the closed regular neighbourhood of a bouquet $B_g = S^1 \vee \ldots \vee S^1$ of g circles in S^3. This bouquet is a strong deformation retract of the handlebody, i.e. there is a deformation of id_{H_g} to a mapping of H_g onto B_g such that during the deformation the points of B_g remain fixed. In particular, it follows that the embedding $i : B_g \hookrightarrow H_g$ induces an isomorphism $\pi_1(B_g) \xrightarrow{\cong} \pi_1(H_g)$. Now taking any triangulation of an orientable closed 3-manifold M^3 a regular neighbourhood of the 1-skeleton is a handlebody H_g of genus g has is its complement: $H'_g = \overline{M^3 \setminus H_g} \approx H_g$. The pair (H_g, H'_g) is called a *Heegaard decomposition of* M^3. A consequence is that every path can be deformed into ∂H_g and that the g free generators of $\pi_1(H_g)$ deliver a generating system for $\pi_1(M^3)$. The attaching of H'_g to H_g can be done in such a way that first g disks (and regular neighbourhoods of them) are adjoined and finally a 3-ball. The first step corresponds to introducing relations, the last one has no effect on the fundamental group. Together this gives a *balanced presentation* with as many defining relators as generators. A similar construction can be done for non-orientable closed 3-manifolds, using non-orientable handlebodies. For compact manifolds, perhaps with boundary, the method can also be used; however now the number of defining relations will be smaller than the number of generators. For the sketch above we made the assumption that the manifold is triangulated. But this is no restriction according to the deep theorem of Moise [Moise 1977] that every 3-manifold can be triangulated and that the "Haupvermutung" is true, i.e. any two triangulations of the same manifold admit isomorphic subdivisions or, in other words, if two PL-3-manifolds M_1^3, M_2^3 are homeomorphic then there is a PL-homeomorphism $M_1^3 \to M_2^3$.

5.1.2. Theorem. *The fundamental group of a closed* 3*−manifold admits a balanced presentation. More precisely, if the manifold possesses a Heegaard decomposition of genus g then the fundamental group can be presented by g generators and g defining relators.* □

As an application we consider abelian fundamental groups of compact 3-manifolds. Clearly, if a group admits a presentation with less defining relators than generators then it also has a balanced presentation. The following groups obviously admit balanced presentations: $1 = \langle - \mid - \rangle$, $\mathbb{Z} = \langle s \mid - \rangle$, $\mathbb{Z}_n = \langle s \mid s^n \rangle$ where $n \geq 1$, $\mathbb{Z}^3 = \langle a, b, c \mid [a, b],\ [a, c],\ [b, c] \rangle$ and these groups are the fundamental groups of the 3-manifolds S^3, $S^1 \times S^2$, of lens spaces (see 5.2.2), and of $S^1 \times S^1 \times S^1$. Further abelian groups with balanced presentation are: $\mathbb{Z}^2 = \langle a, b \mid [a, b], [a, b] \rangle$ and $\mathbb{Z} \oplus \mathbb{Z}_n = \langle a, b \mid [a, b], b^n \rangle$. But except for $\mathbb{Z} \oplus \mathbb{Z}_2$

none of them is the fundamental group of a closed 3-manifold; of course, $\mathbb{Z}^2 \cong \pi_1(S^1 \times S^1 \times [0,1])$.

5.1.3. Theorem [Reidemeister 1936]. *The following is complete a list of abelian groups which can be fundamental groups of closed* 3*–manifolds:* $\mathbb{Z}_n$, $\mathbb{Z}$, $\mathbb{Z} \oplus \mathbb{Z} \oplus \mathbb{Z}$, $\mathbb{Z} \oplus \mathbb{Z}_2$.

To prove this Epstein [Epstein 1961] studies the deficiency of a group. The *deficiency* of a presentation $\langle \mathcal{X} \mid \mathcal{R} \rangle$ with m generators and q defining relators is $m-q$ and the *deficiency* def G *of the group* G is the maximum over the deficiencies of all finite presentations of G. To determine the homology groups $H_k(G)$ one has to construct a $K(G,1)$–space. Up to dimension 2 one can take the complex corresponding to any presentation of G and then add 3–cells to kill π_2, 4–cells to kill π_3 etc. By a result of [Hopf 1943], $H_2(G) \cong [F,F] \cap N/[F,N]$ where $F = \langle \mathcal{X} \mid - \rangle$; here N denotes the smallest normal subgroup of G containing R. A quite easy consequence is

5.1.4 $$\text{def}\,\langle \mathcal{X} \mid \mathcal{R} \rangle \;\le\; p_1(G) - \text{rank}(H_2(G)),$$

where $p_1(G)$ is the Betti number of G^{ab}, see 1.1.12 (a). A group G is called *efficient* if equality holds for some presentation $\langle \mathcal{X} \mid \mathcal{R} \rangle$ and in this case def G = def $\langle \mathcal{X} \mid R \rangle$. Using the canonical presentation of an abelian group it follows that finitely generated abelian groups are efficient. For a 3-manifold M^3

$$\text{def}\,\pi_1(M^3) \begin{cases} \ge 1 - \chi(M) & \text{if } \partial M^3 \ne \emptyset, \\ \ge 0 & \text{if } M^3 \text{ is closed.} \end{cases}$$

In general, for an n–manifold M^n there is at most one torsion element in $H_{n-1}(M^n)$ and this is of order 2 if it exists. This property will eliminate the groups $\mathbb{Z} \oplus \mathbb{Z}_r$, $r > 2$.

Important tools for the study of 3-manifolds are the following theorems of [Papakyriakopoulos 1957a, 1957b].

5.1.5. Loop Theorem and Dehn's Lemma. *Let* M^3 *be a* 3*–manifold and* S *a component of* ∂M^3*; let* N *be a normal subgroup of* $\pi_1(S)$ *such that*

$$(\pi_1(S) \setminus N) \cap \ker(\pi_1(S) \to \pi_1(M^3)) \ne \emptyset.$$

Then there is a 2*–cell* $D \subset M^3$ *such that* $\partial D \subset S$ *and represents an element of* $\pi_1(S) \setminus N$. □

5.1.6. Sphere Theorem. *Let* M^3 *be a* 3*-manifold and let* A *be a* $\pi_1(M^3)$*-submodule of* $\pi_2(M^3)$ *such that* $\pi_2(M^3) \setminus A \ne \emptyset$*. Then there is* $X \subset M^3$*, homeomorphic to the* 2*-sphere or to the real projective plane, and such that* X *has a neighbourhood in* M^3 *homeomorphic to* $X \times [-1,+1]$ *and such that a generating element of* $\pi_2(X) \cong \mathbb{Z}$ *represents an element of* $\pi_2(M^3) \setminus A$. □

In particular, when $\pi_2(M^3) \ne 0$ then there exists an embedded 2-sphere which is not contractible. There are close connections between these theorems and properties of the fundamental groups of 3-manifolds.

5.1.7. Corollary. *Let M^3 be an orientable 3-manifold with boundary ∂M^3, no component of which is a 2-sphere. If $\pi_2(M^3, \partial M^3) \neq 0$ then $\pi_1(M^3)$ is either infinite cyclic or a free product.* □

If we omit the assumption that M^3 is compact but postulate only that the fundamental group is finitely presentable then this will not give anything new for the groups:

5.1.8. Theorem [Scott 1973]. *If G is the fundamental group of a 3-manifold then every finitely generated subgroup of G is finitely presentable. If the fundamental group of a 3-manifold M^3 is finitely presentable then M^3 contains a compact submanifold N^3 such that the inclusion induces an isomorphism $\pi_1(N^3) \to \pi_1(M^3)$.* □

Although the homology is determined by the fundamental group let us collect some general important results from the homology theory for 3-manifolds which all follow from the fact that the Euler characteristic vanishes.

5.1.9. Theorem

(a) *The Euler characteristic of a closed 3-manifold vanishes.*

(b) *For the Betti numbers of a closed connected 3-manifold M^3 the following relations hold:*

(i) $p_0 = p_3 = 1,\ p_1 = p_2$ *if M^3 is orientable;*

(ii) $p_0 = 1,\ p_3 = 0,\ p_2 = p_1 - 1$ *if M^3 is non-orientable.*

(c) *For a non-orientable closed 3-manifold* $|H_1(M^3, \mathbb{Z})| = \infty$, *and thus,* $|\pi_1(M^3)| = \infty$.

(d) *Let M^3 be a compact manifold with boundary. Then $\chi(\partial M^3) = 2\chi(M^3)$. The first Betti number p_1 of an orientable compact manifold M^3 with boundary is at least as large as the total number of handles of ∂M^3.* □

§ 5.2. Haken Manifolds

Surfaces can be classified by their homology groups, say, together with the classes corresponding to the boundary components. A similar result does not hold for 3-manifolds. The Poincaré dodecahedron space has the same homology groups as S^3 but the fundamental group is the group from 1.1.4 (1) and has order 120. Poincaré [Poincaré 1904] introduced the concept of fundamental groups in order to show that the dodecahedron space is different from the sphere although they have the same homology. In this context he pointed out the problem, later called the Poincaré conjecture, of whether a closed 3-manifold with trivial fundamental group is homeomorphic to S^3. Whether he expected a positive answer or not cannot be decided from his published work. This question is not yet decided and it is reasonable to avoid so-called Poincaré spheres, that is 3-manifolds of the same homotopy type as S^3, by postulating that every 2-sphere within the 3-manifold under consideration bounds a ball. Moreover, for the classification problem it seems reasonable to take account not only the fundamental group but also its subgroups corresponding to the boundary.

5.2.1. Definition. Let M^3 be a 3-manifold and $S_1, \ldots, S_r$ its boundary components. The embedding of $i_j : S_j \hookrightarrow M^3$ defines a homomorphism $i_{j\#} : \pi_1(S_j, v_\#) \to \pi_1(M^3, v_\#)$ if $v_\# \in S_j$. The *peripheral system of* M^3 consists of $\pi_1(M^3)$ and the r conjugacy classes of the images of the $\pi_1(S_j)$ in $\pi_1(M^3)$.

If two manifolds are homeomorphic then there is an isomorphism between the fundamental groups sending one peripheral system one to the other. However it has long been known that the fundamental group and the peripheral system do not classify compact 3-manifolds. Counterexamples are given by the lens spaces.

5.2.2. Lens Spaces. Consider on $S^3 = \{(z_1, z_2) \in \mathbb{C}^2 : |z_1|^2 + |z_2|^2 = 1\}$ the group of order p generated by the transformation $\tau : S^3 \to S^3$, $(z_1, z_2) \mapsto (e^{\frac{2\pi i}{p}} z_1, e^{\frac{2\pi i q}{p}} z_2)$; here $p \geq 1$, $\gcd(p,q) = 1$. Then the space $L(p,q) = S^3/\langle\tau\rangle$ is called a *lens space*. Since $S^3 \to L(p,q)$ is a covering it follows that S^3 is the universal cover of $L(p,q)$ and $\pi_1(L(p,q)) \cong \mathbb{Z}_p$. Hence *$L(p,q)$ and $L(p',q')$ are homeomorphic if $p = p'$ and $q \equiv \pm q'$* mod p *or* $qq' \equiv \pm 1$ mod p. It can be shown using the Reidemeister-Franz torsion that *these conditions are also necessary* [Reidemeister 1936].

Since the universal cover of a lens space is the 3-sphere it follows that $\pi_2(L(p,q)) = 0$, $\pi_3(L(p,q)) = \mathbb{Z}$ and that there are infinitely many n with $\pi_n(L(p,q)) \neq 0$, see [Novikov 1936, p. 39]. The situation is quite different for 3-manifolds with infinite fundamental groups.

5.2.3. Theorem. *Let M^3 be a compact orientable 3–manifold such that $\pi_2(M^3) = 0$ and $|\pi_1(M^3)| = \infty$. Then M^3 is aspherical, that is $\pi_n(M^3) = 0$ for $n \geq 2$. In other words, M^3 is a $K(\pi, 1)$–space, where $\pi = \pi_1(M^3)$.*

Proof. Consider the universal cover $\tilde{p} : \tilde{M}^3 \to M^3$. Then $\pi_1(\tilde{M}^3) = 1$, thus by the Hurewicz theorem, see [Spanier 1966, 7.5.5], and the relationship between homotopy groups of total and base space of a covering, see [Novikov 1986, p. 33]: $H_2(\tilde{M}^3) = \pi_2(\tilde{M}^3) \cong \pi_2(M^3) = 0$. Moreover $\pi_3(M^3) \cong \pi_3(\tilde{M}^3) \cong H_3(\tilde{M}^3) = 0$ since $|\pi_1(M^3)| = \infty$ and therefore $\tilde{M}^3$ is not a closed 3-manifold. By iterative application of the Hurewicz theorem, $\pi_n(M^3) \cong \pi_n(\tilde{M}^3) \cong H_n(\tilde{M}^3) = 0$. □

If M^3 is a $K(\pi,1)$–space then $\pi_1(M^3)$ does not have torsion. Otherwise there would be a covering p: $\bar{M}^3 \to M^3$ where $\pi_1(\bar{M}^3) \cong \mathbb{Z}_m$ for some $m \geq 2$ and $\pi_n(\bar{M}^3) \cong \pi_n(M^3)$. However $\bar{M}^3$ is a $K(\mathbb{Z}_m, 1)$–space, hence we get the contradiction $\mathbb{Z}_m \cong \pi_k(\bar{M}) \cong \pi_k(M^3) = 0$ for $k = 3, 5, 7, \ldots$. This proves

5.2.4. Corollary. *Under the assumptions from Theorem* 5.2.3, *$\pi_1(M^3)$ is torsionfree.* □

There is a class of 3-manifolds, namely the Haken manifolds, for which the fundamental group contains enough information to classify them, and to indicate their main topological properties, as was the case for 2-manifolds.

5.2.5. Definitions. Let M^3 be a 3-manifold and S a surface which is either *properly embedded in* M^3, that is $S \cap \partial M^3 = \partial S$, or $S \subset \partial M^3$. The surface S may have several connected components.

(a) The surface S is called *compressible* in M^3 if one of the following conditions (i) - (iii) is fulfilled. Otherwise S is called *incompressible.*

(i) S is a 2-sphere which bounds a homotopy 3-cell in M^3;

(ii) S is a disc and either $S \subset \partial M^3$ or there is a homotopy 3-cell $X \subset M^3$ with $\partial X \subset S \cup \partial M^3$;

(iii) there is a 2-cell $D \subset M^3$ with $D \cap S = \partial D$ and with ∂D not contractible on S.

(b) S is $2-$*sided in* M^3 if there is an embedding $h : S \times [-1,1] \to M^3$ with $h(x,0) = x$ for all $x \in S$ and $h(S \times [-1,1]) \cap \partial M^3 = h(\partial S \times [-1,1])$.

(c) M^3 is called *irreducible* if every 2-sphere S^2 embedded in M^3 bounds a 3-ball in M^3, $\mathbb{P}^2-$*irreducible* if M^3 is irreducible and does not contain 2-sided projective planes, and *boundary irreducible* if ∂M^3 is incompressible. M^3 is *sufficiently large* if it contains a properly embedded 2-sided incompressible surface. A sufficiently large irreducible and boundary irreducible 3$-$manifold which does not contain 2$-$sided projective planes is called a *Haken manifold.*

By simple arguments using the theorems 5.1.5-6 of Papkyriakopoulos one proves the following statements: (a) A system of surfaces in M^3 or ∂M^3 is incompressible if and only if every component is incompressible. (b) A 2-sided surface S in M^3 which is not a 2-sphere is incompressible if and only if $i_{\#} : \pi_1(S) \to \pi_1(M^3)$ is injective where $i : S \hookrightarrow M^3$. (c) For a system S of 2-sided incompressible surfaces in M^3 take a regular neighbourhood $U(S)$ and define $M'^3 = \overline{M^3 \setminus U(S)}$. Then M^3 is irreducible or $\mathbb{P}^2-$irreducible if and only if M'^3 is irreducible or $\mathbb{P}^2-$irreducible, respectively. Moreover the embedding $M'^3 \hookrightarrow M^3$ induces monomorphisms of the fundamental groups of every component of M'^3.

There is an algebraic topological criterion for the existence of an incompressible surface:

5.2.6. Proposition [Waldhausen 1968]. *Let M^3 be a $\mathbb{P}^2-$irreducible surface. Then M^3 is sufficiently large if and only if one of the following conditions is fulfilled.*

(a) *$|H_1(M^3)| = \infty$ and, hence, $\pi_1(M^3)$ is an HNN-extension.*

(b) *$\pi_1(M^3) = A *_C B$ where $A \neq C \neq B$.*

If $\partial M^3 \neq \emptyset$ and if there is no $2-$sphere in ∂M^3 then $|H_1(M^3)| = \infty$, see 5.1.9 (d). □

The existence of an incompressible surfaces allows one to construct a so-called *(Haken) hierarchy*, that is, to find a finite collection of incompressible surfaces such that finally the manifold M^3 is decomposed into balls. To recover M^3 boundary surfaces have to be glued together and here the fact that for surfaces homeomorphisms are determined up to isotopy by the induced isomorphisms of the fundamental groups can be extended to the 3-dimensional case. This gives the following important result of Waldhausen.

5.2.7. Theorem [Waldhausen 1968], [Hempel 1976, 13.7]. *Let M^3, N^3 be two Haken manifolds and $f_\# : \pi_1(M^3) \to \pi_1(N^3)$ an isomorphism between the peripheral systems. Then there is a boundary preserving map $f : (M^3, \partial M^3) \to (N^3, \partial N^3)$ inducing $f_\#$. Either f is homotopic to a homeomorphism of M^3 to N^3 or M^3 is a twisted I-bundle (that is there is a fibration of M^3 over a surface S which is not trivial) over a closed surface and N^3 is the product bundle over a homeomorphic surface.* □

For Haken manifolds there is also a theorem like the Baer Theorem; thus the homeotopy group corresponds to the outer automorphism group, see [Waldhausen 1968].

5.2.8. Fibred 3-Manifolds. Assume that S is a compact surface and $h : S \to S$ a self-homeomorphism. Denote by $M^3 = S \times I/h$ the 3-manifold obtained from $S \times I$ by identifying $(x, 0)$ and $(h(x), 1)$, and let $i : S \to M^3$, $x \mapsto (x, 1)$. Then M^3 is called a *fibred* $3-$*manifold* since there is a locally trivial fibration $p : M^3 \to S^1$ with fibre S. From the long exact homotopy sequence, see [Novikov 1986, p. 32], it follows that $p_\# : \pi_1(M^3) \to \mathbb{Z} = \pi_1(S^1)$ is surjective with kernel $\pi_1(S)$ and that $i_\# : \pi_n(S) \to \pi_n(M^3)$ for $i \geq 2$ is an isomorphism. In particular, if S is not a 2-sphere or projective plane, then M^3 as well as S is a $K(\pi, 1)-$space; hence $\pi_2(M^3) = 0$. Moreover, by geometric arguments one can show that an embedded 2-sphere bounds a ball.

Conversely, Waldhausen's Theorem 5.2.7 implies an earlier theorem proved in [Stallings 1962] which *characterizes fibred 3-manifolds as those irreducible 3-manifolds whose fundamental group contains a finitely generated subgroup with quotient $\mathbb{Z}$.* Many properties of fibred 3-manifolds correspond to properties of the surface S and, hence, can be obtained from the fundamental group.

5.2.9. Seifert Manifolds. An orientable compact 3-manifold M^3 together with an effective action of the group S^1 such that no point of M^3 is fixed for all transformations of S^1 is called a *Seifert fibre space* or *Seifert manifold.* Such manifolds were introduced in [Siefert 1933] and classified in [Waldhausen 1967]. Denote by B the space of orbits and by $p : M^3 \to B$ the projection; give B the quotient topology. Then B has the topological type of a surface. (*In the following we will restrict ourselves to the case when this surface is orientable.*) An arbitrary point $x \in B$ has a disc D as neighbourhood such that $p^{-1}(D)$ is homeomorphic to $D \times S^1$. However the fibration is not trivial in the sense that $p^{-1}(y) = y \times S^1 \to M^3$ is injective for $y \in D$, since the action of $\varphi_t : D \times S^1$, $0 \leq t \leq 1$, in general, has the form $(z, w) \mapsto (z \cdot e^{2\pi i p t}, w \cdot e^{2\pi i \lambda t})$ for $z \in D = \{\zeta \in \mathbb{C} : |\zeta| \leq 1\}$, $w \in \mathbb{C}$, $|w| = 1$, $\gcd(p, \lambda) = 1$. If $p > 1$ then the orbit of a point (z, w) with $z \neq 0$ consists of p segments every point of which is the image of (z, w) for just one value of t, while the orbit of $(0, w)$ consists of only of the segment $0 \times S^1$ where every point is image of $(0, w)$ p–times, see Fig. 5.2.1; now all fibres except the central one have a neighbourhood which is trivially fibred. The central fibre is called *exceptional*. Compactness arguments show that the number m of exceptional fibres is finite. To find a presentation of the fundamental group consider first the space obtained by removing solid torus neighbourhoods of the exceptional fibres and one normal fibre. The space obtained is of the form $M' = F' \times S^1$ where F' is a surface with $m+1$ boundary components and, thus, $\pi_1(M') \cong \pi_1(F') \oplus \pi_1(S^1)$. The Seifert manifold M^3 is obtained from M' by pasting solid tori to boundary tori and this gives for every solid torus one relation $s_i^{p_i} f^{q_i}$ corresponding to the meridian of the solid torus; here $\lambda_i q_i \equiv 1 \bmod p_i$. For the exceptional fibres $p_i \geq 2$ and one may normalize so that $0 < q_i < p_i$; for the normal fibre $p_0 = 1$ and $e = q_0 \in \mathbb{Z}$ is arbitrary. The first normalization corresponds to a choice of a section on the boundary components belonging to exceptional fibres and the number e is the obstruction to extending this section to all of M'. Define the *rational Euler number* $e_0 = e - \sum_{i=1}^m q_i/p_i$. We denote the Seifert manifold by $S(g; e_0; q_1/p_1, \ldots, q_m/p_m)$. This is part (a) of the following theorem.

5.2.10. Theorem

(a) *The fundamental group of the Seifert manifold* $SF = S(g; e_0; q_1/p_1, \ldots, q_m/p_m)$ *has the following presentation:*

$$\pi_1(SF) = \langle\, s_1, \ldots, s_m, t_1, u_1, \ldots, t_g, u_g, f \mid [s_i, f],\ [t_j, f],\ [u_j, f],\ s_i^{p_i} f^{q_i},\ \prod_{i=1}^m s_i \prod_{j=1}^g [t_j, u_j] f^e \,\rangle \text{ where } e = e_0 + \sum_{i=1}^m q_i/p_i \in \mathbb{Z}.$$

(b) *If* $2g + \sum_{i=1}^m (1 - q_i/p_i) \geq 2$, *in particular, if* $2g + m \geq 4$ *then* f *has infinite order. Moreover,* f *generates the centre of* $\pi_1(SF)$ *if the inequality is strict. In the latter case* SF *is called* sufficiently complicated.

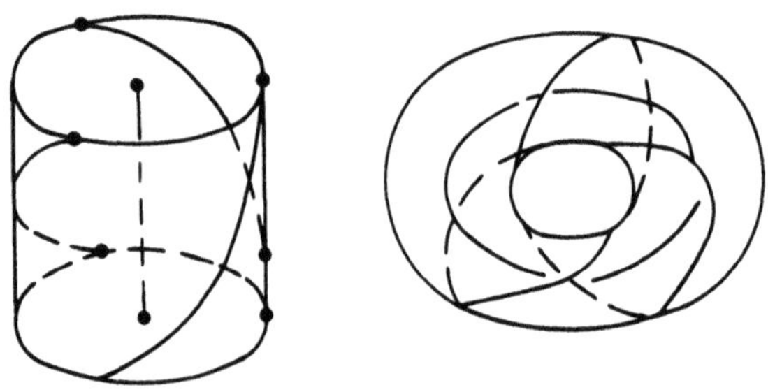

Fig. 5.2.1

(c) *Two sufficiently complicated Seifert manifolds* $S(g; e_0; q_1/p_1, \dots, q_m/p_m)$ *and* $S(g'; e'_0; q'_1/p'_1, \dots, q'_{m'}/p'_{m'})$ *where* $2g + \sum_{i=1}^{m}(1 - q_i/p_i) \geq 2$ *are homeomorphic if and only if* $g = g'$, $m = m'$ *and, after a suitable permutation of the subscripts, either* $q'_i/p'_i \equiv q_i/p_i \bmod 1$ *for* $1 \leq i \leq m$, $e_0 = e'_0$ *or* $-q'_i/p'_i \equiv q_i/p_i \bmod 1$ *for* $1 \leq i \leq m$, $e_0 = e'_0$. *This condition is also necessary and sufficient for the fundamental groups to be isomorphic.*

Proof. The arguments for (b) and (c) are similar to those for 3.2.15 and 3.4.7, respectively, see [Orlik-Vogt-Zieschang 1967]. □

5.2.11. Corollary. *If two Seifert manifolds with infinite fundamental group are homeomorphic then there is a fibre preserving homeomorphism.* □

A closed orientable 3-manifold M^3 admits infinitely many types of Heegaard decompositions, see text before 5.1.2. For, given one such decomposition, one may add handles by deleting an unknotted cube in a small ball in one of the handlebodies (stabilization procedure). The minimal genus of all Heegaard decompositions of M^3 is called the *Heegaard genus of* M^3 and is denoted by $h(M^3)$. Let $d(M^3) = d(\pi_1(M^3))$ be the rank of the fundamental group. As remarked above, $d(M^3) \leq h(M^3)$. A question of Waldhausen [Waldhausen 1978] is whether equality must hold. For the special case $d(M^3) = 0$ this reduces to the problem whether a simply connected 3-manifold has a Heegaard decomposition of genus 0; if so the manifold is a 3-sphere, and thus this is the Poincaré problem. However there are Seifert manifolds for which the question of Waldhausen has a negative answer:

5.2.12. Theorem

(a) *Let* $SF = S(0; e_0; 1/2, \dots, 1/2, q/(2\ell + 1))$, $\ell \geq 1$, $\gcd(q, 2\ell + 1) = 1$ *with an even number* m *of exceptional fibres. Then* $m - 2 = d(SF) \leq h(SF) \leq m - 1$. *If, in addition,* $e_0 = \pm 1/2(2\ell + 1)$, *then* $d(SF) = h(SF) = m - 2$. *If* $m = 4$ *and* $e_0 \neq \pm 1/2(2\ell + 1)$ *then* $2 = d(SF) < h(SF) = 3$. *For* $m > 4$ *the question of Waldhausen remains open.*

(b) *For all other Seifert manifolds Heegaard genus and rank coincide (and equals* $2g + m - 1$ *in most cases; here* g *is the genus).*

Proof. Let us consider the case $g = 0$, $m \geq 4$. Take $m-1$ exceptional fibres, connect them by simple arcs which project to simple arcs on the base surface, and take a regular neighbourhood H_{m-1} of the graph obtained. Then H_{m-1} and $\overline{SF \setminus H_{m-1}}$ are handlebodies of genus $m - 1$ as is easily seen, proving $h(SF) \leq m - 1$. Factoring out the centre $\langle f \rangle$, a Fuchsian group arises the rank of which is $m - 1$ (hence $d(SF) \geq m - 1$) except in the case of an even m and $p_1 = \dots = p_{m-1} = 2$, p_m odd. In this case one can refine the proof of 3.2.18 and prove that $d(SF) = m - 2$. (In fact, by a more careful calculation one obtains that *there is a balanced presentation for* G *with two generators and two defining relations*). It remains to show that $h(SF) = 3$ and this is done using theorems of 3-dimensional topology. For details see [Boileau-Zieschang 1984]. □

§ 5.3. On Knots and Their Groups

A knot is an isotopy class of simple closed curves in S^3. A general theory would be rather complicated because of *wild* behavior and therefore one restricts oneself to *tame knots* that are piecewise linear or smooth knots.

5.3.1. Definition. A *knot* is a simple closed polyhedral path k in some triangulation of S^3 or $\mathbb{R}^3$). *Two knots* are called *equivalent* if there is an isotopy of S^3 mapping one of them to the other. (This is called an *ambient isotopy.*) A knot is called *trivial* if it is equivalent to a triangle (or circle). The equivalence class with respect to this equivalence is also called a knot. A consequence of the assumption that the knot is tame is the existence of a regular neighbourhood $U(\mathrm{k})$ which is a solid torus $D^2 \times S^1$. Now $C = \overline{S^3 \setminus U(\mathrm{k})}$ is called the *complement of* k and $\pi_1(C)$ is called the *group of* k or, briefly, a *knot group.*

The corresponding concept is used for embeddings of several disjoint circles; then it is called a *link*, the number of components is the *multiplicity of the link.* We will mostly restrict to knots = links of multiplicity 1.

By declaring one point of the complement C to be the infinite one the theory of knots in S^3 corresponds to the theory of knots in $\mathbb{R}^3$ and one chooses the form which is more convenient for the problem considered. Mostly a knot k in $\mathbb{R}^3$ is described by an orthogonal projection to a plane. By general position arguments one proves that there is a projection such that there are only finitely many multiple points, all of them have order two, that is they are *double points*, and no vertex of the polygon k is mapped to a double point. At every double point it is marked which arc is the upper one. Homotopic deformations can be replaced by sequences of elementary alterations, namely by replacing one side of a (geometrical) triangle by the two other ones or vice versa. The projection of such an alteration is shown in Fig. 5.3.1. This allows a combinatorial theory of knots which was basic for knot theory and is still strongly in use.

From the Alexander duality theorem, see [Novikov 1986, p. 52]), it follows that $H_0(C) \cong H_1(C) \cong \mathbb{Z}$ and that all other homology groups are trivial; hence, the homology groups of the complement cannot distinguish different knots. However the knot groups are strong invariants of knots. To get a presentation of the group we consider an orthogonal projection of the knot k into the plane $z = 0$ and the mapping cylinder $Z = \{(x, y, z) : -\infty < z \leq z_0\}$ if $(x, y, z_0) \in \mathrm{k}$. The projecting cylinder Z has self-intersections in n projecting rays a_i corresponding to the n double points of the projection. The rays a_i

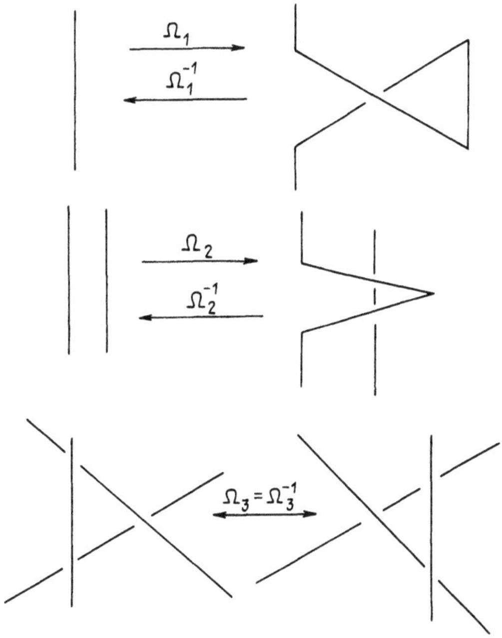

Fig. 5.3.1

decompose Z into n 2-cells Z_i where Z_i is bounded by a_{i-1}, a_i and the overcrossing arc σ_i of k. The complement of Z can be retracted parallel to the rays onto a halfspace above the knot and, thus, is contractible.

For the computation of $\pi_1(C, v)$ for some basepoint $v \in C$ observe that there is (up to a homotopy fixing v) exactly one polygonal path in general position relative to Z which intersects a given Z_i with intersection number 1 and which does not intersect the other Z_j. Paths of this type, taken for $i = 1, \ldots, n$, represent, by 1.2.19, a system $s_1, \ldots, s_n$ of generators for $\pi_1(C, v)$. In other words, every arc σ_i corresponds to a generator s_i and a word for the path ω is obtained by examining its projection and writing s_i (or s_i^{-1}) when ω undercrosses the arc σ_i from right to left (or left to right, respectively). To obtain relators consider a small path ρ_j in C encircling the ray a_j and joined to v by an arc λ_j. Then $\lambda_j \rho_j \lambda_j^{-1}$ is contractible and the corresponding word $l_j r(s_i) l_j^{-1}$ is a relator. The word $l_j r(s_i) l_j^{-1}$ can easily be read off from the knot projection; it has the form $r_j = s_j s_i^{-\eta_j} s_k^{-1} s_i^{\eta_j}$; see Fig. 5.3.2. It is easily verified that these form a system of defining relations and we get the following fundamental theorem.

5.3.2. Theorem on the Wirtinger Presentations. *Let σ_i, $i = 1, \ldots, n$ be the overcrossings of a regular projection of a knot (or link)* k. *Then the knot group admits the following so-called Wirtinger presentation:* $\pi_1(C) =$

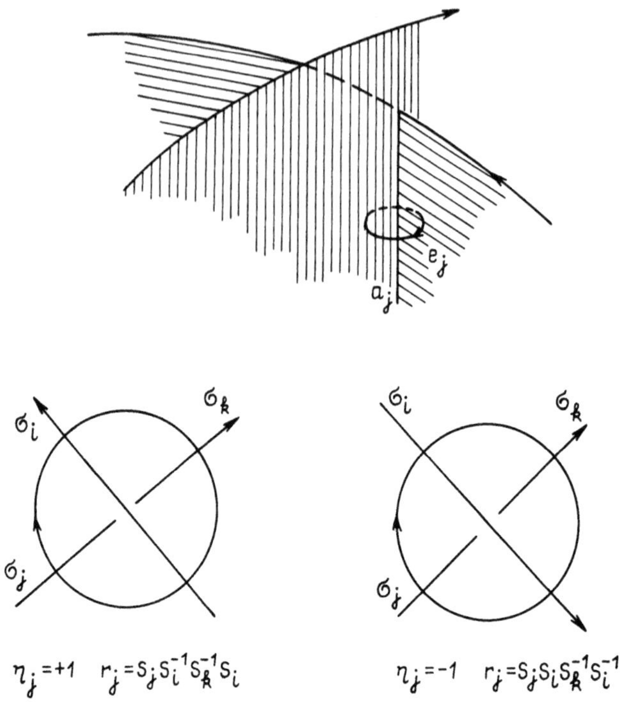

Fig. 5.3.2

$\langle s_1, \ldots, s_n \mid r_1, \ldots, r_n \rangle$. *The arc* σ_i *corresponds to the generator* s_i*; a crossing of characteristic* η_j *as in* Fig. 5.3.2 *gives rise to the defining relator* $r_j = s_j s_i^{-\eta_j} s_k^{-1} s_i^{\eta_j}$. □

A path with a projection enclosing the knot projection is contractible in C; on the other hand it is the product of n conjugates of the defining relators r_j from above. Hence:

5.3.3. Corollary. *Each of the defining relators* r_k *from* 5.3.2 *is a consequence of the other defining relators* r_j, $j \neq k$. □

Let us illustrate matters with some examples.

5.3.4. Examples

(a) **Trefoil knot:** From Fig. 5.3.3 we obtain the Wirtinger generators s_1, s_2, s_3 and defining relators $s_1s_2s_3^{-1}s_2^{-1}$ at the vertex A, $s_2s_3s_1^{-1}s_3^{-1}$ at B, $s_3s_1s_2^{-1}s_1^{-1}$ at C. One of the defining relations is a consequence of the other two and one generator and one relator can be dropped giving the presentation

$$\langle s_1, s_2 \mid s_1s_2s_1s_2^{-1}s_1^{-1}s_2^{-1} \rangle = \langle x, y \mid x^3y^2 \rangle = \langle x \mid - \rangle *_{\langle x^3 \rangle = \langle y^{-2} \rangle} \langle y \mid - \rangle$$

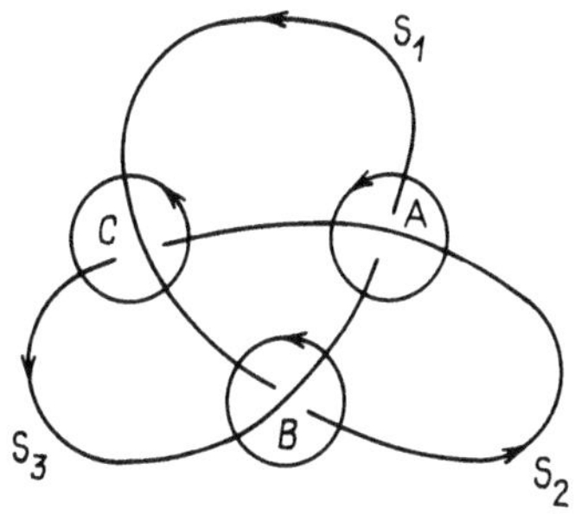

Fig. 5.3.3

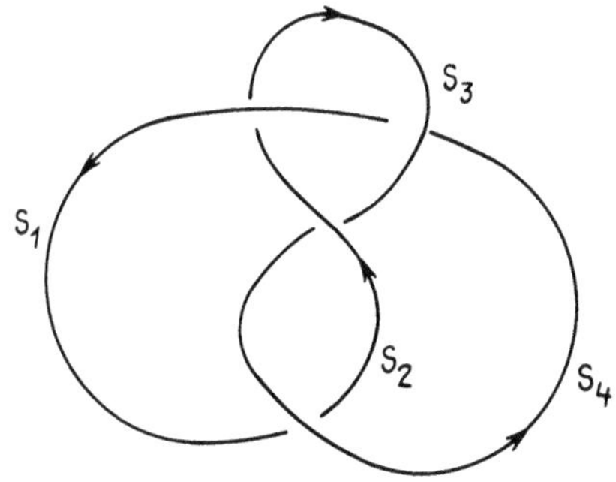

Fig. 5.3.4

where $y = s_2^{-1}s_1^{-1}s_2^{-1}$, $x = s_1s_2$. The presentation as an amalgamated free product is also of geometric origin: realize the knot on the canonical torus in S^3 by a curve running 3 times along the longitude and twice along the meridian gives this presentation by applying the Seifert-van Kampen Theorem 1.2.18 to the two solid tori. From 2.2.14 it follows that the group is neither cyclic nor abelian and that the centre is the infinite cyclic group generated by x^3. In particular this shows that the trefoil is not the trivial knot.

The second presentation can be generalized to arbitrary *torus knots* $\mathrm{t}(p,q)$, i.e. knots represented by simple closed curves lying on the standard torus in S^3 and it follows that their groups have the presentations $\langle x, y \mid x^p y^{-q}\rangle$ with $2 \le p, q,\ \gcd(p,q) = 1$. (For further properties of torus knots see 5.3.8, 10.)

(b) **Figure eight knot,** see Fig. 5.3.4. By dropping one of the Wirtinger relations we obtain the first of the following presentations; the second is obtained by expressing s_2, s_4 in terms of s_1, s_3 and then defining $s = s_1$ and $u = s_1^{-1}s_3$:

$$\begin{aligned} G &= \langle s_1, s_2, s_3, s_4 \mid s_3s_4^{-1}s_3^{-1}s_1, s_1s_2^{-1}s_1^{-1}s_3, s_4s_2^{-1}s_3^{-1}s_2\rangle \\ &= \langle s, u \mid u^{-1}sus^{-1}u^{-2}s^{-1}us\rangle\ . \end{aligned}$$

Abelianized the group becomes infinite cyclic, in accordance with the facts about the homology groups stated above; here s is mapped to a generator of $\mathbb{Z}$ while u is mapped to 0. Hence, $\{s^i : i \in \mathbb{Z}\}$ is a Schreier system of coset representatives and $\{x_i = s^ius^{-i} : i \in \mathbb{Z}\}$ the corresponding system of generators for the commutator subgroup G', see 1.3.7. The defining relations are

$$r_n = s^n(u^{-1}sus^{-1}u^{-2}s^{-1}us)s^{-n} = x_n^{-1}x_{n+1}x_n^{-2}x_{n-1}, \quad n \in \mathbb{Z}.$$

By successively dropping generators and defining relators it turns out that $G' = \langle x_0, x_1 \mid -\rangle$ is a free group of rank 2. This shows, in particular, that the figure eight knot is not trivial.

(c) **The 2-bridge knot** $\mathrm{b}(7,3)$. From Fig. 5.3.5 we determine generators and relators as before. It suffices to use the Wirtinger generators v, w which

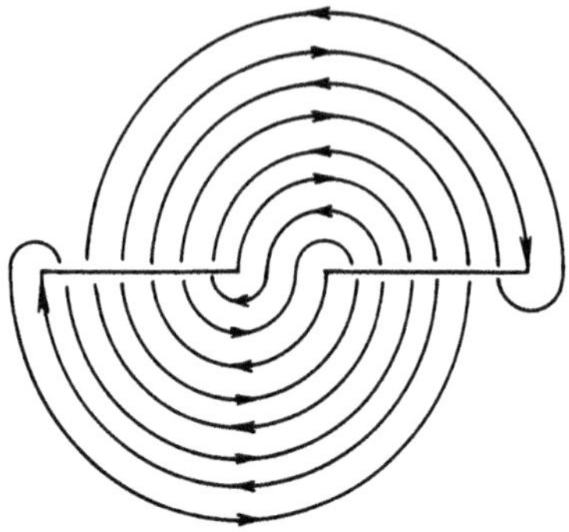

Fig. 5.3.5

correspond to the bridges, i.e. the segments overcrossing the curved arcs. One obtains the presentation

$$\begin{aligned} G &= \langle v, w \mid vwvw^{-1}v^{-1}wvw^{-1}v^{-1}w^{-1}vwv^{-1}w^{-1}\rangle \\ &= \langle s, u \mid susu^{-1}s^{-1}usu^{-1}s^{-2}u^{-1}sus^{-1}u^{-1}\rangle \end{aligned}$$

where $s = v,\ u = wv^{-1}$. A system of coset representatives for G' is $\{s^i : i \in \mathbb{Z}$ and these lead to the generators $\{x_i = s^i u s^{-i} : i \in \mathbb{Z}\}$ and the defining relations

$$x_{n+1}x_{n+2}^{-1}x_{n+1}x_{n+2}^{-1}x_n^{-1}x_{n+1}x_n^{-1}, \quad n \in \mathbb{Z}.$$

By abelianizing we obtain the relations $x_n^{-2}x_{n+1}^3x_{n+2}^{-2} = 1$. Thus the abelianized group G'/G'' and, hence, G' is not finitely generated. In fact, G' is an infinite free product $G' = \ldots *_{B_{-2}} A_{-1} *_{B_{-1}} A_0 *_{B_0} A_1 *_{B_1} \ldots$, where $A_n = \langle x_n, x_{n+1}, x_{n+2}\rangle$ is a 1-relator group and hence $B_n = \langle x_n, x_{n+1}\rangle$ is free group of rank 2 by the Freiheitssatz 2.4.1.

5.3.5. 2-Bridge Knots. More generally, a knot b in $\mathbb{R}^3$ is called a 2-*bridge knot* if it meets a plane $E \subset \mathbb{R}^3$ in 4 points A, B, C, D such that the two arcs of b in each halfspace defined by E possess orthogonal projections onto E which are simple and disjoint. Assume that the projection of the arcs from one side are line segments $w_1 = AB,\ w_2 = CD$; the other pair of arcs are projected onto disjoint simple curves v_1 (from B to C) and v_2 (from D to A). The arcs v_1, v_2 can be deformed so that their projections traverse alternatingly w_1, w_2 and v_i runs initially to $w_j, i \neq j$, comp. Fig 5.3.5. Then the number of intersection points is the same on both "bridges" w_1, w_2; denote it by $\alpha - 1$. Number the double points on each bridge successively by $1, \ldots, \alpha - 1$ in the order they occur when going from B to A or D to C, respectively. Let $|\beta|$ be the number of the intersection point where v_1 first meets w_2 and take β positive if v_1 crosses from above. It turns out that β is odd and that the number of components is 1 if α is odd and 2 otherwise. The knot described above is denoted by $\mathrm{b}(\alpha, \beta)$. It is easy to prove that *groups of* 2*-bridge knots admit* 1*-relator presentations.* Moreover, the *knots* $\mathrm{b}(\alpha, \beta)$ *and* $\mathrm{b}(\alpha', \beta')$ *are equivalent (as oriented knots as defined by the paths) if and only if* $\alpha = \alpha'$ *and* $\beta^{\pm 1} \equiv \beta' \bmod 2\alpha$; *if the orientation condition is dropped (that is one considers only the point set given by the path) then the second condition is weakened to* $\beta^{\pm 1} \equiv \beta' \bmod \alpha$. *This is also the necessary and sufficient condition for the knot groups to be isomorphic.* The classification of the knots [Schubert 1956] is done by quite difficult topological arguments; for the weaker statement there is a nice geometric proof of Seifert using a twofold branched covering, but it can also be proved purely algebraically that the groups are not isomorphic [Funcke 1975]. See also 5.3.10.

By a theorem of Alexander every piecewise linearly embedded S^2 in S^3 separates S^3 into two 3-balls and from the Sphere Theorem 5.1.6 it follows that for the complement C of a knot $\pi_2(C) = 0$. The Loop Theorem 5.1.5 implies that for a non-trivial knot the inclusion $i : \partial C \hookrightarrow C$ induces a monomorphism $i_\# : \pi_1(\partial C) \to \pi_1(C)$. Since $H_1(C) = \mathbb{Z}$, 5.2.6 and 5.2.7 yield:

5.3.6. Proposition. *The complement C of a non-trivial knot is a Haken manifold and is determined by its fundamental group together with its peripheral system* (*see* 5.2.1). □

Take a Seifert surface S, i.e. a compact connected orientable surface in S^3 bounded by the knot k, of minimal genus g and "bisect" the complement C along it. Denote the space obtained by C^* and the two copies of S by S^- and S^+. The Loop Theorem 5.1.5 implies that the embeddings $i_\pm : S^\pm \hookrightarrow C^*$ induce monomorphisms $i_{\pm *} : \pi_1(S^\pm) \to \pi_1(C^*)$. Consider now the covering $p_\infty : C_\infty \to C$ of the knot complement corresponding to the commutator subgroup G'; it consists of countably many copies C_j^* of C^* where the "upper side" S_j^+ of C_j^* is identified with the "lower side" S_{j+1}^- of C_{j+1}^*. The Seifert-van Kampen Theorem implies that

$$G' = \pi_1(C_\infty) = \ldots *_{B_{-2}} A_{-1} *_{B_{-1}} A_0 *_{B_0} A_1 *_{B_1} \cdots$$

where $A_j = \pi_1(C_j^*)$ and $B_j = \pi_1(S_j^+) = \pi_1(S_{j+1}^-)$ and the identifications and embeddings are obtained from the inclusions. In particular, since S is a compact orientable surface of genus g with one boundary curve, it follows that B_j is a free group of rank $2g$. Simple arguments, using the solution of the word problem for amalgamated free products, see 2.2.4, show that G' is finitely generated if and only if the inclusions $B_j \to A_j$ and $B_j \to A_{j+1}$ are surjective, i.e. isomorphisms. Now Stallings' Theorem 5.2.8 implies that $C^* \cong S \times [0, 1]$ and that C is fibred over S^1 with fibre S. By geometric arguments it can be shown that either both inclusions are surjective or neither is [Brown-Crowell 1965]. Let us collect the results in a theorem.

5.3.7. Theorem. *For the notation see the preceding text.*

(a) *If the commutator subgroup G' of a knot group G is finitely generated then G' is a free group of rank $2g$ where g is the genus of the knot. The knot complement admits a fibration over S^1 with fibre a Seifert surface of genus g. The knot is called* fibred.

(b) *If G' is not finitely generated then*

$$G' = \pi_1(C_\infty) = \ldots *_{B_{-2}} A_{-1} *_{B_{-1}} A_0 *_{B_0} A_1 *_{B_1} \cdots$$

and the generator t of the group of covering transformations of $p_\infty : C_\infty \to C$ induces an automorphism τ of G' such that $\tau(A_j) = A_{j+1}, \tau(B_j) = B_{j+1}$. Here $A_j \cong \pi_1(C^*)$, $B_j \cong \pi_1(S) \cong F_{2g}$, g the genus of the knot, and B_j is a proper subgroup of A_j and A_{j+1}. The subgroups B_j and B_{j+1} do not coincide. □

This throws some light on the calculations for the trefoil, the figure eight knot and the 2-bridge knot b(7, 3) in 5.3.4: the complements of the first two knots can be fibred over S^1 with fibre a torus with a hole, the complement of b(7, 3) cannot be fibred. The genera of the trefoil and the figure eight knot are 1; it turns out that those are the only fibred knots with genus 1, see [Burde-Zieschang 1985, 6.1, 15.8]. Moreover, b(7, 3) is a non-trivial knot, but at this stage it is not clear that its genus is 1 (since we did not obtain the presentation of the commutator group from a Seifert surface) which is in fact the case, see [Funcke 1978]. The commutator subgroup can be used to study geometric properties of the knot, particularly when it is finitely generated, see [Burde-Zieschang 1985, Chap. 4–6]. Using the solution of the word problem for amalgamated free products and elementary group theory one obtains the first part of the following proposition.

5.3.8. Proposition

(a) *The centre of the commutator subgroup of a knot group G is trivial. If the centre $\mathcal{Z}(G)$ of G is non-trivial then G' is finitely generated and $\mathcal{Z}(G)$ is infinite cyclic generated by an element $t^n \cdot u$, $n > 1, u \in G'$.*

(b) *The group $G(p, q) = \langle x, y \mid x^p y^{-q} \rangle$ of the torus knot* t(p, q), *where $p, q \geq 2$,* gcd$(p, q) = 1$, *has finitely generated commutator subgroup, which is a free group of rank $2g(p, q) = (p-1) \cdot (q-1)$ where $g(p, q)$ is the genus of* t(p, q).

(c) *If the fundamental group of a knot has non-trivial centre then the knot is a torus knot.*

Proof. For (b) it only remains to prove the rank formula. This can be done using the Reidemeister-Schreier method. It is simpler, though, to prove that the natural projection $G(p, q) \to \mathbb{Z}_p * \mathbb{Z}_q$ induces an isomorphism $G' \to (\mathbb{Z}_p * \mathbb{Z}_q)'$. Consider $\mathbb{Z}_p * \mathbb{Z}_q$ as the fundamental group of the 2-complex C^2 consisting of one vertex, two edges ξ, η and two faces with boundaries ξ^p, η^q and use the covering related to the commutator subgroup. For the proof of

(c) one needs topological arguments either from surface topology or from the theory of Seifert fibre spaces. (It is also part of a general result of Waldhausen that if the fundamental group of a 3-manifold M^3 has a centre then M^3 is a Seifert manifold, see 5.2.9.) □

The knot group is a strong invariant but it is in general too difficult to handle. The most efficient methods for explicit calculations use the first homology group of the infinite cyclic covering in the theory of Alexander modules and Alexander polynomials. Although this shows nicely the interaction of geometric arguments with combinatorial group theory we omit discussion. For this topic see [Burde-Zieschang 1985, Chap. 8+9].

The notion of a Heegaard decomposition, see 5.1.2 can be extended to compact 3-manifolds with boundary.

5.3.9. Heegaard Decompositions and Tunnels of Knot Exteriors. We restrict to the case where M^3 is a compact orientable 3-manifold with ∂M^3 a torus $S^1 \times S^1$. A *Heegaard decomposition of* M^3 *of genus* g is obtained from a handlebody H_g of genus g by attaching $g-1$ disjoint 2*–handles* $D_i^2 \times [0,1]$; more precisely:

$$M^3 = H_g \cup K_g \quad \text{with } K_g = (S_g \times [0,1]) \cup \bigcup_{i=1}^{g-1} (D_i^2 \times [0,1]) ,$$

$$H_g \cap K_g = \partial H_g = S_g \times 0, \; K_g \cap (D_i^2 \times [0,1]) = \partial D_i^2 \times [0,1] \subset S_g \times 1 .$$

Two Heegaard decompositions (H_g, K_g), $(\tilde{H}_g, \tilde{K}_g)$ of manifolds $M^3, \tilde{M}^3$ are called *homeomorphic* if there is a homeomorphism $M^3 \to \tilde{M}^3$ mapping H_g to $\tilde{H}_g$ and K_g to $\tilde{K}_g$. It is easy to prove that every compact orientable 3-manifold with boundary a torus admits a Heegaard decomposition of some genus; clearly, the fundamental group then has a presentation with g generators corresponding to longitudes of H_g and $g-1$ relators corresponding to meridians $\partial D_i^2 \times \frac{1}{2}$ of K_g.

In particular, the exterior of an arbitrary knot $\mathrm{k} \in S^3$ has a Heegaard decomposition. The minimal genus g of a Heegaard decomposition of the exterior of k is called the *Heegaard genus of* k.

Dual to attaching 2-handles is digging tunnels. The *tunnel number of a knot* k is the minimal number of simple arcs (*tunnels*) which must be attached to k in order that the complement of an open neighbourhood of the resulting complex is a handlebody. Two systems of tunnels are called homeomorphic if there exists a homeomorphism of S^3 preserving the knot and sending one system of tunnels to the other. Since a regular neighbourhood of an arc is a 2-handle the tunnel number of k equals the Heegaard genus of k minus 1. Therefore *knots with Heegaard genus* 2 *are the knots with tunnel number one.* The group of such a knot has a 1-relator presentation. There arise the following questions:

1) Is every knot with a 1-relator group a one tunnel knot (i.e. a Heegaard genus 2 knot)? Or stronger: Is every 1-relator presentation obtained from a Heegaard decomposition of genus 2?

2) How many different tunnels does a one tunnel knot posses? We consider these problems now for torus knots.

5.3.10. Genus 2 Heegaard Decompositions of Torus Knots. We construct a Heegaard decomposition of genus 2 for the torus knot $\mathrm{t}(p,q)$, $\gcd(p,q) = 1$. Choose $a, b \in \mathbb{Z}$ such that $pb - qa = 1$, $0 < a < p$, $0 < b < q$. Consider the standard Heegaard decomposition (H_2, H_2') of S^3 and let $\pi_1(H_2) = \langle s, t \mid - \rangle$. Then there exists a system of discs μ_1, μ_2 of H_2' such that $\overline{H_2' \setminus (U(\mu_1) \cup U(\mu_2))}$ is a 3-ball and $\partial\mu_1 \in s^p t^{-q}$, $\partial\mu_2 \in s^a t^b$ (considered in H_2). Then $\overline{S^3 \setminus (H_2 \cup U(\mu_1))}$ is a solid torus in S^3 knotted like the torus knot $\mathrm{t}(p,q)$. This gives the standard presentation $G(p,q) = \langle s, t \mid s^p t^{-q} \rangle$ of the knot group and we say that this presentation is *geometric*. In 2.2.28 (c) we have seen that the group $G(p,q)$ admits infinitely many non-Nielsen-equivalent pairs of generators, namely the generating pairs s^α, t^β where $0 < 2\alpha \leq p\beta$, $0 < 2\beta \leq q\alpha$, $\gcd(\alpha, \beta) = 1$ and they belong to 1-relator presentations if and only if $\alpha = 1$ or $\beta = 1$. It can be shown by geometric arguments that the 1-relator presentations belonging to the generating pairs (s, t^b), (s^a, t), for a, b see above, are also geometric, that is, they result from Heegaard decompositions of genus 2 of the exterior of $\mathrm{t}(p,q)$. No other classes of generating pairs are geometric. In most cases the three generating pairs (s,t), (s, t^b), (s^a, t) correspond to non-homeomorphic Heegaard decompositions or tunnels. But exceptions do occur. For details see [Boileau-Rost-Zieschang 1988].

General information about the problems 1) and 2) is quite meagre. For instance, a 2-bridge knot $\mathrm{b}(a,b)$ has Heegaard genus 2 and admits at least one Heegaard decomposition of genus 2. If $b^2 \not\equiv \pm 1 \bmod a$ then there are at least two non-homeomorphic Heegaard decompositions of genus 2; in some special cases 4 different ones are known. There are also known 1-relator presentations which are not geometric, in fact, "most" are not. However, for these knots neither all 1-relator presentations of the group nor all Heegaard decompositions of genus 2 are known nor is it known whether there are only finitely many such presentations and decompositions.

5.3.11. Braids. Place on opposite sides of a rectangular frame R in $\mathbb{R}^3$ equidistant points p_i, q_i, $1 \leq i \leq n$. Let f_i, $i = 1, \ldots, n$ be n pairwise disjoint polygonal simple strings with f_i starting at p_i and ending at $q_{\pi(i)}$, where $i \mapsto \pi(i)$ is a permutation on $\{1, \ldots, n\}$. The f_i are required to run "strictly downwards", that is, each f_i meets any plane perpendicular to the lateral edges of the rectangle at most once. The strings f_i constitute an $(n-)$*braid* z, see Fig. 5.3.6. Two braids are *equivalent* or *isotopic* if one can be moved into the other by a "level preserving" isotopy which does not move the points p_i, q_i. The equivalence class of the braid z is also called a braid and denoted by z. A braid

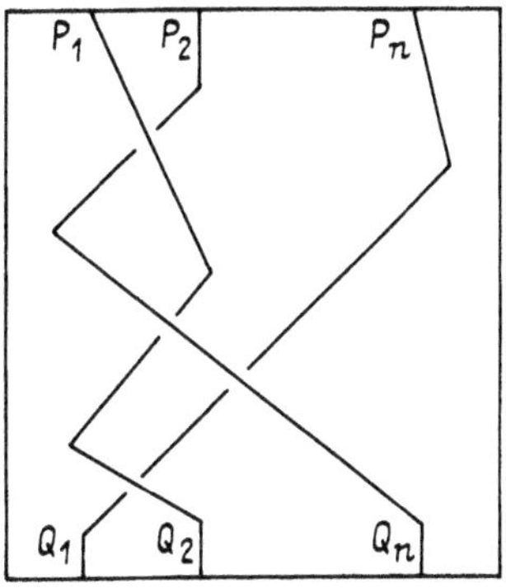

Fig. 5.3.6

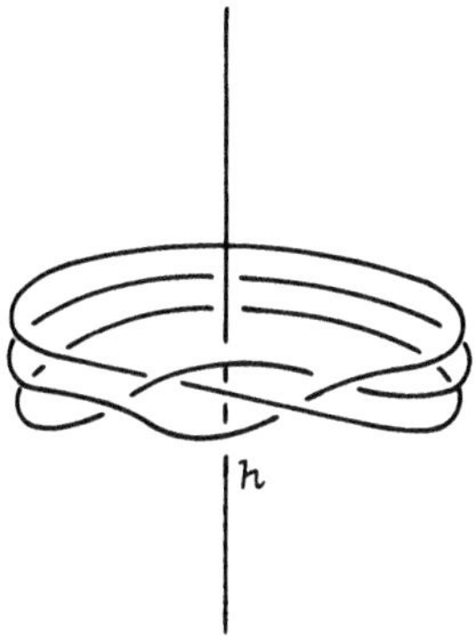

Fig. 5.3.7

can be closed with respect to an *axis* h by identifying the endpoints p_i and q_i as shown in Fig. 5.3.7. Every braid defines a *closed braid.* For closed braids one introduces a similar equivalence, postulating that at every intermediate stage of the isotopy there occurs a closed braid with respect to the axis h. A theorem of Alexander states that *every knot or link in* $\mathbb{R}^3$ *can be deformed into a closed braid.*

There is an obvious composition of two braids z, z′ by identifying the ends q_i of z with initial points p_i of z′, see Fig. 5.3.8. The braid consisting of n strings parallel to the lateral edges serves as identity and an inverse z^{-1} is obtained from z by a reflection in a plane perpendicular to the braid.

5.3.12. Proposition and Definition. *The isotopy classes of n-braids form a group called the* braid group B_n. □

Denote by σ_i the braid where the i-th string overcrosses the $(i+1)$-th and all strings except the i-th and $(i+1)$-th run parallel to the lateral edges, see Fig. 5.3.9. It is easy to see that the *elementary braids* $\sigma_1, \ldots, \sigma_{n-1}$ generate B_n. By simple geometric arguments one proves that the relations given in 1.1.4 (k) and repeated in the following proposition are defining relations.

5.3.13. Proposition

(a) *The braid group* B_n *has the following presentation:*

$$\begin{aligned} B_n = &\langle \sigma_1, \ldots, \sigma_{n-1} \mid \sigma_j \sigma_{j+1} \sigma_j \sigma_{j+1}^{-1} \sigma_j^{-1} \sigma_{j+1}^{-1},\ 1 \le j \le n-2, \\ &[\sigma_j, \sigma_k],\ 1 \le j < k-1 \le n-2 \rangle. \end{aligned}$$

(b) *Two n-braids define the same closed braid if and only if they are conjugate in* B_n. □

There are two other ways to approach the braid groups. To describe the first one we place the frame R used to define braids into a cube Q the axes of

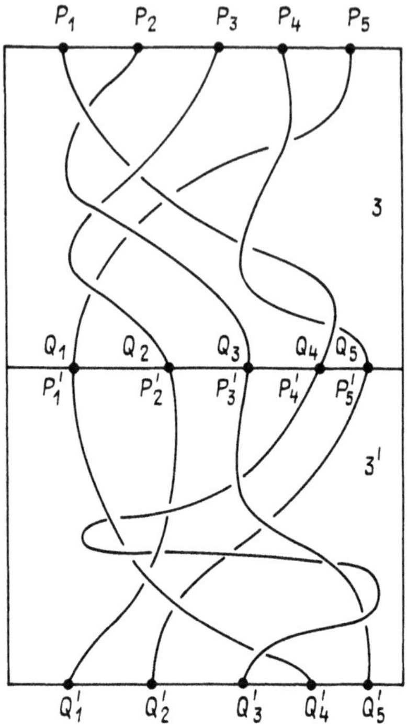

Fig. 5.3.8

which are parallel to the coordinate axes. The upper side of the frame which carries the points p_i coincides with the upper back edge of Q parallel to the x-axis and the opposite side which contains the q_i is assumed to bisect the base-face D of Q, see Fig. 5.3.10. Then one can isotopically deform any braid within Q so that the z-value of any point of the braid is constant during the deformation and the z-projection of the result consists of disjoint simple arcs. These start at the images p_i' of the p_i and end at the q_i; call such a system of arcs a *normal dissection.* Equivalent braids define isotopic normal dissections. Let $D_n = D \setminus \{q_1, \ldots, q_n\}$. For any normal dissection, there is a self-homeomorphism of D_n which maps the standard dissection consisting of the straight segments $p_i'q_i$ into the given one; it is uniquely determined up to isotopy. Thus this gives an isomorphism between the braid group B_n and the isotopy classes of self-homeomorphisms of D_n, that is the mapping class group of D_n.

The fundamental group of D_n is a free group $F_n = \langle s_1, \ldots, s_n \mid - \rangle$ where the generators correspond to circles around the holes. A homeomorphism of D_n "maps holes into holes and preserves the boundary of D_n" and, hence, induces an automorphism $\beta : F_n \to F_n$ with $\beta(s_j) = L_j s_{\pi(j)}^{\varepsilon} L_j^{-1}$, $1 \leq j \leq n$, and $\beta(\prod_{i=1}^n s_i) = (\prod_{i=1}^n s_i)^{\varepsilon}$, $\varepsilon \in \{1, -1\}$ where π is a permutation of $\{1, \ldots, n\}$.

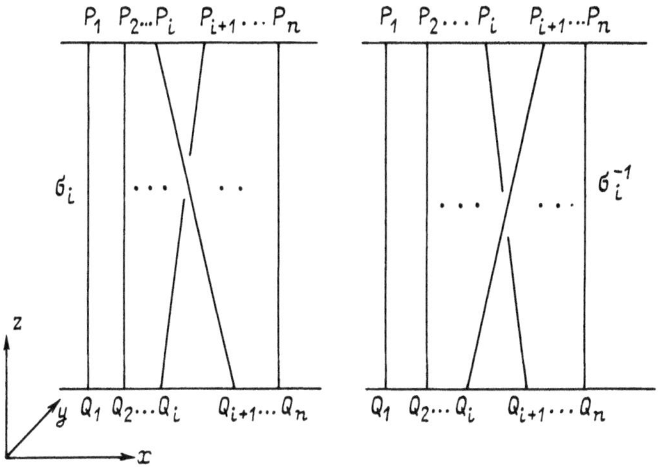

Fig. 5.3.9

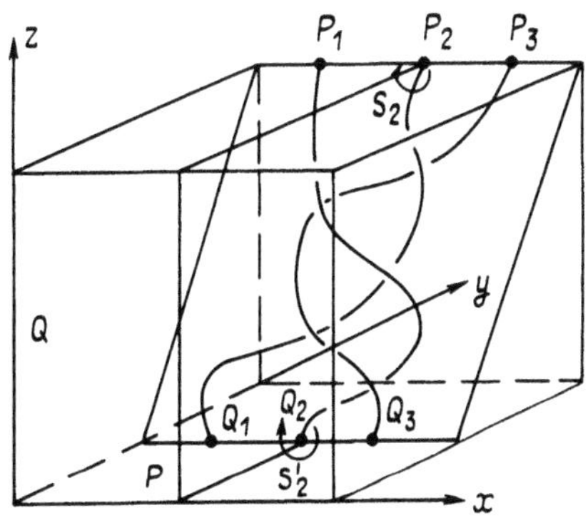

Fig. 5.3.10

If the homeomorphism corresponds to a braid then $\varepsilon = 1$. Automorphisms of this type are called *braid automorphisms*. For instance, the elementary braid σ_i induces the elementary braid automorphism $s_i \mapsto s_i s_{i+1} s_i^{-1}$, $s_{i+1} \mapsto s_i$, $s_j \mapsto s_j$ for $j \neq i, i+1$ which is also denoted by σ_i. By the Dehn-Nielsen and Baer theorems 3.4.6, 16 the mapping class group of orientation preserving self-homeomorphisms of D_n is isomorphic to the group of braid automorphisms modulo the cyclic subgroup generated by the inner automorphism defined by $\prod_{i=1}^{n} s_i$. The above interpretation of the braid group, of course, gives an immediate solution of the word problem, since one has only to check whether the induced braid automorphism, i.e. an automorphism of the free group, is the identity or not. Moreover, one can define normal forms for braids or braid automorphisms, respectively, see e.g. [Burde-Zieschang 1985, 10 B]). By purely combinatorial group theoretical arguments one proves easily that the elementary braid automorphisms σ_i, $1 \leq i \leq n-1$ generate the group of braid automorphisms, and this can be used to prove the Dehn-Nielsen Theorem 3.4.6 for the sphere with holes, see 3.4.7.

For braid groups the conjugacy problem turns out to be much more complicated than the word problem. It has been solved in [Garside 1969] and [Makanin 1968]. Since there is no easily described solution known we omit further comments. The conjugacy problem is equivalent to the classification of closed $n-$braids (see 5.3.11) which can be used as one step in the classification of arbitrary links. The second step is a stabilization procedure, i.e. adding a new trivial string to a closed braid and deciding which closed $n-$braids become equivalent $(n+1)-$braids. This shows the importance of the solution of the conjugacy problem for braid groups.

5.3.14. Configuration Spaces and Braid Groups. Let us now describe the third approach to the braid groups. We start with the definition of braids in 5.3.11. A braid z meets a plane $z = c$ in n points $(z_1, \ldots, z_n)$ for $0 \leq c \leq 1$; here $z = 1$ and $z = 0$ contain the initial points p_i and endpoints q_i, respectively. Therefore z can be interpreted as a simultaneous motion of n points in a plane E^2, $\{(z_1(t), \ldots, z_n(t)) : 0 \leq t \leq 1\}$. We construct a $2n-$manifold such that $(z_1, \ldots, z_n)$ represents a point and $(z_1(t), \ldots, z_n(t))$ a loop such that B_n becomes its fundamental group. Every $n-$tuple $(z_1, \ldots, z_n)$ represents a point $z = (x_1, y_1, x_2, y_2, \ldots, x_n, y_n)$ in the euclidean space E^{2n}, where (x_i, y_i) are the coordinates of z_i in E^2. Let Λ denote the subspace E^{2n} of points $(x_1, y_1, x_2, y_2, \ldots, x_n, y_n)$ with $(x_i, y_i) = (x_j, y_j)$ for at least one pair $i < j$. The symmetric group $\mathrm{Sym}(n)$ operates on E^{2n} by permuting the coordinate pairs (x_j, y_j), maps Λ into Λ, and operates freely on the *configuration space* $E^{2n} \setminus \Lambda$. The projection $q : E^{2n} \to \hat{E}^{2n} = E^{2n}/\mathrm{Sym}(n)$ maps Λ onto $\hat{\Lambda}$ and $q : E^{2n} \setminus \Lambda \to \hat{E}^{2n} \setminus \hat{\Lambda}$ is a regular covering of an open $2n-$manifold with $\mathrm{Sym}(n)$ as group of covering transformations.

5.3.15. Proposition. $\pi_1(\hat{E}^{2n} \setminus \hat{\Lambda}) \cong B_n$, $\pi_1(E^{2n} \setminus \Lambda) \cong I_n = ker(B_n \to \mathrm{Sym}(n))$. □

If one performs the construction above more carefully one gets a $2n-$dimensional cell complex on E^{2n} where Λ is a $(2n-2)-$dimensional subcomplex such that the action of $\mathrm{Sym}(n)$ is cellular. It turns out that $\hat{E}^{2n} \setminus \hat{\Lambda}$ is aspherical. This implies the following result, which can also be proved using arguments of $3-$dimensional topology.

5.3.16. Proposition. *The braid group B_n is torsion-free.* □

For another proof use the result of the work of [Baumslag-Taylor 1968] that for the free group F_n the group $IA(F_n) = \mathrm{Ker}(\mathrm{Aut}(F_n) \to GL(n, \mathbb{Z}))$ is torsion-free and the fact that $IA(F_n)$ contains the braid automorphisms.

The approach to braid groups via configuration spaces was used in [Arnol'd 1969] to calculate the cohomology groups of braid groups.

Chapter 6
Cohomological Methods and Ends

§ 6.1. Group Extensions and Cohomology

In this section we recall basic facts of the general theory of group extensions.

6.1.1. On the Presentation of an Extension. Consider groups $A = \langle S \mid R \rangle$ and $G = \langle T \mid Q \rangle$ and a short exact sequence $1 \to A \xrightarrow{i} E \xrightarrow{p} G \to 1$. Then E is called an *extension of* A *by* G. To get a presentation of E choose a mapping, not necessarily a homomorphism, $\alpha : G \to E$ with $p \circ \alpha = id_G$. Then $i(S) \cup \alpha(T)$ generates E and we may consider $S \cup T^\alpha$ as a set of generators of E; we write T^α instead of T to underline that α is not "natural", in contrast to i. Of course, R is a set of relators of E. Other relations arise from the action of G on A: $a \mapsto \alpha(g)^{-1} a \alpha(g)$ defines an automorphism $\gamma_g : A \to A$. This gives the relations $\alpha(t)^{-1} s \alpha(t) = \gamma_t(s)$, $s \in S$, $t \in T$ where $\gamma_t(s)$ is a word over S. Via these relations each element of E can be brought into the form $w(s)v(\alpha(t))$. This element is mapped by p to $v(t)$; hence, if it is trivial in E then $v(t)$ is a relation of G, and $v(\alpha(t)) \in i(A)$. Now $v(t)$ is a product of conjugates of the defining relations Q of G and their inverses. Therefore the value of $v(\alpha(t))$ can be calculated if the values $q(\alpha(t))$ for $q \in Q$ are known and $w(s)v(\alpha(t))$ is a relation if and only if $w(s)^{-1}$ equals the value of $v(\alpha(t))$. For each $q \in Q$ we choose a suitable word w_q. The relations described above are defining and this gives the following presentation for E:

6.1.2

$$\langle S \cup T^\alpha \mid R \cup \{\alpha(t)^{-1} s \alpha(t) \gamma_t(s)^{-1} : s \in S, t \in T\} \cup \{w_q(s) q(\alpha(t)) : q \in Q\}\rangle.$$

This suggests the following approach to the theory of extensions of a group A by a group G: Take the generators in 1-1 correspondence with the generators $S \cup T$ of A and G. As defining relations use first those of A. Next fix the action of G on A by some mapping $\alpha : T \to \text{Aut } A$ and introduce the relations corresponding to the action. Finally, replace the defining relations of G by equations of the type $w_q(s)q(\alpha(t)) = 1$ with $w_q(s) \in A$. But the following questions arise:

Is the group with the presentation 6.1.2 *an extension of* A *by* G*?*

Which presentations 6.1.2 *determine the "same" extension?*

Let us first make precise the expression "same extension". By definition two *extensions* $1 \to A \xrightarrow{i} E \xrightarrow{j} G \to 1$ and $1 \to A \xrightarrow{i'} E' \xrightarrow{j'} G \to 1$ are *equivalent* if there is an isomorphism ψ such that the diagram

$$\begin{array}{ccccccccc} 1 & \longrightarrow & A & \xrightarrow{i} & E & \xrightarrow{j} & G & \longrightarrow & 1 \\ & & \| & & \downarrow{\scriptstyle\varphi} & & \| & & \\ 1 & \longrightarrow & A & \xrightarrow{i'} & E' & \xrightarrow{j'} & G & \longrightarrow & 1 \end{array}$$

is commutative. Let $\alpha : G \to E$ and $\alpha' : G \to E'$ be mappings with $j \circ \alpha = j' \circ \alpha' = \mathrm{id}_G$ and γ_g, γ'_g the automorphisms of A corresponding to $g \in G$, that is, $\gamma_g(a) = \alpha(g)^{-1} a \alpha(g)$, $\gamma'_g(a) = \alpha'(g)^{-1} a \alpha'(g)$. Now $\alpha'(g) = \varphi\alpha(g) \cdot i'(a_g)$ for some $a_g \in A$. By a simple calculation we deduce that γ_g and γ'_g differ by the inner automorphism with factor a_g. Of course, each automorphism γ_g can be altered by an inner automorphism with no essential effect. Factor out the group of inner automorphisms and consider $\beta : G \to \mathrm{Out}\ A = \mathrm{Aut}\ A/\mathrm{Inn}\ A$, $g \mapsto \beta_g$. For $g, h \in G$ we have $\beta_{gh} = \beta_h \circ \beta_g$, since $\alpha(gh)^{-1}\alpha(g)\alpha(h) \in i(H)$; hence $\beta : G \to \mathrm{Out}\ A$ is a homomorphism. Equivalent extensions define the same homomorphism.

6.1.3. Definition. A system (G, A, β) of groups G and A and a homomorphism
$\beta : G \to \mathrm{Out}\ A$ is called an *abstract kernel.* A group E together with an exact sequence $1 \to A \xrightarrow{i} E \xrightarrow{j} G \to 1$ is called an *extension realizing the kernel* (G, A, β) if, for $e \in j^{-1}(g)$, $g \in G$, the automorphism of A defined by $a \mapsto i^{-1}[e^{-1}i(a)e]$ belongs to the class β_g.

The main problems of extension theory now are:

6.1.4. Existence Problem. *Given an abstract kernel, is there a corresponding extension?*

6.1.5. Uniqueness Problem. *Classify, up to equivalence, the extensions realizing the same kernel.*

Let us first deal with the existence problem. The following example represents an abstract kernel that cannot be realized by an extension. The next propositions show that some abstract kernels which are important for topology can be realized by extensions. In the following we use the letter $\mathcal{Z}$ for a cyclic group in the role of G.

6.1.6. Example. Consider $\pi_1(S_{2p+1}) = \langle t_1, u_1, \dots, t_{2p+1}, u_{2p+1} \mid \prod_{i=1}^{2p+1} [t_i, u_i] \rangle$ and put $A = \pi_1(S_{2p+1}) \oplus \langle z \mid - \rangle$. By $t_1 \mapsto t_1 z$, $u_1 \mapsto u_2 t_2 u_2^{-1} t_2^{-1} u_1$, $t_i \mapsto t_{i+1}$, $u_i \mapsto u_{i+1}$ for $2 \le i \le 2p$, $t_{2p+1} \mapsto t_1 t_2 t_1^{-1}$, $u_{2p+1} \mapsto t_1 u_2 t_1^{-1}$, $z \mapsto z^{-1}$ we define an automorphism $\varphi : A \to A$, and φ^{2p} is the inner automorphism $a \mapsto t_1 a t_1^{-1}$. Let $G = \mathcal{Z}_{2p}$ and assume that there is an extension E for the abstract kernel (G, A, β) where β maps the generator of $G = \mathcal{Z}_{2p}$ to the class of φ in Out A. Choose $x \in E$ such that conjugation by x induces the automorphism φ of A. Now $x^{2p} = t_1 z^k$ for a suitable k, and

$$t_1 z^k = x^{2p} = x^{-1} x^{2p} x = x^{-1} t_1 z^k x = x^{-1} t_1 x \cdot x^{-1} z^k x = t_1 z \cdot z^{-k} ;$$

this implies the contradiction $k = \frac{1}{2}$. The geometrical background of this example is that there is a Seifert manifold with a mapping class of order $2p$ which cannot be realized by a mapping of order $2p$. (See [Zieschang 1981, 62.1], [ZVC 1988, 12.2.1].)

Let (G, A, β) be an abstract kernel. If the group A is abelian the *semi-direct product* $G \ltimes A = \{(g, a) : g \in G, a \in A\}$ with the product rule $(g', a') \cdot (g, a) = (g'g, \gamma_g(a') \cdot a)$ defines an extension as desired. On the other hand, if the centre of A is trivial then the homomorphism $A \to \text{Aut } A$ adjoining to $a \in A$ the inner automorphism $x \mapsto a^{-1} x a$ is injective. If, moreover, β is injective and E is the inverse image in Aut A of $\beta(G)$, then the unique extension realizing (G, A, β) is $1 \to A \to E \to \beta(G) \to 1$. Hence:

6.1.7. Proposition

(a) *Each abstract kernel (G, A, β) where A is abelian can be realized by an extension.*

(b) *If the centre of A is trivial and β is injective then (G, A, β) can be realized. Any two extensions realizing the kernel are equivalent.* □

By 6.1.7 (b), the problem of extension for a group A with trivial centre is reduced to the, in general, very difficult problem of determining the subgroups of the automorphism group. When the centre is not trivial, in particular, when A is abelian the situation becomes more interesting since then there may be non-equivalent extensions realizing the same kernel. The best known examples are the extensions of $\mathbb{Z}$ by a finite cyclic group $\mathcal{Z}_n, n \geq 2$ with trivial action, i.e. $\beta(\mathcal{Z}_n) = 1$: the groups $\mathcal{Z}_n \oplus \mathbb{Z}$ and $\mathbb{Z}$ with the given group as the subgroup $n\mathbb{Z}$ solve the problem.

Next we consider the abelian case and show the relationship with cohomology theory. Let us remark here that Schreier [Schreier 1926] gave a theorem which solves in an abstract sense both Problems 6.1.4,5. For more recent, more general results than mentioned in the following see [Cartan-Eilenberg 1956, 1960]. In the following we write the operation in A additively. The mapping α, see 6.1.1, amounts to the choice of a representative of each coset of A in E; we assume that the subgroup A is represented by 1.

6.1.8. Definition and Simple Properties. For $g, h \in G$ define $(g, h) \in A$ by $\alpha(g)\alpha(h) = \alpha(gh) \cdot (g, h)$. The function $(-,-) : G \times G \to A$ is called a *factor set.* By simple calculations one obtains:

(1) $\quad (gh, k) \cdot \gamma_k((g, h)) = (g, hk) \cdot (h, k) \quad$ (associative law of E) ,

(2) $\quad (g, 1) = (1, g) = 1 \quad$ ($1 \in G$ represented by 1) ,

(3) $\quad \gamma_h \gamma_g(a) = (g, h)^{-1} \gamma_{gh}(a)(g, h) \quad$ for $a \in A,\ g, h \in G$.

That the equations (1) and (2) are sufficient to construct an extension E for the abstract kernel (G, A, β) is a theorem of [Schreier 1926]. In case of an abelian A the equations (3) disappear and $\gamma = \beta$. This gives (a) and (b) of the following proposition; (c) follows by direct calculations.

6.1.9. Proposition

(a) *Let E be a group with abelian normal subgroup A and factor group $G = E/A$. Then E defines an action β of G on A and a factor system in A with respect to G that solves the equations* 6.1.8 (1), (2).

(b) *Conversely, let β be an action of a group G on an abelian group A and let $\{(g,h) \in A : g, h \in G\}$ be a factor set that solves* 6.1.8 (1), (2). *Let E be the set of pairs $[g, a]$ with $a \in A$, $g \in G$ and define the multiplication $[g,a][h,b] = [gh, (g,h)\beta_h(a)b]$. Then E forms a group with normal subgroup A and $E/A \cong G$. The extension $1 \to A \to E \to G \to 1$ is determined (up to equivalence) by the operation of G on A and the factor set.*

(c) *Let E and E' be extensions realizing the abstract kernel (G, A, β) which have the factor sets $\{(g,h)\}$ and $\{(g,h)'\}$, respectively. Then the extensions E and E' are equivalent if and only if there exists a mapping $\lambda : G \to A$ with the properties that $\lambda(1) = 1$ and $(g,h)' = \lambda(gh)^{-1}(g,h)\ [\beta_h(\lambda(g))]\lambda(h)$ for all $g, h \in A$.* □

An important tool for extension theory is the cohomology theory of groups and we will briefly describe it here.

6.1.10. Definition of Cohomology Groups of a Group

(a) Consider a group G, written multiplicatively, that operates on an abelian group A, written additively. For $a \in A$, $g \in G$ the operation will be written in the form $a \cdot g$ instead of $\beta_g(a)$. This action extends to the group ring $\mathbb{Z}G$ by $a \cdot \sum_i n_i g_i = \sum_i n_i(a \cdot g_i)$ and this turns A into a $\mathbb{Z}G$-module, sometimes just called a $G-$module.

(b) Let $C^n(G, A)$, $n \leq 1$, be the group of all functions $f : G^n \to A$ with the property that $f(g_1, \ldots, g_n) = 0$ if some g_i equals 1. Then f is called a *$n-$dimensional cochain.* The usual addition of functions makes $C^n(G, A)$ an abelian group. Let $C^0(G, A) = A$ and $C^n(G, A) = 0$ for $n < 0$.

(c) The *coboundary operator* $\delta^n : C^n(G, A) \to C^{n+1}(G, A)$ is defined by

$$\begin{aligned}
&(\delta^n f)(g_0, \ldots, g_n) \\
&\quad = f(g_1, \ldots, g_n) + \sum_{j=1}^{n} (-1)^j f(g_0, \ldots, g_{j-2}, g_{j-1}g_j, g_{j+1}, \ldots, g_n) \\
&\qquad + (-1)^{n+1} f(g_0, \ldots, g_{n-1}) \cdot g_n \quad \text{for } n \geq 1; \\
&(\delta^0 a)(g_1) = a - a \cdot g_1; \\
&\delta^n = 0 \quad \text{for } n < 0.
\end{aligned}$$

(d) $\left(C^n(G,A),\delta^n\right)_{n\in\mathbb{Z}}$ *is a cochain complex, i.e.* $\delta^{n+1}\delta^n = 0$. *The* n*–th cohomology group* $\ker \delta^n/\delta^{n-1}(C^{n-1}(G,A))$ *is called the* n*-th cohomology group of* G *with coefficients in* A*, and is denoted by* $H^n(G,A)$.

6.1.11. Examples. (a) *Let* $G = \mathcal{Z}_2 = \{1,-1\}$ *and* $A = \mathbb{Z}$ *and let* G *act trivially on* A*. Then* $H^0(\mathcal{Z}_2,\mathbb{Z}) = \mathbb{Z}$, $H^1(\mathcal{Z}_2,\mathbb{Z}) = 0$ *and* $H^2(\mathcal{Z}_2,\mathbb{Z}) = \mathbb{Z}_2$.

Proof. $H^0(G,A) = \ker\delta^0$. Now $(\delta^0 a)(-1) = a - a\cdot(-1) = a - a = 0$ for $a \in A = \mathbb{Z}$; hence $\ker\delta^0 \cong \mathbb{Z}$. If $h \in C^1(\mathcal{Z}_2,\mathbb{Z})$ then $(\delta^1 h)(-1,-1) = h(-1) - h(1) + h(-1)\cdot(-1) = 2h(-1)$; hence $\ker\delta^1 = 0$ and $H^1(\mathcal{Z}_2,\mathbb{Z}) = 0$. If $f \in C^2(\mathcal{Z}_2,\mathbb{Z})$ then $(\delta^2 f)(g_0,g_1,g_2) = 0$ if some $g_i = 0$. Hence $\delta^2 f$ is determined by the value of $(-1,-1,-1))$ which vanishes. Thus $\ker\delta^2 = C^2(\mathcal{Z}_2,\mathbb{Z}) = \mathbb{Z}$. From $(\delta^1 h)(-1,-1) = 2h(-1)$ it follows that $\delta^2(C^2(\mathcal{Z}_2,\mathbb{Z})) = 2\mathbb{Z}$ and $H^2(\mathcal{Z}_2,\mathbb{Z}) = \mathbb{Z}_2$.

(b) *Now let* G *and* A *be as above, but assume that the operation of* $\mathcal{Z}_2$ *on* $\mathbb{Z}$ *is non-trivial:* $1\cdot(-1) = -1$. *Then* $H^0(\mathcal{Z}_2,\mathbb{Z}) = 0$, $H^1(\mathcal{Z}_2,\mathbb{Z}) = \mathbb{Z}_2$ *and* $H^2(\mathcal{Z}_2,\mathbb{Z}) = 0$. (Proof as exercise.)

For another example see 6.1.14, 16.

Let (G,A,β) be an abstract kernel and $(-,-) : G\times G \to A$ a corresponding factor set satisfying 6.1.8. Now the equations 6.1.8 correspond to the vanishing assumptions in 6.1.10 (b). When we write $\beta_g(a) = a\cdot g$ the equation 6.1.8 (1) turns into

$$0 = (h,k) - (gh,k) + (g,hk) - (g,h)\cdot k = (\delta^2 f)(g,h,k)\ .$$

Hence each extension defines a 2-cocycle, and vice versa. The semi-direct product is obtained from the trivial factor set. If the factor sets $(-,-), (-,-)'$ define equivalent extensions then there is a function $f : G \to A$ which solves the equations

$$(g,h)' - (g,h) = f(h) - f(gh) + f(g)\cdot h = (\delta^1 f)(g,h)\ ,$$

see 6.1.9 (c) and 6.1.10 (c). Hence:

6.1.12. Theorem. *Let* (G,A,β) *be an abstract kernel where* A *is abelian. Then each factor set is a 2-cocycle; two factor sets define equivalent extensions if and only if they differ by a coboundary. Hence, the set of extensions realizing the abstract kernel* (G,A,β) *and the cohomology group* $H^2(G,A)$ *are in one-to-one correspondence where the semi-direct product* $G \ltimes A$ *correponds to* $0 \in H^2(G,A)$. *Thus a group structure can be defined on the set of extensions realizing an abstract kernel.* □

The group structure on the extensions can, of course, be defined directly as is done in books on group theory, see [Hall 1959, Chap. 15], [Kurosh 1967]. The following statements are direct consequences of the definitions; they offer tools for general calculations.

6.1.13. Proposition. *Let G act on the abelian groups A, A' by $a \mapsto a \cdot g$ and $a' \mapsto a' \cdot g$, respectively, and let $\varphi : A \to A'$ be a homomorphism compatible with the action of G, i.e. $\varphi(a \cdot g) = \varphi(a) \cdot g$ for $a \in A$, $g \in G$. Then:*

(a) *φ induces homomorphisms $\varphi_* : H^n(G, A) \to H^n(G, A')$, $f \mapsto \varphi \circ f$, $n \in \mathbb{Z}$.*

(b) *Assume that φ is a monomorphism. Let the extensions*

$$1 \to A \hookrightarrow E \xrightarrow{j} G \to 1 \quad \text{and} \quad 1 \to A' \hookrightarrow E' \xrightarrow{j'} G \to 1$$

correspond to elements $\xi \in H^2(G, A)$ and $\xi' \in H^2(G, A')$, respectively. If $\varphi_*(\xi) = \xi'$ then there is a monomorphism $\Phi : E \to E'$ with the properties $\Phi(a) = \varphi(a)$, for $a \in A$, and $j'\Phi = j$. Hence, E can be considered as subgroup of E' and $[E' : E] = [A' : A]$. □

6.1.14. Proposition

(a) *Let A, G be as above and assume that the order m of G is finite. Then the homomorphisms $\mu : C^q(G, A) \to C^q(G, A)$, $f \mapsto m \cdot f$ induce the trivial homomorphisms $\mu_* : H^q(G, A) \to H^q(G, A)$ for $q \geq 1$, i.e. $\mu_*(H^q(G, A)) = 0$.*

(b) *$H^q(\mathbb{Z}_m, \mathbb{Z}_n) = 0$ for $q \geq 1$ if $\gcd(m, n) = 1$.*

Proof. Let $g_1, \dots, g_q \in G$ and $f \in \ker \delta^q$ a cocycle. Then

$$\begin{aligned} 0 &= \sum_{h \in G} (\delta^q f)(a, g_1, \dots, g_q) \\ &= \sum_{h \in G} f(g_1, \dots, g_q) - \sum_{h \in G} f(hg_1, g_2, \dots, g_q) \\ &\quad + \sum_{i=2}^{q} \sum_{h \in G} f(h, g_1, \dots, g_{i-2}, g_{i-1}g_i, g_{i+1}, \dots, g_q) \\ &\quad + (-1)^{q+1} \sum_{h \in G} f(h, g_1, \dots, g_{q-1}) \cdot g_q \,. \end{aligned}$$

This implies for $K(x_1, \dots, x_{q-1}) = \sum_{h \in G} f(h, x_1, \dots, x_{q-1})$:

$$\begin{aligned} mf(g_1, \dots, g_q) = K(g_2, \dots, g_q) & \\ &- \sum_{i=2}^{q} (-1)^i K(g_1, \dots, g_{i-2}, g_{i-1}g_i, g_{i+1}, \dots, g_q) \\ &- (-1)^{q+1} K(g_1, \dots, g_{q-1}) \cdot g_q \,. \end{aligned}$$

K is a cochain since it vanishes if some $g_i = 1$. From the last equation and the definition of δ^{q-1} it follows that

$$\mu(f)(g_1, \dots, g_q) = m \cdot f(g_1, \dots, g_q) = (\delta^{q-1} K)(g_1, \dots, g_q) \,. \quad \square$$

6.1.15. An Alternative Approach to Cohomology of Groups. Explicit calculations of cohomology groups following the definition above are quite messy. Another way to obtain these groups which is closer to the construction known in algebraic topology is, briefly, as follows. See [Gruenberg 1970], [Hilton-Stammbach 1971].

We consider $\mathbb{Z}$ as a G–module with trivial action of G and take a *projective* $(G-)$*resolution of* $\mathbb{Z}$, i.e. an exact sequence

$$\ldots \to X_{n+1} \xrightarrow{\partial_{n+1}} X_n \xrightarrow{\partial_n} X_{n-1} \to \ldots \xrightarrow{\partial_3} X_2 \xrightarrow{\partial_2} X_1 \xrightarrow{\partial_1} X_0 \xrightarrow{\varepsilon} \mathbb{Z} \to 0$$

where the X_i are projective (for instance, free) G-modules. Define

$$\underline{C}^n(G,A) = \mathrm{Hom}_{\mathbb{Z}G}(X_n, A) \quad \text{for } 0 \le n, \qquad \underline{C}^n(G,A) = 0 \quad \text{for } n < 0;$$
$$\underline{\delta}^n : \underline{C}^n(G,A) \to \underline{C}^{n+1}(G,A),\ f \mapsto f\partial_{n+1} \text{ for } n \ge 0, \qquad \underline{\delta}^n = 0 \quad \text{for } n < 0;$$
$$\underline{H}^n(G,A) = \ker \underline{\delta}^n / \underline{\delta}^{n-1}(\underline{C}^{n-1}(G,A)) \ .$$

It is easily checked that $\underline{\delta}^{n+1}\underline{\delta}^n = 0$, and hence the cohomology groups $\underline{H}^n(G,A)$ are defined. From general theorems it follows that the cohomology groups thus defined do not depend on the special choice of the projective resolution of $\mathbb{Z}$. In particular, the groups $H^q(G,A)$ and $\underline{H}^q(G,A)$ are isomorphic, see [Hilton-Stammbach 1971, p. 184]; we will use the first notation. The following example throws some light on the above construction.

6.1.16. Cohomology of Cyclic Groups. Let $G = \mathcal{Z}_m = \{g^i : 0 \le i \le m-1\}$. There is the free $\mathcal{Z}_m$–resolution for $\mathbb{Z}$ (with trivial $\mathcal{Z}_m$–action):

$$\ldots \to \mathbb{Z}\mathcal{Z}_m \xrightarrow{\partial_{2i}} \mathbb{Z}\mathcal{Z}_m \xrightarrow{\partial_{2i-1}} \ldots \xrightarrow{\partial_2} \mathbb{Z}\mathcal{Z}_m \xrightarrow{\partial_1} \mathbb{Z}\mathcal{Z}_m \xrightarrow{\varepsilon} \mathbb{Z} \to 0$$

where ∂_{2i-1} is multiplication by $(g-1)$ and ∂_{2i} multiplication by $(g^{m-1} + g^{m-2} + \ldots + g + 1)$ for $i \ge 1$ and $\varepsilon(\sum_{i=0}^{m-1} n_i g^i) = \sum_{i=0}^{m-1} n_i$. The sequence is exact and the cohomology groups are obtained from the cochain complex

$$0 \to \mathrm{Hom}_{\mathbb{Z}\mathcal{Z}}(\mathbb{Z}\mathcal{Z}, A) \to \mathrm{Hom}_{\mathbb{Z}\mathcal{Z}}(\mathbb{Z}\mathcal{Z}, A) \to \mathrm{Hom}_{\mathbb{Z}\mathcal{Z}}(\mathbb{Z}\mathcal{Z}, A) \to \ldots \ .$$

Since $f \in \mathrm{Hom}_{\mathbb{Z}\mathcal{Z}}(\mathbb{Z}\mathcal{Z}, A)$ is defined by the value $f(g)$ one can identify A and $\mathrm{Hom}_{\mathbb{Z}\mathcal{Z}}(\mathbb{Z}\mathcal{Z}, A)$ and the sequence becomes

$$0 \to A \xrightarrow{\tau} A \xrightarrow{\sigma} A \xrightarrow{\tau} A \xrightarrow{\sigma} \ldots, \quad \tau(a) = a \cdot (g-1),\ \sigma(a) = a \cdot \sum_{i=0}^{m-1} g^i \ .$$

Let $A^{\mathcal{Z}_m} = \{a \in A : a \cdot g = a\}$, the fixed point set of $\mathcal{Z}_m$. Then we obtain from the sequence above:

$$H^0(\mathcal{Z}_m, A) = A^{\mathcal{Z}_m},\ H^{2i-1}(\mathcal{Z}_m, A) = \ker \sigma / A \cdot (g-1),$$
$$H^{2i}(\mathcal{Z}_m, A) = A^{\mathcal{Z}_m} / A \cdot \sum_{i=0}^{m-1} g^i,\ i \ge 1,$$

and this gives the following corollary.

6.1.17. Corollary. *If the action of* $G = \mathcal{Z}_m$ *on* $A = \mathbb{Z}^n$ *is trivial then* $H^0(\mathcal{Z}_m, \mathbb{Z}^n) = \mathbb{Z}^n$, $H^{2i-1}(\mathcal{Z}_m, \mathbb{Z}^n) = 0$, $H^{2i}(\mathcal{Z}_m, \mathbb{Z}^n) = (\mathbb{Z}_m)^n$, $i \geq 1$. □

Like the uniqueness problem for extensions, the existence problem can be treated with the help of the cohomology of groups. To an abstract kernel (G, A, β) one defines an obstruction which is an element in $H^3(G, Z(A))$, where $Z(A)$ denotes the centre of A. *An extension exists if and only if this obstruction element vanishes.* Next we consider an important tool for combinatorial group theory: cohomological dimension.

6.1.18. Definition and Proposition

(a) *The* cohomological dimension $\text{cd}(G)$ *of the group* G *is* n *if* $H^n(G, A_0)$ *is non-zero for some* A_0 *and* $H^m(G, A) = 0$ *for all* A *and all* $m > n$.

(b) $\text{cd}(G) \leq n$ *if and only if there is a projective resolution of* $\mathbb{Z}$ *which is zero after the n-th term.* □

6.1.19. Corollary

(a) $\text{cd}(F) = 1$ *if* F *is free and* $F \neq 1$.

(b) *If* $H \subset G$, *then* $\text{cd}(H) \leq \text{cd}(G)$.

(c) $\text{cd}(G)$ *finite implies* G *torsion-free.* □

6.1.20. Theorem

(a) *If* $H < G$ *then* $\text{cd}(H) \leq \text{cd}(G)$ *and if* $\text{cd}(G) < \infty$ *and* $[G : H] < \infty$ *then one has equality.*

(b) *If* G *is torsion-free and* $[G : H] < \infty$ *then* $\text{cd}(H) = \text{cd}(G)$.

Proof. The first claim is a simple consequence of Shapiro's Lemma. The second is a difficult result of Serre. (See [Cohen 1972, p. 9].) □

This theorem permits the introduction of the notion of *virtual cohomological dimension.* For if G has torsion-free subgroups H and K then $\text{cd}(H) = \text{cd}(H \cap K) = \text{cd}(K)$. Then define $\text{vcd}(G) = \text{cd}(H)$.

Groups of finite actual or virtual cohomological dimension arise naturally in topological and geometric contexts. The finiteness involved is usually derived by applying the following result or some variant thereof.

6.1.21. Proposition. *Let the torsion-free group* G *act freely and cellularly on the contractible* $n-$*dimensional CW-complex* X. *Then* $\text{cd}(G) \leq n$.

Proof. This follows from the Definition 6.1.18 (a) since the cellular chain groups form a free resolution of $\mathbb{Z}$ of length n. □

Important examples of groups of finite virtual cohomological dimension are given by the following two results due to [Borel-Serre 1974] and [Harer 1986], respectively.

6.1.22. Proposition. $\text{vcd}(SL(n, \mathbb{Z})) = \binom{n}{2}$. □

6.1.23. Proposition. *Let* G *be the mapping class group of a closed orientable surface of genus* $g \geq 2$. *Then* $\text{vcd}(G) = 4g - 5$. □

There is a substantial theory of groups of finite cohomological dimension. We quote only some well-known results.

6.1.24. Proposition. *Let G be a free abelian group of rank n. Then* $\mathrm{cd}(G) = n$.

Proof. It follows easily from 6.1.18 (b) or 6.1.21 that $\mathrm{cd}(G) \leq n$ since the "cubical" tesselation of $\mathbb{R}^n$ with vertices at integer points is an $n-$dimensional contractible CW-complex on which G acts freely and cellularly. A more delicate argument is needed to establish the equality, see [Gruenberg 1970, p. 150]. □

6.1.25. Proposition

(a) *Let G be a free product of A and B amalgamating C. Then* $\max\{\mathrm{cd}(A), \mathrm{cd}(B)\} \leq \mathrm{cd}(G) \leq \max\{\mathrm{cd}(A), \mathrm{cd}(B)\} + 1$.

(b) *Let G be an HNN-extension with base group A. Then* $\mathrm{cd}(A) \leq \mathrm{cd}(G) \leq \mathrm{cd}(A) + 1$.

Proof. In both cases the result is achieved by constructing a Mayer-Vietoris sequence along the lines of that which can be constructed to give the cohomology of a space to which the Seifert-van Kampen Theorem applies [Massey 1967, Chap. 4]. □

6.1.26. Proposition. *Let $1 \to N \to E \to G \to 1$ be a group extension. Then* $\mathrm{cd}(E) \leq \mathrm{cd}(G) + \mathrm{cd}(N)$.

The last result requires the notion of spectral sequence, developed in algebraic topology, and in particular the so-called Lyndon-Hochschild-Serre sequence [Hilton-Stammbach 1971]. □

To conclude this section we state a result of Serre [Huebschmann 1979] that has interesting applications to the question of torsion in groups.

6.1.27. Proposition. *Let G be a group and suppose that there is a family of subgroups $(G_i)_{i \in I}$ and an integer $n \geq 1$ such that for every $\mathbb{Z}G-$module A the cohomology group $H^n(G, A)$ is isomorphic to the direct sum of the cohomology groups $H^n(G_i, A)$. Then any finite subgroup of G is conjugate, in an essentially unique way, to a subgroup of some unique G_i.* □

§ 6.2. Ends of Groups

In an obvious but imprecise sense the real line $\mathbb{R}$ has two ends. Similarly a twice punctured sphere, which is obviously homomorphic to an infinite cylinder, has two ends and, more generally a sphere with n punctures can be regarded as having n ends. Then the real plane $\mathbb{R}^2$ and, more generally, real n-dimensional space should be regarded as having just one end. The study of ends of topological spaces has led to some very fruitful interactions between topology and group theory. Detailed accounts of the theory of ends and applications are to be found in [Cohen 1972], [Scott-Wall 1979].

Formally an *end of a (non-compact) space* X is an equivalence class of descending chains of subsets $\mathcal{D} : D_1 \supset D_2 \supset D_3 \supset \ldots$, where D_n is open with compact boundary and $\bigcap_{n=1}^{\infty} D_n$ is empty. Two such chains $\mathcal{D}$ and D' are *equivalent* if for every m there exists n such that $D_m \supset D'_n$ and $D'_m \supset D_n$. This formal concept of end was introduced in [Freudenthal 1931] as a natural way of compactifying topological spaces. Provided that the space in question satisfies some mild conditions (if, for instance, it is locally compact and connected) a theory of ends can be defined. The focus of [Freudenthal 1931] was topological groups and the two main results obtained are:

6.2.1. Theorem

(a) *A topological group has at most two ends.*

(b) *The direct product of two non-compact spaces has one end and the ends of a direct product of a compact and a non-compact space are in one-to-one correspondence with the ends of the non-compact factor.* □

Discrete groups first enter the picture in [Hopf 1942]. The typical example considered is that of a group of covering transformations of a regular covering of a compact space.

6.2.2. Theorem. *Let the space X have a compactification by ends and suppose that the discrete group G operates properly discontinuously on X with compact fundamental domain. Then the cardinality of the set of ends of X is either* 1, 2 *or* $2^{\aleph_0}$.

For the case of a finitely generated group of covering transformations, it is also shown in [Hopf 1943] that the set of ends is essentially independent of the covering and so the set of ends can be regarded as the set of ends of the abstract group involved.

6.2.3. Theorem

(a) *If the finitely generated group G has two ends, then G has an infinite cyclic subgroup of finite index.*

(b) *A direct product of two finitely generated infinite groups has one end.*

Part (b) of the theorem is an application of Theorem 6.2.1(b). □

An approach to the theory of ends that is convenient for dealing with groups is the following, which is derived from ideas in [Freudenthal 1944]. Let X be a locally finite, infinite connected graph with vertex set V and edge set E. For any subset Y of E let $n(Y)$ be the number of infinite components of the graph $X \setminus Y$ obtained by deleting the edges in Y. Then $e(X) = \sup\{n(Y) : Y \text{ is finite}\}$ is called the *number of ends* of X. (There is a discrepancy, which is usually ignored, between the definition of the number of ends and the cardinality of the set of ends when the latter is uncountable). If X is the Cayley graph of a group relative to some finite generating set then we define the *number of ends* of G to be $e(G) = e(X)$. An indication is given below of why this is well-defined, i.e. independent of the particular generating set. With this definition, one can prove part of Theorem 6.2.2 for finitely generated groups as follows.

Proof of 6.2.2. Let X be the Cayley graph of G relative to the generating set S. Suppose that $e(G)$ is the positive integer n. Then there is a finite connected subgraph L of X such that the graph $X \setminus L$ obtained by deleting the edges of L and any resulting isolated vertices consists of exactly n infinite components. Since G is infinite there exists $g \in G$ such that $gL \cap L$ is empty and so gL lies within some component Y of $X \setminus L$. Now exactly one of the components of $Y \setminus gL$ is infinite and $L \cup (X \setminus Y)$ is connected whence $X \setminus gL$ has at most two infinite components. Since the operation of g is an isomorphism of X it follows that $n \leq 2$. □

6.2.4. Example. If F is a free group then $e(F)$ is infinite unless F has rank 1 in which case $e(F) = 2$. This is clear from the standard Cayley graph for F associated to a basis of F.

Again let X be a graph with vertex set V and edge set E, and let k denote a field or the ring $\mathbb{Z}$ of integers. Let C^0 be the set of all maps $c : V \to k$ and C^1 the set of all maps $b : E \to k$. (A little care is needed in the definition of C^1 — if edges are regarded as coming in inverse pairs, then either one edge only should be chosen from each pair or only maps satisfying $b(\sigma^{-1}) = -b(\sigma)$ allowed.) Then C^0 and C^1 are groups under pointwise addition and the *coboundary map* $\delta : C^0 \to C^1$ given by $\delta(c)(\sigma) = c(t(\sigma) - s(\sigma))$, is a homomorphism, where, as usual, t and s map an edge to its terminal and initial vertices. An element of C^0 or C^1 has *finite support* if it assumes non-zero values at only finitely many places. Directly from the definitions we have :

6.2.5. Proposition. *Let* $C^0_{\delta f} = \{c \in C^0 : \delta(c)$ *has finite support*$\}$ *and let* $C^0_f = \{c \in C^0 : c$ *has finite support*$\}$. *Then* $e(X) = \dim_k C^0_{\delta f}/C^0_f$. □

For further developments it is convenient to give a more algebraic definition of the number of ends of a group. Let G be an infinite group. Let $\mathcal{P}G$ be the set of all subsets of G and $\mathcal{F}G$ the set of all finite subsets. Under the operation of symmetric difference, denoted by "+", $\mathcal{P}G$ is an abelian group of exponent 2 and $\mathcal{F}G$ is a subgroup. Now G acts by right multiplication on $\mathcal{P}G$ and $\mathcal{F}G$ and hence on $\mathcal{P}G/\mathcal{F}G$. The set of fixed points under this last action is denoted by $\mathcal{Q}G/\mathcal{F}G$ where $\mathcal{Q}G = \{A \subset G : A + Ag$ is finite for all $g \in G\}$. Elements of $\mathcal{Q}G$ are called *almost invariant* subsets of G. Let $\mathbb{F}_2$ be the field of two elements.

6.2.6. Proposition. $e(G) = \dim_{\mathbb{F}_2}(\mathcal{Q}G/\mathcal{F}G)$.

Proof. When $k = \mathbb{F}_2$, an element of C^0 is just the characteristic function of a subset and it is easy to see that if X is the Cayley graph of a group relative to some finite generating set, then $C^0_{\delta f}$ can be identified with $\mathcal{Q}G$ and C^0_f with $\mathcal{F}G$ whence $e(G) = e(X)$. □

Since $\mathcal{F}G$ can be identified with the group algebra $\mathbb{F}_2G$, an easy argument shows that also $e(G) = 1 + \dim H^1(G, \mathbb{F}_2G)$. With this algebraic definition, the following properties are easy to verify.

6.2.7. Proposition

(a) *Let H be a subgroup of finite index in G. Then $e(H) = e(G)$.*
(b) *Let K be a finite normal subgroup of G. Then $e(G/K) = e(G)$.* □

We now describe the structure of groups with two ends. The key technical result (see [Scott-Wall 1979, p. 178]), which we shall not prove is:

6.2.8. Proposition. *Let G be a finitely generated group and let A be an almost invariant subset such that both A and its complement A' are infinite. If $H = \{g \in G : gA + A \in \mathcal{F}G\}$ is infinite, then G has an infinite cyclic subgroup of finite index.* □

Now suppose that G has two ends. Then an almost invariant subset A satisfying the hypotheses of 6.2.8 will exist. The group H is the stabiliser of the class of A in the action of G on the finite set $\mathcal{Q}G/\mathcal{F}G$ and so has finite index in G. Since G is infinite, then H is also infinite and therefore, by Proposition 6.2.8, G has an infinite cyclic subgroup K of finite index which may be assumed to be normal. The centraliser C of K in G has index at most two and so by an old theorem of Schur [Robinson 1972, p. 102], the commutator subgroup of C is finite. Now there must be an epimorphism $\psi : C \to \mathbb{Z}$ with finite kernel L. If $C = G$ then G is an HNN–extension $G = \langle\, L, t \mid t^{-1}Lt = L \,\rangle$. Otherwise G/L is the free product $\mathbb{Z}_2 * \mathbb{Z}_2$ of two cyclic groups of order 2 and then G/L is an amalgamated free product $A *_L B$ with $[A : L] = 2 = [B : L]$.

The upshot of this discussion is that a group with two ends decomposes, in a particular way, either as an HNN–extension or as an amalgamated free product, over a finite subgroup. The remarkable work of Stallings in [Stallings 1971] extends this kind of characterisation to finitely generated groups with infinitely many ends, as we now explain. We say a group G *splits* over a subgroup L if G can be expressed as an amalgamated free product $G = A *_L B$ with $A \neq L \neq B$ or an HNN–extension $G = \langle A, t \mid t^{-1}Lt = L' \rangle$.

6.2.9. Theorem [Stallings 1971]. *A finitely generated group has at least two ends if and only if it splits over a finite subgroup.*

The discussion prior to the statement of the theorem shows that a group has exactly two ends if and only if it either is the semidirect product of a finite normal subgroup by an infinite cyclic group, and thus an HNN–extension with a finite base group, or is an amalgamated free product where the amalgamated subgroup is finite and of index two in each factor. The essential content of

Theorem 6.2.9 beyond what has already been discussed is thus the case of infinitely many ends.

Before discussing the proof of Theorem 6.2.9, something should be said of its background. The theorem is closely tied to the theory of 3–manifolds and it was in the course of a study of 3–manifold theory that Stallings was led to this result. The particular aspect of 3–manifold theory concerned is the Sphere Theorem 5.1.6 to the effect that an orientable 3-manifold M whose second homotopy group $\pi_2(M)$ is non–trivial must contain an embedded sphere representing a non-trivial element of $\pi_2(M)$. The connection with the theory of ends arises from the fact that if G is the fundamental group of M then, by [Hopf 1943], $e(G) = e(\tilde{M})$ where $\tilde{M}$ is the universal cover of M and a relatively easy argument using Poincaré duality [Novikov 1986, p. 52] for $\tilde{M}$ shows that $\pi_2(M) \neq 0$ if and only if $e(\tilde{M}) \geq 2$.

The easy part of the *proof* of 6.2.9 is to show that if G splits over a finite subgroup then $e(G) \geq 2$. Suppose, for instance, that $G = H *_L K$ with L finite. By Theorem 2.2.4 every element of G has a unique normal form $ac_1 \cdots c_n$ where $a \in L$ and the terms c_i come alternately from transversals for L in H and K. If A is the set of all elements whose normal form ends with an element of H, then the finiteness of L shows that A is almost invariant and neither A nor its complement is finite. Thus A defines a non-trivial element of $\mathcal{Q}G/\mathcal{F}G$.

The converse half of the proof is anything but easy. The most elegant argument is based on a result from [Dunwoody 1979].

6.2.10. Theorem. *Let E be a partially ordered set equipped with an involution $\sigma \mapsto \sigma^{-1}$ satisfying the following conditions :*

(a) *if $\sigma \leq \tau$, then $\tau^{-1} \leq \sigma^{-1}$;*

(b) *for any $\sigma, \tau \in E$ the interval $\{\rho \in E : \sigma \leq \rho \leq \tau\}$ is finite;*

(c) *for any $\sigma, \tau \in E$ at least one of $\sigma \leq \tau$, $\sigma \leq \tau^{-1}$, $\sigma^{-1} \leq \tau$, $\sigma^{-1} \leq \tau^{-1}$ holds;*

(d) *for any $\sigma, \tau \in E$ one of $\sigma \leq \tau$, $\sigma \leq \tau^{-1}$ fails to hold.*

Then there exists a tree with E as edge set and σ^{-1} the inverse edge for σ such that $\sigma \leq \tau$ if and only if there is a path in the tree with σ as first edge and τ as last edge. □

This theorem is applied to $\mathcal{Q}G/\mathcal{F}G$ — with the involution induced by taking complements and the partial order induced by almost containment where A is *almost contained* in B if the complement of B in A is finite – to produce a tree upon which the group acts so that the quotient under the action has just one edge and the stabiliser of an edge is finite. The proof is completed by an application of Theorems 2.2.21*-22* and their analogues for HNN–extensions. □

Theorem 6.2.9 has a number of significant applications.

6.2.11. Theorem [Stallings 1971]. *A finitely generated torsion–free group which contains a free subgroup of finite index is free.*

Proof. We argue by induction on the number $d(G)$ of generators of G. There is nothing to prove if $d(G) = 0$. From the hypotheses, 6.2.4 and Proposition 6.2.7, $e(G) \geq 2$ and so G splits over a finite subgroup which must be trivial since G is torsion–free. So either G is infinite cyclic or G is a non-trivial free product. Grushko's Theorem 2.2.27 allows us to apply the induction hypothesis, recalling that a subgroup of a free group is free. □

It is perhaps surprising that no more direct way has been found to prove this result. One can assume, by Theorem 2.2.23 that the given free subgroup acts freely on a tree and all that is then required is construct a tree on which the whole group acts freely. Theorem 6.2.11 was strengthened in [Karrass-Pietrowski-Solitar 1972] to the case when torsion is allowed.

6.2.12. Theorem. *A finitely generated group G has a free subgroup of finite index if and only if G is the fundamental group of a finite graph of finite groups.*

Proof. Let H be a free subgroup of finite index in G. Then H is of finite rank and so to prove that G is the fundamental group of a graph of groups as claimed, we can argue by induction on the rank of H. If rank $H = 1$, then the analysis after Proposition 6.2.8 gives the required form for G. If rank $H > 1$, then G has infinitely many ends and therefore, by 6.2.9, splits over a finite subgroup. Suppose, for example, that $G = A *_L B$. The free subgroup H is, by Theorem 2.2.20, expressible as the fundamental group of a finite graph of groups whose vertex groups are of the form $H \cap g^{-1}Ag$ or $H \cap g^{-1}Bg$ and whose edge groups must be trivial since L is finite and H is torsion–free. Now if the vertex groups are all trivial, then, since H has finite index in G, A and B must be finite and there is nothing further to prove. Otherwise some vertex group must be non–trivial and the finiteness of the index $[G : H]$ implies that at least two vertex groups are non–trivial. A typical vertex group is a free group $H \cap g^{-1}Ag$ which is of finite index in $g^{-1}Ag$ and of smaller rank than H. Hence the induction hypothesis may be applied to A and B whence G can be expressed in the desired manner.

The converse half of the argument relies on an argument involving permutation groups, see [Dicks 1980] for example, which shows that for any finite graph of finite groups there is a homomorphism to a finite group which is injective on the vertex groups. This means that the kernel of this homomorphism does not meet any conjugate of a vertex group and hence, by Theorem 2.2.23, must be free and the proof is complete. □

A natural question that arises from Theorem 6.2.9 is how many times a group can be succesively split over a finite subgroup. Part of the content of Theorem 6.2.12 is that for a finitely generated group that has a free subgroup of finite index, this can happen only finitely often. To formalise this, call a group *accessible* if it can be represented as the fundamental group of a graph of groups in which the vertex groups have at most one end and the edge groups are finite. Not all groups are accessible.

Example 6.2.13. Let $G = \langle a_0, a_1, \ldots, b_1, b_2, \ldots \mid a_{n-1} = [a_n, b_n],\ n \geq 1 \rangle$. Then G is torsion-free since it is the fundamental group of a graph of groups with vertex groups free of rank 2 (the groups $\langle\, a_n, b_n \,\rangle$) and hence is accessible if and only if it is a free product of freely indecomposable groups. However it turns out (see [Scott-Wall 1979, p. 163]) that this is impossible — it should be observed that every finitely generated subgroup of G is free but G is not itself free. In this instance the failure of accessibility is linked to the fact that G is not finitely generated and it has been conjectured that every finitely generated group is accessible. The general question still remains open but we quote two very substantial results in this direction from [Dunwoody 1985] and [Linnell 1983] respectively.

6.2.14. Theorem. *A finitely presented group is accessible.* □

6.2.15. Theorem. *A finitely generated group in which there is a bound on the order of any finite subgroup is accessible.* □

Of interest is the fact that the argument in [Dunwoody 1985] is modelled on one in the paper of Kneser [Kneser 1929] which shows that in a compact 3-manifold there is a bound on the number of disjoint embedded 2-spheres none of which bounds a 3-ball and no two of which bound a region homeomorphic to the product of a 2-sphere with an interval. In almost complete contrast, the argument in [Linnell 1983] is based on a result from [Kaplansky 1972] on group algebras — and this result is effectively a theorem in functional analysis.

Further applications of Theorem 6.2.9 involve cohomology.

6.2.16. Theorem. *A finitely generated group has cohomological dimension* 1 *if and only if it is free (and non-trivial).* □

Proof. By 6.1.19 (a), any non-trivial free group has cohomological dimension 1. Conversely suppose that $\mathrm{cd}(G) = 1$. By 6.1.19 (c), G must be torsion-free. Furthermore a straightforward argument in cohomology theory shows that $H^1(G, \mathbb{Z}G) \neq 0$ and then in turn, see [Swan 1969, p. 595], $H^2(G, \mathbb{F}_2 G) \neq 0$ whence $e(G) \geq 2$. Then G splits over the trivial subgroup and induction on the number of generators shows that G must be free. □

So far we have usually dealt with finitely generated groups. If we examine the question of ends of infinitely generated groups it turns out that there is one additional class of groups that we need to consider. Already it was shown in [Freudenthal 1944] that a finitely generated *periodic* group, that is, one in which all elements are of finite order has at most one end and it is not too hard to show that unless a periodic group G is *locally finite*, that is, every finite subset generates a finite subgroup, then G has at most one end. A countably infinite, locally finite group has infinitely many ends whereas an uncountable locally finite group has one end [Holt 1981]. The most general form of Theorem 6.2.9 reads:

6.2.17. Theorem. *A group G has at least two ends if and only if either G splits over a finite subgroup or G is countably infinite and locally finite. Furthermore if $e(G) > 2$ then $e(G)$ is infinite.* □

In contrast to the result on ends, Theorem 6.2.16 does generalise directly – see [Swan 1969].

6.2.18. Theorem. *A group has cohomological dimension one if and only it is free (and non-trivial).* □

The arguments in [Swan 1969] are algebraic in character and involve examining summands of direct sums of countably generated modules and we make no further comment. Theorem 6.2.16 combines with Theorem 6.1.27 to yield a generalisation of Theorem 6.2.11.

6.2.19. Theorem. *If the torsion-free group G has a free subgroup H of finite index, then G is free.*

Proof. Since H is free and of finite index in G then $\mathrm{cd}(G) = \mathrm{cd}(H) = 1$ and so G is free. □

Theorem 6.2.12 also generalises [Cohen 1973], [Scott 1974]:

6.2.20. Theorem. *A group G has a free subgroup of finite index if and only if G is the fundamental group of a graph of finite groups.* □

The theory of ends of groups was generalized in [Houghton 1974], [Scott 1977] to a theory of ends of pairs of groups. The motivation for Theorem 6.2.9 lay originally in the study of embeddings of a 2-sphere in an orientable 3-manifold. However one can examine embeddings of other surfaces and hope to obtain a connection with the theory of ends. In particular, the quantity $e(G)$ is replaced by a number $e(G, S)$, where S is a subgroup of G, with $e(G, S) = e(G)$ when $S = 1$.

Much of the work in this area centres around the Torus Theorem for 3-manifolds.

6.2.21. Theorem. *Let M^3 be a compact orientable irreducible 3-manifold which admits an essential torus, i.e. its fundamental group is embedded into $\pi_1(M^3)$. Then either M^3 admits an essential embedded torus or the fundamental group of M^3 contains an infinite cyclic normal subgroup.* □

A comparatively algebraic proof of this result is given by Scott [Scott 1980]. The crucial point is that if one can show that the fundamental group splits over a free abelian group of rank two, i.e. a subgroup isomorphic to the fundamental group of a torus, then the manifold must contain an embedded torus.

Chapter 7
Decision Problems

§ 7.1. Decision Problems and Algorithms

In a decision problem, a class of mathematical entities is partitioned into two subclasses by some defining condition, and a solution to the problem consists of an effective procedure or *algorithm* which specifies, in a finite number of steps, which subclass an arbitrarily given entity lies in. For the problem to be well-posed, each entity must be specified by a finite description and it must be clear whether a putative description actually describes an entity in the given class.

When Dehn first formulated the word problem (see 1.1.9) there was no precise concept of algorithm. Many examples of algorithms had been given and were accepted as procedures which, for any input, could be effectively carried through in a finite number of steps. The standard Euclidean algorithm for calculating the greatest common divisor of two integers is the best-known example. Dehn's solution to the word problem for surface groups of genus greater than one (see 4.1.1) constitutes another simple procedure.

Interest in giving a precise notion of algorithm developed among logicians in the 1930's and several formulations were put forward. Among the best known are those formulated in [Church 1941], [Markov 1954] and [Turing 1936]. These different formulations are all equivalent and this is generally accepted as evidence for what is known as *Church's thesis* or *Markov's Normalisation Principle* which sets out the belief that the intuitive notion of algorithm is precisely captured by these equivalent definitions. With a precise concept to work with, it is possible to contemplate the idea of a decision problem for which no algorithmic solution is possible and examples of such *unsolvable* problems were soon found in mathematical logic. Much subsequent effort was needed before unsolvable decision problems were found in group theory but ultimately it has turned out that, for general decision problems involving group presentations, the existence of instances where the problem is unsolvable is the rule rather than the exception. A more detailed account is given in 7.2.

With the advent of theoretical computer science, more restricted notions of algorithm have been developed. Of interest to group theory are those kinds of algorithm which are embodied in the notions of finite and pushdown automata. We shall discuss these in 7.3.

In this section we give a precise notion of algorithm.

7.1.1. Turing Machine. We begin with a loose description of a Turing machine which will then be formalised into a definition. One imagines a tape attached to a control device. The tape is subdivided into squares on which a letter taken from a finite *tape alphabet* may be printed, and which is potentially infinite in the sense that additional squares may be added at either end.

Attached to the tape is a *reading/writing head* through which the symbol printed on a single square may be *scanned*, and the control device displays one of a specified finite number of symbols to indicate its *internal state*. The machine performs computations in accordance with a finite set of rules which permit it to perform the following types of operation:

(a) replace the scanned tape symbol by another (in both instances regarding a blank as a symbol);

(b) move the reading head one square to the right or to the left, attaching an additional blank square if necessary.

In both cases the control device may either enter a new internal state or return to the previous state.

The operating instructions are such that, at any given moment when the machine is running, the next operation to be applied is completely determined by the current scanned symbol and the internal state. In some cases the combination of scanned symbol and internal state will dictate that the machine should halt. This means that the operating instructions can be represented by a finite set of quadruples of the form

(a) $qSS'q'$,

(b) $qSLq'$ or $qSRq'$ where

(1) q and q' denote internal configurations, which may be the same, S and S' are tape letters, again possibly the same, and L and R stand for left and right, respectively;

(2) there is at most one quadruple beginning with a given pair qS.

The quadruples are to be interpreted as meaning : when in internal state q and scanning S, (a) print S' to replace S and enter internal state q', and (b) move the reading head one square to the left or right, respectively, and enter internal state q'.

The condition that there is at most one quadruple beginning with a given qS ensures that the machine makes no arbitrary choices and will halt if no such quadruple exists.

7.1.2. Example. We give a simple example of a Turing machine which distinguishes between words of odd and even length. Consider the Turing machine specified by :

tape alphabet : $X, \bar{X}, Y, \bar{Y}, E$ and B (for blank)

internal states : q_0 , q_1

rules: q_0XRq_1, q_0YRq_1, $q_0\bar{X}Rq_1, q_0\bar{Y}Rq_1$

q_1XRq_0, q_1YRq_0, $q_1\bar{X}Rq_0$, $q_1\bar{Y}Rq_0$, q_0BEq_0.

Given a word W in the letters X, $\bar{X}$, Y and $\bar{Y}$ of the tape alphabet, if the machine is started with W printed on the tape, the control device is in internal state q_0 and the reading head is scanning the leftmost symbol of W, then the machine will halt scanning the symbol E if and only if the word has even length. To see this think of the internal state q_0 as standing for 'even' and q_1 as standing for 'odd'. (If W has odd length, then the machine will halt but will scan a blank.)

7.1.3. The formal definition of a *Turing machine* is then simply a set of two alphabets — tape symbols, including the blank B and internal states — with two additional symbols (L and R) together with a set of quadruples of the form 7.1.1 (a) or (b) above which satisfy 7.1.1 (1) and (2).

The application of this to decision problems is as follows. An algorithm to solve the decision problem for the subclass P of the class C is defined to exist if there exists a Turing machine with the following properties:
Each entity in the class C is specified by a unique finite sequence of tape symbols. To determine whether a given entity belongs to the subclass P or its complementary subclass P' the Turing machine begins a computation in a chosen *initial* internal state, with the sequence specifying the entity printed on the tape and the reading head scanning, say, the leftmost symbol of the sequence. In all cases the computation must come to a halt after a finite number of steps. Furthermore, when the Turing machine halts, the reading head scans a particular tape symbol, specified in advance, if and only if the entity lies in the subclass P.

The Turing machine described in Example 7.1.2 solves the word problem for the presentation $\langle X, Y \mid X^2 = XY = 1\rangle$ of the cyclic group of order two since, taking $\bar{X}$ to denote X^{-1} and $\bar{Y}$ to denote Y^{-1}, a word W represents the identity element if and only if it has even length.

7.1.4. Example. We indicate a Turing machine that solves the word problem for the natural presentation $\langle X, Y \mid X^2 = Y^2 = [X, Y] = 1\rangle$. The strategy for this machine is that it counts the occurrences of X and Y modulo 2, treating $\bar{X}$ and $\bar{Y}$ as if they were X and Y. The tape alphabet is again $\{X, Y, \bar{X}, \bar{Y}, E, B\}$ but this time there are four internal states, namely $q_{0,0}$,$q_{0,1}$,$q_{1,0}$ and $q_{1,1}$. The two subscripts will encode the number, modulo 2, of occurrences of X and Y encountered as the reading head traverses the word from left to right. The rules needed to achieve this are the sixteen rules of the form

$$q_{0,0}XRq_{1,0}\ ,\ q_{0,0}\bar{X}Rq_{1,0}\ ,\ q_{0,0}YRq_{0,1}\ ,\ q_{0,0}\bar{Y}Rq_{0,1}\quad \text{etc.}$$

plus the single rule $q_{0,0}BEq_{0,0}$. The machine will always halt after it has scanned the complete word but this final rule guarantees that the symbol E is printed on the tape if and only if W contains an even number of occurrences of both X and Y respectively.

To construct a Turing machine which solves the word problem for an infinite group requires more effort because of the fact that the basic operations used in a Turing machine are of such a primitive character. Despite their seemingly elementary nature, Turing machines are as powerful as any other form of idealised computer that has yet been conceived.

§ 7.2. Unsolvable Decision Problems

Unsolvable decision problems are first to be found in the foundations of the concept of algorithm. The most usual basic decision problem is the *halting problem* for a Turing machine. This asks for an algorithm to decide, for any word in the tape alphabet of the Turing machine, whether or not the machine will halt when the Turing machine is started running with the given word printed on the tape and the reading head scanning the leftmost symbol.

7.2.1. Theorem. *There exists a Turing machine with unsolvable halting problem.* □

We shall not attempt to explain the proof of this result beyond saying that it employs self-reference in somewhat the same kind of way as the diagonalisation argument used to show the uncountability of the real numbers. A full account may be found in, for example, [Rotman 1973]. The classic and fundamental decision problem for a group presentation is the word problem, see 1.1.9: is there an algorithm to determine of an arbitrary word in the generators of the presentation whether, as a consequence of the relators of the presentation, the word defines the identity element? The answer to this can be negative, that is, there exists a presentation whose word problem is not algorithmically solvable - moreover it is possible to write down an example of such a presentation in a comparatively short time.

7.2.2. Theorem. *There is no algorithm to solve the word problem for the group presentation B given by:*
generators: $a, b, c, d, e, p, q, r, t, k$.
relations: $p^{10}a = ap$, $p^{10}b = bp$, $p^{10}c = cp$, $p^{10}d = dp$, $p^{10}e = ep$, $qa = aq^{10}$, $qb = bq^{10}$, $qc = cq^{10}$, $qd = dq^{10}$, $qe = eq^{10}$, $ra = ar$, $rb = br$, $rc = cr$, $rd = dr$, $re = er$, $pacqr = rpcaq$, $p^2adq^2r = rp^2daq^2$, $p^3bcq^3r = rp^3cbq^3$, $p^4bdq^4r = rp^4dbq^4$, $p^5ceq^5r = rp^5ecaq^5$, $p^6deq^6r = rp^6edbq^6$, $p^7cdcq^7r = rp^7cdceq^7$, $p^8caaaq^8r = rp^8aaaq^8$, $p^9daaaq^9r = rp^9aaaq^9$, $pt = tp$, $qt = tq$, $pk = kp$, $qk = kq$, $k(aaa)^{-1}t(aaa) = (aaa)^{-1}t(aaa)k$.

The presentation has 29 relations among 10 generators which require 427 occurrences of a generator. It is obtained by applying a construction of [Borisov 1969] to the following semigroup presentation C introduced in [Tsejtin 1958].
generators: a, b, c, d, e
relations: $ac = ca$, $ad = da$, $bc = cb$, $bd = db$, $ce = eca$, $de = edb$, $cdca = cdcae$, $caaa = aaa$, $daaa = aaa$.

It should be explained that a semigroup presentation defines a semigroup in a manner that parallels the way in which a group presentation defines a group. Specifically one considers the set of all (positive) words in the generating symbols — not allowing inverse symbols. Regarding the relations of the presentation as rules for replacing one subword by another induces an equivalence relation that is compatible with the multiplication of words defined by juxtaposition. The resulting equivalence classes form a semigroup with the class of the empty word as identity element.

7.2.3. Theorem. *There is no algorithm to determine of an arbitrary word of the semigroup presentation C whether or not $W = aaa$ in C, i.e. represents the same element of the semigroup defined by C.* □

The transition to the presentation of Borisov is as follows.

7.2.4. Proposition. *For any positive word W in the generators of C,*

$$kW^{-1}tW = W^{-1}tWk$$

in B if and only if $W = aaa$ in C.

The *proof* of this proposition is rather technical and relies on the fact that Borisov's presentation is constructed from the free group on the letters p and q by successive formation of HNN-extensions. The normal form theorem 2.2.5 is the tool that provides the necessary understanding of how words can be equal in B. A very rough description of the argument is as follows. If $W = aaa$ in C then a sequence of applications of the relations of C transforms W into aaa. A parallel calculation in B establishes an equality of the form $W = UaaaV$ where U is a word on r and p and V is a word on r and q which provide a record of the calculation in C. The first group of relations serve to pass to the two ends of the word being operated on the symbols recording which relation of C was applied. The relations involving k and t then provide a means of eliminating this record and thereby establish the implication that if $W = aaa$ in C then k and $W^{-1}tW$ commute in B.

We illustrate with a sample calculation. Clearly, if $W = cdaaa$ then $W = aaa$ in C. The following equalities then hold in B :

$$p^{90}cdaaaq^9r = cp^9daaaq^9r = crp^9aaaq^9 = rp^{90}caaaq^9.$$

Thus we obtain

$$cdaaa = p^{-90}rp^{90}caaaq^9r^{-1}q^{-9}.$$

Combining this with a similar calculation involving the relation of B parallel to the relation $caaa = aaa$ of C gives an equality $W = UaaaV$ of the required type. Then the relations $pt = tp$ and $pr = rp$ show that

$$W^{-1}tW = V^{-1}(aaa)^{-1}U^{-1}tU(aaa)V = V^{-1}(aaa)^{-1}t(aaa)V$$

whence it follows that k and $W^{-1}tW$ commute.

To establish the converse implication one has to show that the only means by which k and $W^{-1}tW$ can commute is by the above process. This involves showing that the presence of inverse symbols has no significant effect. The key is the use of relations of the type $p^{10}a = ap$ and $qa = aq^{10}$ which control how the record letters p and q and the letters of C may move across one another. We shall not attempt to describe this in detail for this presentation but instead will illustrate the principle with a very elementary example.

7.2.5. Example. Let

$$G = \langle x, s, t \mid xs = sx^2, xt = tx^2 \rangle$$

and let W be a word in s and t, possibly involving s^{-1} and t^{-1}. If n is the length of W then an easy inductive argument shows that

$$W^{-1}xW = x^{2^n}$$

in G if and only if W is in fact a positive word not involving inverse symbols. Thus the question of whether a word is positive can be specified in terms of an equality within the group G.

This particular idea can be found in the earliest examples of group presentations with unsolvable word problem, namely those due to Novikov [Novikov 1955] and then Boone [Boone 1957], whose constructions also involved a transition from an unsolvable semigroup problem. Construction of a semigroup presentation with unsolvable word problem was first achieved by Post [Post 1947] and Markov [Markov 1947] whose technique was to mimic the operation rules of a Turing machine or Markov algorithm by the relations of a semigroup.

Before turning to applications of the unsolvability of the word problem we describe a different approach due to [Higman 1961], to the construction of a group with unsolvable word problem which indicates a profound connection between computational ideas like that of algorithm and finite presentability. Let $\mathcal{X}$ be an alphabet. A set A of words over $\mathcal{X}$ is called *recursively enumerable* if there is some effective procedure which enumerates the elements of A. It should be emphasised that it is not assumed that the elements of A are enumerated in a way that enables one to determine whether or not an arbitrarily given word lies in A. We leave the notion of effective procedure imprecise and merely remark that a formal definition must be given in terms of something like a Turing machine. A good example of such a procedure is that of enumerating the set of all consequences of a finite set of defining relations. For this one one must order products of conjugates of the relators in some effective and systematic way and then the enumeration is carried out by successively calculating each product. The existence of a finitely presented group with unsolvable word problem gives an example of a set of defining relations whose consequences are recursively enumerable but where the existence of the effective enumeration does not provide an algorithm to determine when a word is a consequence of the relations.

7.2.6. *Theorem. A finitely generated group G can be embedded in a finitely presented group if and only if it has a presentation (with a finite generating* *set) for which the set of defining relations is a recursively enumerable set of words.* □

7.2.7. Theorem. *There is a finitely presented group which contains an isomorphic copy of every finitely presented group.*

Proof. The set of all finitely presented groups is countable and therefore there is a countable group H, for instance the direct product of all finitely presented groups, containing an isomorphic copy of every finitely presented group. Now by Theorem 2.2.6, this countable group H can be embedded in a two generator group G and it is not difficult to see, at least in principle, that the group G is given by a recursively enumerable set of defining relations. The corollary now follows directly from the theorem. □

To derive a finitely presented group with unsolvable word problem from Theorem 7.2.6 one proceeds as follows. The existence of a Turing machine problem with unsolvable halting problem is, by standard theorems in logic, equivalent to the existence of a recursively enumerable set A of positive integers with the property that there is no algorithm to determine whether or not an arbitrary positive integer lies in A. With this result to hand it is then very easy to construct an example of a finitely generated group with unsolvable word problem.

7.2.8. Example. Let $G = \langle a, b, c, d \mid a^n b a^n = c^n d c^n : n \in A \rangle$ Now an easy argument, using for instance the method of 2.1.7, shows that $a^m b a^m$ lies in the subgroup of the free group on a and b generated by $\{a^n b a^n : n \in A\}$ if and only if $m \in A$. Since then $a^m b a^m = c^m d c^m$ if and only if $m \in A$, any algorithm which solved the word problem for G would also determine whether or not an arbitrary integer m lies in A.

According to Theorem 7.2.6 G can be embedded in a finitely presented group which must necessarily have unsolvable word problem.

Once a presentation with unsolvable word problem had been obtained, many other decision problems were shown to algorithmically unsolvable. The most notable was Dehn's isomorphism problem which was shown to be unsolvable by Adyan [Adyan 1955] and then Rabin [Rabin 1958] in the following way.

Let $G = \langle \mathcal{X} \mid \mathcal{R} \rangle$ be a group presentation with algorithmically unsolvable word problem. Construct, in a uniform manner, for each word W over $\mathcal{X}$, a presentation P_W with the property that P_W defines the trivial group if and only if $W = 1$ in G. Then in the class $\Pi = \{P_W : W \text{ is a word over } \mathcal{X}\}$, the presentations P_W and P_1 are isomorphic if and only if $W = 1$ in G. Any algorithm that could decide whether two presentations in Π defined isomorphic groups would also solve the word problem for G. It should be observed that the construction given actually shows that there can be no algorithm to determine whether a presentation defines the trivial group. As with the example with unsolvable word problem, the method of construction is that of amalgamated free products and HNN-extensions.

The full form of the theorem proved by the method of Adyan and Rabin is as follows.

7.2.9. Theorem. *Let $\mathcal{P}$ be a property of finite presentations of groups satisfying the following conditions :*

(a) *if the presentations G and G' define isomorphic groups, then G satisfies $\mathcal{P}$ if and only if G' satisfies $\mathcal{P}$;*

(b) *there is a presentation G_0 satisfying $\mathcal{P}$;*

(c) *there is a presentation G_1 such that if the presentation G defines a group in which the group defined by G_1 can be embedded, then G does not satisfy $\mathcal{P}$.*

Then there is no algorithm to determine of an arbitrary presentation whether it satisfies $\mathcal{P}$. □

This theorem embraces a very large number of properties. The most elementary example of a property $\mathcal{P}$ satisfying (a), (b) and (c) is that of defining the trivial group. Another simple example is that of defining an abelian group. One surprising variation of this theorem [Collins 1970] is that it remains valid even when the kinds of presentations considered are all assumed in advance to have *solvable* word problem. One might think that if one can always solve the word problem then all that one has to do to see if a presentation defines the trivial group is to check whether all the generators represent the identity. However this does not imply that one can decide which presentations in a class define the trivial group since the mapping from a presentation to the algorithm which solves its word problem may not be algorithmic.

As noted in Chapter 5 every finitely presented group can be realised as the fundamental group of a 4-manifold. This and the unsolvability of the isomorphism problem were exploited in [Markov 1958] to show that the problem of homeomorphy for 4-manifolds is algorithmically unsolvable. Markov's work was subsequently extended in [Boone-Haken-Poenaru 1966] to diffeomorphy and combinatorial equivalence. Unlike the situation for groups, some attention must be given to formulating a description of a manifold that meets the criteria for decision problems. Traditional topological "presentations" in terms of adding handles or defining a cell structure are not suitable in this context. The former will not in general yield a finite description while in the latter approach it is not, in general, possible to determine whether a purported presentation actually defines a manifold. The problem is resolved, for example in [Boone-Haken-Poenaru 1966] by working with a rectilinear simplicial complex, equipped with additional structures which make it a combinatorial n-manifold and also provide it with a C^∞-atlas described by algebraic equations. The various equivalence problems for a class of manifold presentations are then translated into equivalence problems about group presentations via the fundamental group. The algorithmic unsolvability of the manifold equivalence problems then follows from the corresponding algorithmic unsolvability of the group-theoretic problems.

§ 7.3. Automata and Groups

In our discussion of decision problems so far, the most general notion of algorithm — a Turing machine or its equivalent — was used. However interesting results also occur when the kind of algorithm considered is subject to restrictions. Perhaps the simplest kind of algorithm that still has significant applications is that embodied in the notion of a

7.3.1. Finite Automaton. Such an automaton also possesses a tape and reading head and has a finite number of internal states. In contrast to a Turing machine, a finite automaton cannot print symbols on its tape. Instead it just reads whatever word is initially printed on the tape, symbol by symbol, at each step moving to a new internal state. The new state is a function of both the symbol read and the existing state. When the whole word on the tape has been read the automaton halts. Some of the internal states are designated as *accepting states* and a word is *accepted* by the automaton if it halts in an accepting state having started with the given word as tape input. The *language* of the automaton is the set of all words it accepts.

Formally a finite automaton $\mathcal{A}$ consists of a finite *tape alphabet*, a finite *state alphabet*, a specified *initial state* and a *transition function* which can be represented symbolically as a set of substitution rules of the form $qS \to q'$ where q and q' are internal states and S is a tape letter. A *computation* on the automaton is then a sequence

$$q_0S_1S_2\dots S_n \to q_1S_2\dots S_n \to \dots \to q_{n-1}S_n \to q_n$$

where q_i, $i = 0, 1, \dots, n$ are internal states, with q_0 initial, and $W = S_1S_2\dots S_n$ is the word printed on the tape. The physical description given above is that of a *deterministic* automaton in which there is exactly one rule for each pair qS. It is also possible to have non-deterministic finite automata where the requirement is that there are finitely many substitution rules for each pair qS and in a computation any of these rules may be applied. A well-known result in automata theory (see [Hopcroft-Ullman 1979, p. 22]) asserts that a language can be accepted by a non-deterministic finite automaton if and only it can be accepted by some deterministic finite automaton — essentially because the computations on any non-deterministic automaton can be copied by a deterministic automaton with a larger number of states.

A group presentation on a finite alphabet is called *regular* if there is a finite automaton whose language is precisely the set of all words in the generators of the presentation which represent the identity. (This terminology is drawn from the theory of formal languages where a language is called regular if it is the language of a finite automaton.) Regular presentations are characterised by the following theorem of Anisimov [Anisimov 1972].

7.3.2. Theorem. *A group has a regular presentation on a finite alphabet if and only if it is finite.*

Proof. If the group G is finite then its multiplication table presentation is regular. For one may construct an automaton $\mathcal{A}$ whose internal states are in bijective correspondence with the elements of G, whose tape alphabet consists of the non–zero elements of G and whose transition function consists of rules $q_x y = q_z$ whenever $xy = z$ in G. The initial state and the only accepting state is q_1.

Conversely suppose that G is infinite and has a regular presentation. Since G is infinite there exists an infinite sequence $(w_n), n \geq 1$, of words such that no subword of any w_n represents the identity element. Let $\mathcal{A}$ accept the language consisting of the words representing 1. Now there exists n which is greater than the number Q of states of $\mathcal{A}$. Further $w_n w_n^{-1}$ represents 1 and so there is an accepting computation with $w_n w_n^{-1}$ as tape input. Since $n > Q$, during the course of reading in w_n some internal state q is repeated. Thus we may write w_n in the form $w_n' u w_n''$ where u is what is read in during the cycle at the state q. However this means that there is an accepting computation with $w_n' w_n'' w_n^{-1}$ as tape input since the cycle at q can be omitted. So $w_n' w_n''$ and w_n represent the same element of G which means that u must represent the trivial element contradicting our assumption about the sequence (w_n). □

7.3.3. Pushdown Automata (PDA). This is a more general kind of automaton, which is also of great importance in the theory of formal languages. The basic method of operation is the same as that of a finite automaton but now some storage and printing capacity is added. Specifically the automaton can store a single finite string of *stack letters* and it reads, say, the rightmost letter in the string. The typical operational step is then a transition

$$ZAqSW \rightarrow ZZ'q'W$$

where A is the stack letter scanned, q is the internal state, S is the tape letter scanned and Z' is an arbitrary word in the stack letters. The transition function thus consists of a finite set of substitution rules of the form

$$AqS \rightarrow Z'q'$$

where, in the non-deterministic case which is the one usually considered, there is at least one such rule for each triple AqS. A computation on a PDA begins in a given initial state, with a word printed on the tape and a single specified initial stack letter scanned. The *language accepted* by the PDA is the set of all words in the tape alphabet for which there is a computation which halts in an *accepting state.* Anisimov [Anisimov 1972] asked : what can be said about the structure of a group having a presentation for which the set of words representing the identity is a *context-free* language, that is a language accepted by a PDA. He also proved:

7.3.4. Theorem. *A finitely generated virtually free group has a presentation for which the set of words representing the identity is a context-free language.*

(A group is *virtually free* if it has a free subgroup of finite index.) We begin the *proof* of Anisimov's theorem by describing a PDA that accepts the words which represent the identity in the standard presentation of a free group of rank two. It is convenient to modify slightly the definition of a pda. Specifically the initial stack symbol is always taken to denote a blank — and is usually denoted by λ. Furthermore, it is convenient to allow the machine to print letters into the stack without affecting the word printed on the tape. In terms of the substitution rules which define the transition function, this amounts to introducing rules of the form $Aq \to Zq'$.

Let the free group have basis $\mathcal{X} = \{X, Y\}$. The automaton $\mathcal{A}$ has two internal states q and q', the tape alphabet is

$$\mathcal{X} \cup \mathcal{X}^{-1} = \{X, Y, X^{-1}, Y^{-1}\}$$

and the stack alphabet is just the tape alphabet with the blank λ adjoined. The transition function for $\mathcal{A}$ is given by the substitution rules

$$\begin{aligned}
&\lambda q S \to S q', \quad \text{for any } S \in \mathcal{X} \cup \mathcal{X}^{-1},\\
&S q' T \to S T q', \quad \text{for } S, T \in \mathcal{X} \cup \mathcal{X}^{-1}, \quad T \neq S^{-1},\\
&S q' S^{-1} \to q', \quad \text{for any } S \in \mathcal{X} \cup \mathcal{X}^{-1},\\
&\lambda q' \to \lambda q.
\end{aligned}$$

Then, when given a word W as tape input, the automaton changes to the state q' and successively reads the letters of W into the stack until it finds a pair that are inverse to one another which it then cancels (if such exist). This process is repeated until the input word is completely absorbed and either the stack contains a non-empty reduced word equal to the original word and the automaton halts in the state q' or the stack contains only the blank and the automaton returns to the state q which is the only accepting state. For example with $W = YXX^{-1}Y^{-1}X$ as input word the automaton performs the following computation :

$$\begin{aligned}
\lambda q YXX^{-1}Y^{-1}X &\to Yq'XX^{-1}Y^{-1}X \to YXq'X^{-1}Y^{-1}X\\
&\to Yq'Y^{-1}X \to \lambda q'X \to \lambda qX \to Xq'.
\end{aligned}$$

(In this computation we have followed the usual convention of explicitly writing the symbol λ which denotes the blank only when no other symbol lies to the left of the state symbol.)

Now let G be a virtually free group. Then without loss of generality we may suppose that G has a free normal subgroup K of finite index. Let $\mathcal{X}$ be a (necessarily) finite basis for K. From the group extension

$$1 \to K \to G \to G/K \to 1$$

we obtain (see 6.1.1) a finite presentation of G of the following form, where $\mathcal{T}$ is a transversal for K in G:

generators : $\mathcal{X} \cup \mathcal{T}$
relations: $TXT^{-1} = W_{T,X}(\mathcal{X})$, for $X \in \mathcal{X}$, $T \in \mathcal{T}$, $T \neq 1$
$ST^{\varepsilon} = W_{S,T,\varepsilon}(\mathcal{X})R$, with $R, S, T \in \mathcal{X}$ and $\varepsilon = \pm 1$.
Now, working stepwise from the left, any word in the above generators can be brought, by means of the above relations, into a word of the form $W(\mathcal{X})$T, for some word W and some $T \in \mathcal{T}$. The aim is to construct a PDA $\mathcal{A}$ that follows this calculation, keeping track of the transversal elements via the internal states and the word W in the stack. The tape alphabet consists of all the letters of $\mathcal{X} \cup \mathcal{T}$ and their inverses and the stack alphabet is just $\mathcal{X} \cup \mathcal{X}^{-1}$ with a blank adjoined. The internal states come in pairs q_T, q'_T , in one-to-one correspondence with the elements of $\mathcal{T}$, together with some additional *working* states that we do not specify precisely.

We provide an illustration of how the automaton works. Suppose that, after several steps, the internal state is q'_S, where $S \neq 1$, the stack contains W and the tape symbol scanned is $X \in \mathcal{X}$. Suppose further that $SXS^{-1} = Y^{-1}X$ is one of the relations of the presentation. The aim is to return to one of the states q_S, q'_S with the reduced form of $WY^{-1}X$ in the stack. This is achieved by using substitution rules

$$\begin{aligned} Zq'_SX &\to ZY^{-1}q'_{S,1}, \text{ for } Z \neq Y, \\ Yq'_SX &\to q'_{S,1}, \\ Zq'_{S,1}X &\to ZXq'_S, \text{ for } Z \neq X^{-1}, \\ X^{-1}q'_{S,1}X &\to q'_S, \\ \lambda q'_S &\to \lambda q_S, \end{aligned}$$

where $q'_{S,1}$ is a supplementary working state. In general if the relation is $SXS^{-1} = V$ and V has length n, then $n-1$ supplementary states are needed. If T^{ε}, $T \in \mathcal{T}$ is scanned and $ST^{\varepsilon} = W_{S,T,\varepsilon}(\mathcal{X})R$ is a relation, then a similar procedure reads $W_{S,T,\varepsilon}(\mathcal{X})$ into the stack and shifts the automaton to internal state q_R or q'_R. Finally if the internal state is q_1, then the rules read letters of $\mathcal{X}$ into the stack, producing a reduced word as they go until a letter of $\mathcal{T} \cup \mathcal{T}^{-1}$ is reached in which case the rules cause the automaton to enter the appropriate internal state q_S.

The main aim of the rest of this section is to sketch the proof of the converse theorem.

7.3.5. Theorem. *If a finitely generated group has a presentation in which the set of words representing the identity is a context-free language, then the group is virtually free.*

The *proof* proceeds in an interesting and roundabout way. The starting point is a fundamental result in the theory of formal languages [Hopcroft-Ullman 1979, pp. 115–117].

7.3.6. Theorem. *A language in a finite alphabet can be accepted by some PDA if and only it can be generated by a context-free grammar.* □

7.3.7. Definition and Example. The meaning of the term *context-free grammar* is as follows. There are given two disjoint alphabets $\mathcal{V}$ and $\mathcal{T}$, a distinguished element $S \in \mathcal{V}$ and a finite set of substitution rules of the form $A \to W$ where $A \in \mathcal{V}$ and W is a word over $\mathcal{V} \cup \mathcal{T}$. A word U lies in the language generated by the grammar if it involves only letters of $\mathcal{T}$ and is derivable from S by a finite number of applications of the substitution rules.

Let $\mathcal{V} = \{S, A, B, C\}$, $\mathcal{T} = \{a, b, c\}$ with the substitution rules

$$S \to ABC, \quad A \to a, \quad B \to aB, \quad B \to b, \quad C \to c.$$

An easy inductive argument shows that the language generated by this grammar is $\{a^n bc \mid n \geq 1\}$.

7.3.8. Theorem (Chomsky normal form). *Every context-free langauge can be generated by a grammar in which the substitution rules are all of the form* $A \to BC$, $A, B, C \in \mathcal{V}$ or $A \to a$, $A \in \mathcal{V}$, $a \in \mathcal{T}$.

We illustrate Theorem 7.3.8 in terms of Example 7.3.7. The changes needed are the adjunction of two addditional letters to the alphabet $\mathcal{V}$, the replacement of the rule $S \to ABC$ by the two rules $S \to AD$ and $D \to BC$ and the replacement of the rule $B \to aB$ by the rules $B \to EB$ and $E \to a$. It is then easy to see that the same language is generated — and that the same technique will apply to any grammar. □

To describe the connection between context–free grammars and PDA's we take this last grammar as an example.

7.3.9. Example. We have a grammar Γ with alphabets

$$\mathcal{V} = \{A, B, C, D, E\} \text{ and } \mathcal{T} = \{a, b, c\}$$

and rules

$$S \to AD, \ B \to EB, \ D \to BC, \ A \to a, \ B \to b, \ C \to c, \ E \to a.$$

Note that *aabc* is generated by the computation

$$S \to AD \to ABC \to ABc \to AEBc \to AEbc \to Aabc \to aabc.$$

The automaton $\mathcal{A}$ which accepts the language generated by the Γ is defined roughly as follows.
Tape alphabet: $\mathcal{T}$;
Stack alphabet: $\mathcal{V} \cup \{\lambda\}$, where λ is the initial stack letter;
Internal states: $\{q_X : X \in \mathcal{V}\} \cup \{q_0, q_h\}$;
Initial state: q_0;
Accepting state: q_h.
The rules which define the transition function of $\mathcal{A}$ are of three types:
(a) Whenever $X \to YZ$ is a rule of Γ then $Yq_Z \to q_X$ is a rule of $\mathcal{A}$;

(b) whenever $X \to x$ is a rule of Γ, then for any $Y, Z \in \mathcal{V}'$, $Yq_Zx \to YZq_X$ is a rule of $\mathcal{A}$ (the possibility that Y is the blank λ is also allowed,

(c) simple rules involving the initial and accepting states.

The aim is to be able to copy in *reverse* computations in Γ. For example

$$\lambda q_0 aabc \to \lambda q_A abc \to Aq_E bc \to AEq_B c \to Aq_B c \to ABq_C \to Aq_D \to \lambda q_S \to \lambda q_h$$

shows that $aabc$ is accepted by $\mathcal{A}$. The first and last steps are achieved via the special rules (c) for the initial and accepting states. The intermediate steps use a rule of type (a) or of type (b) according as the corresponding step in the computation in Γ uses a rule of form $X \to YZ$ or of form $X \to x$. (As before λ is written explicitly only when no other symbol precedes the state symbol.)

The above illustrates why every language generated by a context–free grammar is accepted by a PDA. The converse is more complicated and the most that can be said is that one has to construct a grammar which tracks out all possible ways in which a word might be accepted in order to generate it. □

Two consequences of these theorems are relevant. Firstly if a group G has context-free word problem then G has a finite presentation - essentially since a context-free grammar is a finite system of rules. More significant is the following result, proved in [Muller-Schupp 1983], the proof of which depends heavily on the particular form of the substitution rules given in Theorem 7.3.8.

7.3.10. Proposition. *A finitely presented group G with context-free word problem has more than one end.* □

The proof of Theorem 7.3.5 now follows quickly but makes use of very substantial results. By Stallings' Theorem 6.2.9, G splits over a finite subgroup. If G is actually torsion-free, then the finite subgroup must be trivial and, using Grushko's Theorem 2.2.27, an inductive argument on the number of generators of G shows that G is free of finite rank. When G is not torsion-free, Grushko's theorem no longer applies. Fortunately Theorem 6.2.14 of [Dunwoody 1985] shows that a finitely presented group cannot split infinitely often over finite subgroups. So again an induction argument is possible and the desired result is obtained. □

A very recent connection between groups and the theory of finite automata[1] has been introduced by Thurston who has formulated what are called automatic groups. Roughly a group is *automatic* if there is an efficient method

[1] The term "automatic groups" popular in the west is not very well chosen since in our country it was used for a totally different class of groups, namely the class of those groups whose elements are invertible automata with identical input and output alphabets and in which the group operation is superposition of automata. For the groups referred to in this book, it is useful to employ the term "groups with an automatic structure". In regard to the appearance of this class of groups, the authors refer only to Thurston but the names of Cannon, Epstein, Holt, Patterson and other authors are also linked to the introduction of this line of research and the first publications in the field are by Cannon. (Editor's remark)

for drawing its Cayley graph, efficiency being defined in terms of the existence of a finite number of finite automata which recognise vertices of the graph and when two vertices are to be joined by an edge. So far no group-theoretic characterisation of automatic groups has emerged but many groups including fundamental groups of hyperbolic 3-manifolds have been shown to be automatic.

Bibliography*

Adyan, S.I.: The Burnside Problem and Identities in Groups. Moscow: Nauka 1975. Engl. transl: Ergeb. Math. Grenzgeb. *95*, Berlin-Heidelberg-New York: Springer 1979. Zbl. 417.20001

Adyan, S.I., Novikov, P.S.: Infinite periodic groups. I, II, III. Izv. Akad. Nauk SSSR, Ser. Mat. *32*, 212–244, 251–524, 709–731 (1968). Engl. transl.: Math. USSR, Izv. *2*, 209–236, 241–480, 665–685. Zbl. 194, 33

Adyan, S.I.: Algorithmic unsolvability of certain problems of recognition of certain properties of groups. Dokl. Akad. Nauk SSSR *103*, 533–535 (1955). Zbl. 65, 9

Alperin, R.C., Moss, K.N.: Complete trees for groups with a real-valued length function. J. London Math. Soc., II. Ser. *31*, 55–68 (1985). ZBl. 571.20031

Anisimov, A.V.: Some algorithmic problems for groups and context-free languages. Kibernetika *8*:2, 4–11 (1972). Engl. transl.: Cybernetics *8*:2, 174–182 (1972). Zbl. 241.68035

Arnol'd, V.I.: The cohomology ring of the colored braid group. Mat. Zametki *5*, 227–231 (1969). Engl. transl.: Math. Notes Acad. Sci. USSR *5*, 138–140 (1969). Zbl. 277.55002

Baer, R.: Isotopien von Kurven auf orientierbaren, geschlossenen Flächen und ihr Zusammenhang mit der topologischen Deformation der Flächen. J. Reine Angew. Math. *159*, 101–116 (1928). Jbuch 54, 602

Baer, R., Levi, F.: Freie Produkte und ihre Untergruppen. Compositio Math. *3*, 391–398 (1936). Zbl. 15, 6

Baumslag, G., Taylor, T.: The centre of groups with one defining relator. Math. Ann. *175*, 315–319 (1968). Zbl. 157, 349

Birman, J.S.: Braids, links, and mapping class groups. Ann. Math. Studies *82*. Princeton, N.J.: Princeton Univ. Press 1974. Zbl. 305.57013

Birman, J.S., Series, C.: An algorithm for simple curves on surfaces. J. London, Math. Soc., II. Ser. *29*, 331–342 (1984). Zbl. 507.57006

Boileau, M., Collins, D.J., Zieschang, H.: Scindements de Heegaard des petits variétes de Seifert. C. R. Acad. Sci., Paris, Sér. I, *305*, 557–560 (1987). Zbl. 651.57010

Boileau, M., Rost, M., Zieschang, H.: On Heegaard decompositions of torus knot exteriors and related Seifert fibre spaces. Math. Ann. *279*, 553–581 (1988). Zbl. 616.57008

---

* For the convenience of the reader, references to reviews in Zentralblatt für Mathematik (Zbl.), compiled using the MATH database, and Jahrbuch über die Fortschritte der Mathematik (Jbuch) have, as far as possible, been included in this bibliography.

Boileau, M., Zieschang, H.: Heegaard genus of closed orientable Seifert 3-manifolds. Invent. Math. *76*, 455–468 (1984). Zbl. 538.57004

Boone, W.W.: Certain simple unsolvable problems of group theory V, VI. Nederl. Akad. Wetensch. Proc., Ser. A *60*, 22–27, 227–232 (1957). Zbl. 67, 257

Boone, W.W., Haken,W., Poenaru, V.: On recursively unsolvable problems in topology and their classfication. In: K. Schütte (ed.): Contributions to Mathematical Logic (Colloq. Hannover 1966), 37–74. Amsterdam: North-Holland 1968. Zbl. 246.57015

Borel, A., Serre, J.-P.: Corners and arithmetic groups. Comment. Math. Helvetici *48*, 436–491 (1973). Zbl. 274.22011

Borisov, V.V.: Simple examples of groups with unsolvable word problem. Mat. Zametki *6*, 521–532 (1969). Zbl. 211, 341. Engl. transl.: Math. Notes *6*, 768–775

Brodskij, S.D.: Equations over groups and groups with a single defining relation. Uspekhi Mat. Nauk *35*:4, 183 (1980). Zbl. 447.20024. Engl. transl.: Russian Math. Surveys *35*:4, 165 (1980)

Brodskij, S.D.: Equations over groups, and groups with one defining relation. Sib. Mat. Zh. *25*:2, 84–103 (1984). Engl. transl.: Sib. Math. J. *25*, 235–251 (1984). Zbl. 579.20020

Brown, E.M., Crowell, R.H.: Deformation retractions of 3-manifolds into their boundary. Ann. of Math., II. Ser. *82*, 445–458 (1965). Zbl. 131, 207

Brunner, A.M.: A group with an infinite number of Nielsen inequivalent one-relator presentations. J. Algebra *42*, 81–84 (1976). Zbl. 342.20014

Burde, G.: Zur Theorie der Zöpfe. Math. Ann. *151*, 101–107 (1963). Zbl. 112, 386

Burde, G., Zieschang, H.: Knots. Berlin: de Gruyter 1985. Zbl. 568.57001

Burnside, W.: On an unsettled question in the theory of discontinuous groups. Quart. J. Pure Appl. Math. *33*, 230–238 (1902). Jbuch 33, 149

Cartan, H., Eilenberg, S.: Homological Algebra. Princeton Math. Ser. *19*. Princeton, N.J.: Princeton Univ. Press 1956. Zbl. 75, 243

Cayley, A.: On the theory of groups. Proc. London Math. Soc. *9*, 126–133 (1878). Jbuch 10, 104

Chiswell, I.M.: Abstract length functions in groups. Math. Proc. Camb. Phil. Soc. *80*, 451–463 (1976). Zbl. 351.20024

Chiswell, I.M., Collins, D.J. Huebschmann, J.: Aspherical group presentations. Math. Z. *178*, 1–36 (1981). Zbl. 443.20030

Church, A.: An unsolvable problem of elementary number theory. Amer. J. Math. *58*, 345–363 (1936). Zbl. 14, 98

Church, A.: The calculi of lambda conversion. Princeton, N.J.: Princeton Univ. Press 1941. Zbl. 26, 242

Cohen, D.E.: Groups of Cohomological Dimension One. Lecture Notes in Math. *245*, Berlin-Heidelberg-New York: Springer 1972. Zbl. 231.20018

Cohen, D.E.: Groups with free subgroups of finite index. In: Lecture Notes in Math. *319*, 26–44. Berlin-Heidelberg-New York: Springer 1973. Zbl. 386.20010

Cohen, D.E.: Combinatorial Group Theory. A Topological Approach. Queen Mary College Math. Notes 1978. Zbl. 389.20024; London Math. Soc. Student Texts 1989. Zbl. 697.20001

Cohen, D.E., Lyndon, R.C.: Free bases for normal subgroups of free groups. Trans. Amer. Math. Soc. *108*, 526–537 (1963). Zbl. 115, 251

Collins, D.J.: On recognizing properties of groups which have solvable word problem. Archiv Math. *21*, 31–39 (1970). Zbl. 195, 20

Collins, D.J.: Presentations of the amalgamated free product of two infinite cycles. Math. Ann. *237*, 233–241 (1978). Zbl. 385.20019

Collins, D.J., Zieschang, H.: Rescuing the Whitehead method for free products, I,II. Math. Z. *185*, 487–504 (1984); *186*, 335–361 (1984). Zbl. 537.20010; Zbl. 542.20012

Collins, D.J., Zieschang, H.: A presentation for the stabilizer of an element in a free product. J. Algebra *106*, 53–77 (1987). Zbl. 608.20020

Coxeter, H.S.M., Moser, W.O.J.: Generators and Relations for Discrete Groups. Ergeb. Math. Grenzgeb. *14*. Berlin-Heidelberg-New York: Springer 1972. Zbl. 239.20040

Culler, M., Morgan, J.W.: Group actions on ℝ-trees. Proc. London Math. Soc., III. Ser. *55*, 571–604 (1987). Zbl. 658.20021

Dehn, M.: Über die Topologie des dreidimensionalen Raumes. Math. Ann. *69*, 137–168 (1910)

Dehn, M.: Über unendliche diskontinuierliche Gruppen. Math. Ann. *71*, 116–144 (1912)

Dehn, M.: Transformation der Kurven auf zweiseitigen Flächen. Math. Ann. *72*, 413–421 (1912)

Dicks, W.: Groups, Trees and Projective Modules. Lecture Notes in Math. *790*. Berlin-Heidelberg-New York: Springer 1980. Zbl. 427.20016

Dunwoody, M.J.: Accessibility and groups of cohomological dimension one. Proc. London Math. Soc., III. Ser. *38*, 193–215 (1979). Zbl. 419.20040

Dunwoody, M.J.: The accessibility of finitely presented groups. Invent. Math. *81*, 449–457 (1985). Zbl. 572.20025

v. Dyck, W.: Gruppentheoretische Studien. Math. Ann. *20*, 1–45 (1882). Jbuch 14, 97

Dyer, J.L., Scott, G.P.: Periodic automorphisms of free groups. Commun. Algebra *3*, 195–201 (1975). Zbl. 304.20029

Eckmann, B., Linnell, P.A.: Poincaré duality groups of dimension two. Comment. Math. Helvetici *58*, 111–114 (1983). Zbl. 518.57003

Eckmann, B., Müller, H.: Plane motion groups and virtual Poincaré duality of dimension 2. Invent. Math. *69*, 293–310 (1982). Zbl. 501.20031

Epstein, D.B.A.: Finite presentations of groups and 3-manifolds. Quart. J. Math., Oxf. II. Ser. *12*, 205–212 (1961). Zbl. 231.55003

Epstein, D.B.A.: Curves on 2-manifolds and isotopies. Acta Math. *115*, 83–107 (1966). Zbl. 136, 446

Fine, B., Howie, J., Rosenberger, G.: One-relator quotients and free products of cyclics. Proc. Amer. Math. Soc. *102*, 249–254 (1988). Zbl. 653.20029

Freudenthal, H.: Über die Enden topologischer Räume und Gruppen. Math. Z. *33*, 692–713 (1931). Zbl. 2, 56

Freudenthal, H.: Über die Enden diskreter Räume und Gruppen. Comment. Math. Helvetici *17*, 1–38 (1944). Zbl. 60, 400

Fuks, D.B.: Classical manifolds. In: Itogi Nauki Tekhn., Sovr. Prob. Mat. I, Fundam. Napravleniya *12*, 253–314 (1986). Zbl. 671.57001. Engl. transl. in: Encycl. Math. Sci. *12*, Berlin-Heidelberg-New York: Springer (in preparation)

Funcke, K.: Nicht frei äquivalente Darstellungen von Knotengruppen mit einer definierenden Relation. Math. Z. *141*, 205–217 (1975). Zbl. 283.20025

Funcke, K.: Geschlecht von Knoten mit zwei Brücken und die Faserbarkeit ihrer Außenräume. Math. Z. *159*, 3–24 (1978). Zbl. 734.55001

Garside, F.A.: The braid groups and other groups. Quart. J. Math., Oxf. II. Ser. *20*, 235–254 (1969). Zbl. 194, 33

Gersten, S.M.: On Whitehead's algorithm. Bull. Amer. Math. Soc., New Ser. *10*, 281–284 (1984). Zbl. 537.20015

Gersten, S.M.: On fixed points of automorphisms of finitely generated free groups. Bull. Amer. Math. Soc., New Ser. *8*, 451–454 (1983). Zbl. 512.20014

Gersten, S.M.: Fixed points of automorphisms of free groups. Adv. Math. *64*, 51–85 (1987). Zbl. 616.20014

Gerstenhaber, M., Rothaus, O.S.: The solution of sets of equations in groups. Proc. Natl. Acad. Sci. USA *48*, 1531–1533 (1962). Zbl. 112, 25
Gilman, J.: On the Nielsen type and the classification for the mapping class group. Adv. Math. *40*, 68–96 (1981). Zbl. 474.57005
Goldstein, R.Z., Turner, E.C.: Automorphisms of free groups and their fixed points. Invent. Math. *78*, 1–12 (1984). Zbl. 548.20016
Goldstein, R.Z., Turner, E.C.: Fixed subgroups of homomorphisms of free groups. Bull. London Math. Soc. *18*, 468–470 (1986). Zbl. 576.20016
Greendlinger, M.D.: Dehn's algorithm for the word problem. Commun. Pure Appl. Math. *13*, 67–83 (1960). Zbl. 104, 19
Gromov, M.: Hyperbolic groups. In: S.M. Gersten (ed.): Essays in Group Theory. Math. Sciences Research Institute Publ. *8*, 75–263. Berlin-Heidelberg-NewYork: Springer 1987. Zbl. 634.20015
Gruenberg, K.W.: Cohomological Topics in Group Theory. Lecture Notes in Math. *143*. Berlin-Heidelberg-New York: Springer 1970. Zbl. 205, 327
Gruško, I.A.: Über die Basen eine freien Produktes von Gruppen. Mat. Sb. *8*, 169–182 (1940). Zbl. 23, 301
Gurevich, G.A.: On the conjugacy problem for groups with a single defining relation. Trudy Mat. Inst. Steklova *133*, 109–120 (1973). Zbl. 291.20045. Engl. transl.: Proc. Steklov Inst. Math. *133*, 108–120
Hall, M., jr.: Solution of the Burnside problem for exponent 6. Proc. Natl. Acad. Sci. USA *43*, 751–753 (1957). Zbl. 79, 30
Hall, M., jr.: The Theory of Groups. New York: Macmillan 1959. Zbl. 84, 22
Handel, M., Thurston, W.P.: New proofs of some results of Nielsen. Adv. Math. *56*, 173–191 (1985). Zbl. 584.57007
Harer, J.L.: The virtual cohomological dimension of the mapping class group of an orientable surface. Invent. Math. *84*, 157–176 (1986). Zbl. 592.57009
Harrison, N.: Real length functions in groups. Trans. Amer. Math. Soc. *174*, 77–106 (1972). Zbl. 255.20021
Hatcher, A., Thurston, W.P.: A presentation of the mapping class group of a closed orientable surface. Topology *19*, 221–237 (1980). Zbl. 447.57005
Hempel, J.: 3-manifolds. Ann. Math. Studies *86*. Princeton: Princeton Univ. Press (1976). Zbl. 345.57001
Higgins, P.J., Lyndon, R.C.: Equivalence of elements under automorphisms of a free group. J. London Math. Soc., II. Ser. *8*, 254–258 (1974). Zbl. 288.20027
Higman, G.: The units of group rings. Proc. London Math. Soc., II. Ser. *46*, 231–248 (1940). Zbl. 25, 243
Higman, G.: A finitely related group with an isomorphic proper factor group. J. London Math. Soc. *26*, 59–61 (1951). Zbl. 42, 21
Higman, G.: Subgroups of finitely presented groups. Proc. Royal Soc. London Ser. A *262*, 455–475 (1961). Zbl. 104, 21
Higman, G., Neumann, B.H., Neumann, H.: Embedding theorems for groups. J. London Math. Soc. *24*, 247–254 (1949). Zbl. 34, 301
Hilton, P.J., Stammbach, U.: A Course in Homological Algebra. Grad. Texts in Math. *4*. Berlin-Heidelberg-New York: Springer 1971. Zbl. 238.18006
Holt, D.F.: Uncountable locally finite groups have one end. Bull. London Math. Soc. *13*, 557–560 (1981). Zbl. 477.20015
Hopcroft, J.E., Ullman, J.D.: Introduction to automata theory, languages and computation. Reading, Mass.: Addison-Wesley 1979. Zbl. 426.68001
Hopf, H.: Enden offener Räume und unendliche diskontinuierliche Gruppen. Comment. Math. Helvetici *16*, 81–100 (1943). Zbl. 60, 400
Houghton, C.H.: Ends of locally compact groups and their coset space. J. Austr. Math. Soc. *17*, 274–284 (1974). Zbl. 289.22005

Howie, J.: On pairs of 2-complexes and systems of equations over groups. J. Reine Angew. Math. *324*, 165–174 (1981). Zbl. 447.20032

Howie, J.: How to generalize one-relator group theory. In: S.M. Gersten, J.R. Stallings (ed.): Combinatorial Group Theory and Topology. Ann. of Math: Studies *111* , 53–78. Princeton, N.J.: Princeton Univ. Press 1987. Zbl. 632.20021

Huebschmann, J.: Cohomology theory of aspherical groups and of small cancellation groups. J. Pure Appl. Algebra *14*, 137–143 (1979). Zbl. 396.20021

Hurwitz, A.: Ueber Riemann'sche Flächen mit gegebenen Verzweigungspunkten. Math. Ann. *39*, 1–61 (1891). Jbuch 23, 429

Hurwitz, A.: Ueber algebraische Gebilde mit eindeutigen Transformationen in sich. Math. Ann. *41*, 401–442 (1893)

Imrich, W.: On metric properties of tree-like spaces. In: Beiträge zur Graphentheorie und deren Anwendungen, Sektion MAROK der Techn. Hochschule Ilmenau, Oberhof (DDR), 129–156, 1979. Zbl. 417.05058

Jaco, W., Shalen, P.: Surface homeomorphisms and periodicity. Topology *16*, 347–367 (1977). Zbl. 447.57006

Kaplansky, I.: Fields and Rings. Chicago: Chicago Univ. Press, 2nd ed. 1972. Zbl. 238.16001

Karrass, A., Pietrowski, A., Solitar, D.: Finite and infinite cyclic extensions of free groups. J. Austr. Math. Soc. *16*, 458–466 (1973). Zbl. 299.20024

Karrass, A., Solitar, D.: The subgroups of a free product of two groups with an amalgamated subgroup. Trans. Amer. Math. Soc. *150*, 227–255 (1970). Zbl. 223.20031

Karrass, A., Solitar, D.: The free product of two groups with a malnormal amalgamated subgroup. Canad. J. Math. *23*, 933–959 (1971). Zbl. 247.20028

Kerckhoff, S.P.: The Nielsen realization problem. Ann. of Math. *117*, 235–265 (1983). Zbl. 528.57008

Kneser, H.A.: Geschlossene Flächen in dreidimensionalen Mannigfaltigkeiten. Jahresber. Dtsch. Math.-Ver. *38*, 248–260 (1929). Jbuch 55, 311

Kuroš, A.G. (= Kurosh, A.G.): Die Untergruppen der freien Produkte von beliebigen Gruppen. Math. Ann. *109*, 647–660 (1934). Zbl. 9, 10

Kurosh, A.G.: The Theory of Groups. Moscow: Nauka 1967. Zbl. 189, 308. German transl.: Berlin: Akademie-Verlag 1970, 1972

Lehner, J.: Discontinuous Groups and Automorphic Functions. Providence, R.I.: Amer. Math. Soc. 1964. Zbl. 178, 429

Lewin, J., Lewin, T.: The group algebra of a torsion-free one-relator group can be embedded in a field. Bull. Amer. Math. Soc. *81*, 947–949 (1975). Zbl. 317.16001

Linnell, P.A.: On accessibility of groups. J. Pure Appl. Algebra *30*, 39–46 (1983). Zbl. 545.20020

Lyndon, R.C.: Length functions in groups. Math. Scand. *12*, 209–234 (1963). Zbl. 119, 264

Lyndon, R.C.: On Dehn's algorithm. Math. Ann. *166*, 208–228 (1966). Zbl. 138, 257

Lyndon, R.C., Schupp, P.E.: Combinatorial Group Theory. Ergebn. Math. Grenzgeb. *89*. Berlin-Heidelberg-New York: Springer 1977. Zbl. 368.20023

Macbeath, A.M.: Generators of linear fractinal groups. In: Proc. Symp. Pure Math *12*, 14–32. Providence, R.I.: Amer. Math. Soc. 1969. Zbl. 192, 357

Macbeath, A.M., Hoare, A.H.M.: Groups of hyperbolic crystallography. Math. Proc. Camb. Philos. Soc. *79*, 235–249 (1976). Zbl. 338.20056

Maclachlan, C.: On representations of Artin's braid group. Mich. Math. J. *25*, 235–244 (1978). Zbl. 366.20032

Magnus, W.: Über diskontinuierliche Gruppen mit einer definierenden Relation. J. Reine Angew. Math. *163*, 141–165 (1930). Jbuch 56, 134

Magnus, W.: Untersuchungen über einige unendliche diskontinuierliche Gruppen. Math. Ann. *105*, 52–74 (1931). Zbl. 2, 113

Magnus, W.: Über n-dimensionale Gittertransformationen. Acta Math. *64*, 353–367 (1935). Zbl. 12, 54

Magnus, W.: Braid groups: a survey. In: Proc. Conf. Canberra 1973, Lecture Notes in Math. *372*, 463–487. Berlin-Heidelberg-New York: Springer 1974. Zbl. 286.20039

Magnus, W.: Noneuclidean Tesselations and Their Groups. Pure Appl. Math. *61*. New York-London: Academic Press 1974. Zbl. 293.50002

Magnus, W., Karrass, A., Solitar, D.: Combinatorial Group Theory: Presentations of Groups in Terms of Generators and Relations. New York-London-Sidney: Interscience Publishers, John Wiley and Sons, Inc. 1966. Zbl. 138, 256

Makanin, G.S.: The conjugacy problem in the braid group. Dokl. Akad. Nauk SSSR *182* , 495–496 (1968). Engl. transl.: Sov. Math., Dokl. *9*, 1156–1157. Zbl. 175, 295

Markov, A.A.: On the impossibility of certain algorithms in the theory of associative systems. Dokl. Akad. Nauk SSSR *55*, 587–590 (1947). Zbl. 29, 101

Markov, A.A.: Theory of Algorithms. Trudy Mat. Inst. Steklova *42*,1–374 (1954). Zbl. 58, 5

Markov, A.A.: The insolubility of the problem of homeomorphy. Dokl. Akad. Nauk SSSR *121*, 218–220 (1958). Zbl. 92, 7

Massey, W.S.: Algebraic Topology: An Introduction. New York: Harcourt, Brace, and World 1967. Zbl. 153, 249

McCool, J.: A presentation for the automorphism group of a free group of finite rank. J. London Math. Soc., II. Ser. *8*, 259–266 (1974). Zbl. 296.20010

McCool, J.: On Nielsen's presentation of the automorphism group of a free group. J. London Math. Soc., II. Ser. *10*, 265–270 (1975a). Zbl. 338.20029

McCool, J.: Some finitely presented subgroups of the automorphism group of a free group. J. Algebra *35*, 205–213 (1975b). Zbl. 325.20025

McCool, J., Pietrowski, A.: On free products with amalgamation of two infinite cyclic groups. J. Algebra *18*, 377–383 (1971). Zbl. 232.20054

Miller, R.T.: Geodesic laminations from Nielsen's viewpoint. Adv. Math. *45*, 189–212 (1982). Zbl. 496.57003

Moise, E.E.: Geometric Topology in Dimensions 2 and 3. Berlin-Heidelberg-New York: Springer 1977. Zbl. 349.57001

Moldavanskij, D.I.: Certain subgroups of groups with one defining relation. Sib. Mat. Zh. *8*, 1370–1384 (1967). Engl. transl.: Sib. Mth. J. *8*, 1039–1048. Zbl. 169, 336

Muller, D.E., Schupp, P.E.: Groups, the theory of ends and context-free languages. J. Comput. Syst. Sci. *26*, 295–310 (1983). Zbl. 537.20011

Murasugi, K.: The center of a group with a single defining relation. Math. Ann. *155*, 246–251 (1964). Zbl. 119, 26

Neumann, H.: Generalized free products with amalgamated subgroups, I, II. Amer. J. Math. *70*, 590–625 (1948); *71*, 491–540 (1949). Zbl. 32, 104; Zbl. 33, 99

Newman, B.B.: Some results on one-relator groups. Bull. Amer. Math. Soc. *74*, 568–571 (1968). Zbl. 174, 46

Nielsen, J.: Die Isomorphismen der allgemeinen unendlichen Gruppen mit zwei Erzeugenden. Math. Ann. *78*, 385–397 (1918). Jbuch 46, 175

Nielsen, J.: Über die Isomorphismen unendlicher Gruppen ohne Relation. Math. Ann. *79*, 269–272 (1919). Jbuch 46, 175

Nielsen, J.: Die Isomorphismengruppen der freien Gruppen. Math. Ann. *91*, 169–209 (1924a). Jbuch 50, 78

Nielsen, J.: Die Gruppe der dreidimensionalen Gittertransformationen. Danske Vid. Selsk. Mat.-Fys. Medd. 5, No. *12*, 1–29 (1924b). Jbuch 50, 74

Nielsen, J.: Untersuchungen zur Topologie der geschlossenen zweiseitigen Flächen, I, II, III. Acta Math. *50*, 189–358 (1927); *53*, 1–76 (1929); *58*, 87–167 (1932). Jbuch 53, 545; Jbuch 55, 971; Zbl. 4, 275

Nielsen, J.: Abbildungsklassen endlicher Ordnung. Acta Math. *75*, 23–115 (1942). Zbl. 27, 266

Novikov, P.S.: On the algorithmic unsolvability of the word problem in group theory. Trudy Mat. Inst. Steklova *44*, 1–140 (1955). Zbl. 68, 13

Novikov, S.P.: Topology. In: Itogi Nauki Tekhn., Sovr. Prob. Mat., Fundament. Napravleniya *12*, 5–252. Moscow: VINITI, 1986. Zbl. 668.55001. Engl. transl. in: Encycl. Math. Sci. *12*. Heidelberg-Berlin-New York: Springer (in preparation)

Ol'shanskij, A.Yu.: *Infinite groups with cyclic subgroups*. Dokl. Akad. Nauk SSSR *245*, 785–787 (1979). Engl. transl.: Soviet Math. Dokl. *20*, 343–346 (1979). Zbl. 431.20025

Ol'shanskij, A.Yu.: Groups of bounded exponent with subgroups of prime order. Algebra Logika *21*, 553–618 (1982). Engl. transl.: Algebra Logic *21*, 369–418 (1983). Zbl. 524.20024

Orlik, P., Vogt, E., Zieschang, H.: Zur Topologie gefaserter dreidimensionaler Mannigfaltigkeiten. Topology *6*, 49–64 (1967). Zbl. 147, 235

Papakyriakopoulos, C.D.: On solid tori. Proc. London Math. Soc., II. Ser. *7*, 281–299 (1957a). Zbl. 78, 163

Papakyriakopoulos, C.D.: On Dehn's lemma and the asphericity of knots. Ann. of Math., II. Ser. *66*, 1–26 (1957b). Zbl. 78, 164

Peczynski, N., Rosenberger, G., Zieschang, H.: Über Erzeugende ebener diskontinuierlicher Gruppen. Invent. Math. *29*, 161–180 (1975). Zbl. 311.20031

Poincaré, H.: Théorie des groupes fuchsiens. Acta Math. *1*, 1–62 (1882). Jbuch 14, 338

Poincaré, H.: Cinquième complément à l'analysis situs. Rend. Circ. Mat. Palermo *18*, 45–110 (1904). Jbuch 35, 504

Post, E.L.: Recursive unsolvability of a problem of Thue. J. Symb. Logic *12*, 1–11 (1947)

Pride, S.J.: The isomorphism problem for two-generator one-relator groups with torsion is solvable. Trans. Amer. Math. Soc. *227*, 109–139 (1977). Zbl. 356.20037

Rabin, M.O.: Recursive unsolvability of group theoretic problems. Ann. of Math., II. Ser. *67*, 172–194 (1958). Zbl. 79, 248

Radó, T.: Über den Begriff der Riemannschen Fläche. Acta Univ. Szeged *2*, 101–121 (1924–26). Jbuch 51, 273

Reidemeister, K.: Knoten und Gruppen. Abhandl. Math. Sem. Univ. Hamburg *5*, 8–23 (1927). Jbuch 52, 578

Reidemeister, K.: Fundamentalgruppen und Überlagerungsräume. Nachr. Ges. Wiss. Göttingen, Math. Phys. Kl. 1928, 69–76. Jbuch 54, 603

Reidemeister, K.: Einführung in die kombinatorische Topologie. Braunschweig: Fr. Vieweg u. Sohn 1932. Zbl. 4, 369

Reidemeister, K.: Homotopieringe und Linsenräume. Abhandl. Math. Sem. Univ. Hamburg *11*, 102–109 (1936). Zbl. 11, 324

Robinson, D.J.S.: Finiteness Conditions and Generalized Soluble Groups, 1. Ergeb. Math. Grenzgeb. *62*. Berlin-Heidelberg-New York: Springer 1972. Zbl. 243.20032

Rosenberger, G.: Bemerkungen zu einer Arbeit von H. Zieschang. Archiv Math. *29*, 623–627 (1977). Zbl. 382.20027

Rotman, J.J.: The Theory of Groups. Boston: Allyn and Bacon, Inc. 1973. Zbl. 262.20001

Sanov, I.M.: Solution of Burnside's problem for exponent 4. Uch. Zap. Leningr. Gos. Univ., Ser. Mat. Nauk *10*, 166–170 (1940)

Schiek, H.: Ähnlichkeitsanalyse von Gruppenrelationen. Acta Math. *96*, 157–252 (1956). Zbl. 71, 252

Schreier, O.: Über die Gruppen $A^a B^b = 1$. Abhandl. Math. Sem. Univ. Hamburg *3*, 167–169 (1924). Jbuch 50, 70

Schreier, O.: Über die Erweiterung von Gruppen, I, II. Monatshefte Math. Phys. *34*, 165–180 (1926); Abhandl. Math. Sem. Univ. Hamburg *4*, 321–346 (1926). Jbuch 52, 113
Schreier, O.: Die Untergruppen der freien Gruppen. Abhandl. Math. Sem. Univ. Hamburg *5*, 161–183 (1927). Jbuch 53, 110
Schubert, H.: Knoten mit zwei Brücken. Math. Z. *65*, 133–170 (1956). Zbl. 71, 390
Schupp, P.E.: Small cancellation theory over free products with amalgamation. Math. Ann. *193*, 255–264 (1971). Zbl. 209, 52
Scott, G.P.: Finitely generated 3-manifold groups are finitely presented. J. London Math. Soc., II. Ser. *6*, 437–440 (1973). Zbl. 254.57003
Scott, G.P.: An embedding theorem for groups with a free subgroup of finite index. Bull. London Math. Soc. *6*, 304–306 (1974). Zbl. 288.20043
Scott, G.P.: Ends of pairs of groups. J. Pure Appl. Algebra *11*, 179–198 (1977). Zbl. 368.20021
Scott, G.P.: A new proof of the annulus and torus theorems. Amer. J. Math. *102*, 241–277 (1980). Zbl. 439.57004
Scott, G.P., Wall, C.T.C.: Topological methods in group theory. In: C.T.C. Wall (ed.): Homological Group Theory. London Math. Soc. Lecture Notes Series *36*, 137–203. Cambridge: Cambridge Univ. Press 1979. Zbl. 423.20023
Seifert, H.: Konstruktion dreidimensionaler geschlossener Räume. Ber. Sächs. Akad. Wiss. *83*, 26–66 (1931). Zbl. 2, 160
Seifert, H.: Topologie dreidimensionaler gefaserter Räume. Acta Math. *60*, 147–238 (1933). Zbl. 6, 83
Seifert, H., Threlfall, W.: Lehrbuch der Topologie. Leipzig: Teubner 1934. Zbl. 9, 86
Selberg, A.: On discontinuous groups in higher-dimensional symmetric spaces. Colloquium Function Theory, 147–164. Bombay: TATA Inst. Fund. Res. 1960. Zbl. 201, 366
Serre, J-P.: Arbres, Amalgames, SL_2. Astérisque *46*. Paris: Soc. Math. France 1977. Zbl. 369.20013. Engl. transl.: Trees. Berlin-Heidelberg-New York: Springer 1980
Shafarevich, I.R.: Algebra. Itogi Nauki Tekhn., Sovr. Prob. Mat., Fundam. Napravleniya *11* Moscow: VINITI, 1986. Zbl. 655.00002. Engl. transl.: Encycl. Math. Sci. *11*. Heidelberg-Berlin-New York: Springer (1990). Zbl. 711.16001
Shalen, P.B.: Dendrology of groups. In: S.M. Gersten (ed.): Essays in Group Theory. Math. Sciences Research Institute Publ. *8*, 265–319. Berlin-Heidelberg-New York: Springer 1987. Zbl. 649.20033
Siegel, C.L.: Einführung in die Theorie der Modulfunktionen n-ten Grades. Math. Ann. *116*, 617–657 (1939). Zbl. 21, 203
Spanier, E.H.: Algebraic Topology. New York: McGraw-Hill 1966. Zbl. 145, 433
Stallings, J.R.: On fibering certain 3-manifolds. In: Topology of 3-manifolds, Proc. 1961 Top. Inst. Univ. Georgia (ed. M.K. Fort, jr), 95–100. Englewood Cliffs, N.J.: Prentice Hall 1962. Zbl. 132, 203
Stallings, J.R.: Group Theory and Three-Dimensional Manifolds. Yale Math. Monographs *4*. New Haven: Yale Univ. Press 1971. Zbl. 241.57001
Stillwell, J.: Classical Topology and Combinatorial Group Theory. Grad. Texts in Math. *72*. Berlin-Heidelberg-New York: Springer 1980. Zbl. 453.57001
Swan, R.G.: Groups of cohomological dimension one. J. Algebra *12*, 585–610 (1969). Zbl. 188, 70
Tartakovskij, V.A.: The sieve method in group theory. Mat. Sb., Nov. Ser. *25*, 3–50 (1949). Zbl. 34, 15
Tietze, H.: Über die topologischen Invarianten mehrdimensionaler Mannigfaltigkeiten. Monatshefte Math. Phys. *19*, 1–118 (1908)

Tits, J.: A theorem of Lie-Kolchin for trees. In: Contributions to Algebra: Collect. Papers dedicated to E. Kolchin, 377–388. New York: Academic Press 1977. Zbl. 373.20039

Tsejtin, G.S.: Associative calculations with an unsolvable equivalence problem. Trudy Mat. Inst. Steklova *52*, 172–189 (1958). Zbl. 87, 253

Todd, J.A., Coxeter, H.S.M.: A practical method for enumerating cosets of a finite abstract group. Proc. Edinburgh Math. Soc., II. Ser. *5*, 26–34 (1936). Zbl. 15, 101

Turing, A.M.: On computable numbers with an application to the Entscheidungsproblem. Proc. London Math. Soc., II. Ser. *42*, 230–265 (1937). Zbl. 16, 97

van Kampen, E.R.: On the connection between the fundamental groups of some related spaces. Amer. J. Math. *55*, 261–267 (1933a). Zbl. 6, 415

van Kampen, E.R.: On some lemmas in the theory of groups. Amer. J. Math. *55*, 268–273 (1933b). Zbl. 6, 392

Waldhausen, F.: Eine Klasse von 3-dimensionalen Mannigfaltigkeiten. I, II. Invent. Math. *3*, 308–333 (1967); *4*, 87–117 (1967). Zbl. 168, 445

Waldhausen, F.: On irreducible 3-manifolds which are sufficiently large. Ann. of Math., II. Ser. *87*, 56–88 (1968). Zbl. 157, 306

Waldhausen, F.: Some problems on 3-manifolds. Proc. Symposia in Pure Math. *32*, Part 2, 313–332 (1978). Zbl. 397, 57007

Whitehead, J.H.C.: On certain sets of elements in a free group. Proc. London Math. Soc., II. Ser. *41*, 48–56 (1936a). Zbl. 13, 248

Whitehead, J.H.C.: On equivalent sets of elements in a free group. Ann. of Math., II. Ser. *37*, 782–800 (1936b). Zbl. 15, 248

Wilkie, H.C.: On non-Euclidean crystallographic groups. Math. Z. *91*, 87–102 (1966). Zbl. 166, 26

Zieschang, H.: Alternierende Produkte in freien Gruppen. Abhandl. Math. Sem. Univ. Hamburg *27*, 13–31 (1964). Zbl. 135, 418

Zieschang, H.: Über die Nielsensche Kürzungsmethode in freien Produkten mit Amalgam. Invent. Math. *10*, 4–37 (1970). Zbl. 185, 52

Zieschang, H.: Generators of the free product with amalgamation of two infinite cyclic groups. Math. Ann. *227*, 195–221 (1977). Zbl. 333.20024

Zieschang, H.: Finite Groups of Mapping Classes of Surfaces. Lecture Notes in Math. *875*. Berlin-Heidelberg-New York: Springer 1981. Zbl. 472.57006

Zieschang, H., Vogt, E., Coldewey, H.-D.: Surfaces and Planar Discontinuous Groups. Lecture Notes in Math. *835*. Berlin-Heidelberg-New York: Springer 1980. Zbl. 438.57001. Enlarged edition published in Russian by Nauka 1988

Index of Notation

$ab: G \to G^{ab}$ 12
$\mathrm{Aut}(p)$ 25
$\mathrm{Aut}F$ 47
(A1′), (A2), (A4) 46
B_n 9
$B_1(K)$ 19
$\mathrm{cd}(G)$ 134
C^1 14
$C_0(K)$, $C_1(K)$, $C_2(K)$ 19
C_∞ 102
$C\langle\mathcal{X}|\mathcal{R}\rangle$ 20
$C(6)$ 97
$C'(\lambda)$ 94
$\mathrm{def}\, G$ 107
$d(G)$ 6
$d(g,h)$ 46

$D(2,3,7)$ 77
$\Delta\langle\mathcal{X}|\mathcal{R}\rangle$ 23
$E(C)$ 13
$e(G)$ 137
$e(G,S)$ 142
$e(X)$ 136
$E^+(X)$ 41
$\mathrm{Fix}(\alpha)$ 53
$\mathcal{F}G$ 137
$f_\sharp$ 18
$\mathbb{F}_2$ 137
$F(C)$ 14
G^{ab} 12
(G,A,β) 128
$G=\langle\mathcal{X}|\mathcal{R}\rangle$ 7
$G=\langle(x_j)_{j\in J}|(R_k(\hat{\mathcal{X}}))_{k\in K}\rangle$ 7
$G=\langle S_1,\dots,S_n|R_1,\dots,R_q\rangle$ 7
$G=\langle s_1,\dots,s_n|-\rangle$ 7
$G=*_{i\in I}G_i$ 35
$G=G_1 *_A G_2$ 35
$G=(G_1*G_2:A_1=A_2)$ 35
$(\mathcal{G},X)$ 41
$\langle G_0,t|t^{-1}A_1t=A_2\rangle$ 35
$GL(2,\mathbb{Z})$ 9
$GL(n,\mathbb{Z})$ 126
$\Gamma\langle\mathcal{X}|\mathcal{R}\rangle$ 22
$\mathbb{H}^2$ 75
$H_0(K),H_1(K),H_2(K)$ 19
$H_1(N_g)$ 12
$H_1(S_g)$ 12
$H^n(G,A)$ 131
$\mathrm{Inn}\,(\pi_1(S_g))$ 53
$IA(F)$ 48
$L(p,q)$ 109
$m(\phi,v)$ 23
$\mu(G)$ 70
N_g 16
$N_{g,r}$ 21, 63
$\mathcal{M}(S)$ 84
(N1)–(N3) 32
$\mathrm{Out}\,(\pi_1(S_g))$ 53
$\mathcal{P}G$ 137
$PSL(2,\mathbb{Z})$ 9
$\pi_1(C,v_0),\pi_1(C)$ 18
$\pi_1(\mathcal{G},X,T)$ 41
$\pi_1(\mathcal{G},X,v)$ 41
$\pi_1(N_g)=\langle v_1,\dots,v_g|v_1^2\dots v_g^2\rangle$ 12
$\pi_1(S_g)=\langle t_1,u_1,\dots,t_g,u_g|\prod_{i=1}^g[t_i,u_i]\rangle$ 12
$\pi_2(M)$ 139
P^2 15
$\mathbb{Q}$ 9
$\mathcal{Q}G$ 137
S^1 15
S_g 15
$S_{g,r}$ 21, 63
$SL(2,\mathbb{Z})$ 9
$\mathrm{Stab}_G(\tilde{v})$ 43
$\mathrm{Stab}_G(\tilde{\sigma})$ 43
$\mathrm{Stab}(w)$ 52
T 94
$Tor\,A$ 12
T(1)–T(3) 33
T(0)–T(2) 45
$V(C)$ 13
$\mathrm{vcd}(G)$ 134
$\chi(C)$ 15
$\mathbb{Z}$ 7
$\mathbb{Z}_n$ 7
$\mathbb{Z}^n$ 8
$Z(G)$ 40
$Z_1(K)$ 19
∂ 14
∂_1,∂_2 19
$\equiv$ 6
$\langle\mathcal{X}|\mathcal{R}\rangle\Rightarrow\langle\mathcal{X}'|\mathcal{R}'\rangle$ 10
$s(\sigma),t(\sigma)$ 13
$|g|$ 31
$|W|$ 6
$[G,G]$ 12
$[x,y]$ 12
$[G:H]$ 140
$\bigvee_{i=1}^n S_i^1$ 15
$1\to A\xrightarrow{i}E\xrightarrow{p}G\to 1$ 128

II. Some Questions of Group Theory Related to Geometry

R.I. Grigorchuk, P.F. Kurchanov

Translated from the Russian
by P.M. Cohn

Contents

Introduction 169

Chapter 1. Equations in Groups and Some Related Questions 172

§ 1. Basic Concepts and the Theorem of Makanin 172
§ 2. Solutions of Systems and Homomorphisms 173
§ 3. Fundamental Sequences and Razborov's Theorem 175
§ 4. On the Structure of the Set of Solutions of Quadratic Equations in Free Groups 178
§ 5. Coefficient-Free Quadratic Equations 179
§ 6. The Classification of Epimorphisms from Surface Groups to Free Groups 180
§ 7. On the Minimal Number of Fixed Points in the Homotopy Class of Mappings and the Width of Elements in Free Groups ... 182
§ 8. On Quadratic Equations in Hyperbolic Groups 184

Chapter 2. Splitting Homomorphisms and Some Problems in Topology 187

§ 1. Heegaard Decompositions of 3-Manifolds and their Equivalence .. 187
§ 2. The Poincaré Conjecture and Three Algorithmic Problems Connected with 3-Manifolds 190
§ 3. Information on Aut $\pi_1(T)$ and Some of its Subgroups and Factor Groups 193

§ 4. On the Problem of the Equivalence of Splitting Homomorphisms . 197
§ 5. On an Algebraic Reduction of the Poincaré Conjecture and the Algorithmic Poincaré Problem 200
§ 6. Some Analogues with the Group of Symplectic Matrices and the Torelli Group 202
§ 7. Algebraic Reduction of the Problem of the Equivalence of Links . . 203
§ 8. On the Andrews–Curtis Conjecture 205

Chapter 3. On the Rate of Growth of Groups and Amenable Groups . 208

§ 1. On the Growth of Graphs and of Riemannian Manifolds 208
§ 2. On the Notion of Growth of a Finitely Generated Group 210
§ 3. On the Proof of Gromov's Theorem and Some Related Results . . . 213
§ 4. Example of a Group of Intermediate Growth and the Construction Scheme of such a Group 216
§ 5. On the Structure of the Set of Growth Degrees of Groups that are Residually-p Groups 218
§ 6. On an Application of the Theory of Groups of Polynomial Growth to Geometry 221
§ 7. Regularly Filtered Surfaces and Amenable Groups 223

Bibliography 226

Index of Notation 231

Introduction

In this survey the reader's notice is drawn to three current questions of combinatorial group theory which have relations to geometry.

Equations in groups, and particularly in free groups, have long been studied. For example, the solvability of the classical conjugacy problem in a group G, posed by Dehn, is connected with the question of solving simple quadratic equations of the form $x^{-1}Ax = B$ in this group.

The theory of equations in free groups developed very intensively in the 1970's to 80's, largely in the papers of the Moscow school. The central result obtained during this time, the theorem of Makanin [Makanin 1982] asserts the existence of an algorithm to recognize the solvability of an arbitrary equation in a free group.

Besides the solvability question, the problem of describing the set of solutions of an equation is also important. For free groups this question was answered by Razborov [Razborov 1987], continuing the work of Makanin. This description is very complicated and so far it is not clear to what extent it can be simplified. However, there is an important case where such a simplification is possible – the case of quadratic equations.

An equation is called quadratic if each variable occurs twice. The theory of quadratic equations is very geometric and is closely connected with the theory of surfaces. The study of this important class of equations was begun in the papers of Lyndon [Lyndon 1959] and Mal'tsev [Mal'tsev 1962]. The problem of describing the set of solutions of arbitrary quadratic equations in free groups was solved in the papers of Comerford and Edmunds [Comerford–Edmunds 1989] and Grigorchuk and Kurchanov [Grigorchuk–Kurchanov 1989a, b, d].

A geometric interpretation of the method was given by Ol'shanskij [Ol'shanskij 1989].

The description of the solution set in [Grigorchuk–Kurchanov 1989a, b, d] is somewhat different from that given by [Comerford–Edmunds 1989]. In the first place, in [Grigorchuk–Kurchanov 1989a] the reduced automorphism group is involved in the description and in [Grigorchuk–Kurchanov 1989b] the existence of a polynomial algorithm is asserted which solves the problem of describing the solution sets. With the help of Lysenok the authors have succeeded in combining these two extensions, as also in our presentation in Sect. 4 of Chap. 1, where a corresponding result is formulated.

Many questions of topology have to be reduced to a study of equations in groups that are not free. An important class of such groups are the hyperbolic groups, and the corresponding study was begun in the paper of Gromov [Gromov 1987]. In Sect. 8 of the first chapter we bring a result of Grigorchuk and Lysenok which asserts the existence of a polynomial algorithm allowing one to find a description of the solution set of an arbitrary quadratic equation in a hyperbolic group.

Section 7 is devoted to the concept of width of elements in free groups and its connexion with the question of the minimal number of fixed points in a homotopy class of continuous self-maps of a compact surface. The problem of effectively calculating the width has a positive solution, thanks to the technique developed by the authors for the solution of quadratic equations in groups.

In Sect. 6 the concept of equivalence of homomorphisms from the fundamental group of a surface to a free group is studied, which finds application in Chap. 2.

Chapter 2 is a survey devoted to the general idea of splitting homomorphism, introduced in group theory quite recently in connexion with studies concerning the Poincaré conjecture.

By a splitting homomorphism we understand any homomorphism of the form

$$\varphi \times \psi : G \to K \times K ,$$

where $\varphi, \psi : G \to K$ are epimorphisms. The most interesting case is that of pairs (G, K), where $G = F_{2m}$, $K = F_m$, $m \geq 2$ and $G = \pi_1(T_g)$, $K = T_g$, where T_g is a closed orientable surface of genus $g \geq 2$. The central problem is whether the splitting homomorphism $\varphi \times \psi : G \to K \times K$ for the given pair of groups is unique up to equivalence. The equivalence of two homomorphisms $\varphi_1 \times \psi_1$, $\varphi_2 \times \psi_2$ means that isomorphisms α, β exist such that the diagram

$$\begin{array}{ccc} G & \xrightarrow{\varphi_1 \times \psi_1} & K \times K \\ {\scriptstyle\alpha}\downarrow & & \downarrow{\scriptstyle\beta} \\ G & \xrightarrow{\varphi_2 \times \psi_2} & K \times K \end{array}$$

commutes.

We remark that some interesting topological questions are connected with this theme and with other questions on splitting homomorphisms. For example, the problem of classifying Heegaard decompositions of genus g of closed 3-manifolds is equivalent to the problem of classifying splitting homomorphisms of pairs $(\pi_1(T_g), T_g)$. In Sect. 1 we shall prove this theorem by algebraic methods for the orientable and the non-orientable case at the same time.

Chapter 3 is devoted to questions of growth of finitely generated groups. The concept of growth of a group appeared in the works of Efremovich [Efremovich 1953] and Shvarts [Shvarts 1955], as well as that of Milnor [Milnor 1968a]. The consideration of numerous examples showed that the growth of a group is either polynomial or exponential. Gromov [Gromov 1981] succeeded in proving that groups of polynomial growth include the class of almost nilpotent groups. On the other hand, Grigorchuk [Grigorchuk 1983] constructed the first example of a group of intermediate growth between polynomial and exponential, thus solving negatively the problem of Milnor [Milnor 1968b]. The study of this class of groups undertaken by Grigorchuk [Grigorchuk 1984a, b,

1985a] showed that groups of intermediate growth possess many interesting properties. These results are exposed in shortened form in Sect. 5 of Chap. 3.

The final Sect. 7 is devoted to the concept of amenability, introduced in a classical paper of von Neumann [von Neumann 1929]. It gives an account of two unsuccessful attempts to describe the class of amenable groups – more precisely, the negative solution of two problems of Day [Day 1957].

We give a combinatorial criterion for amenability established by Grigorchuk [Grigorchuk 1978] and related to the notion of co-growth in groups.

For an understanding of the results of this survey, besides mastering the standard concepts of algebra and geometry in a general university course, the reader is required to know the bases of combinatorial group theory (for example, as in the general survey given in Part I) and the basic notions of topology (cf. e.g. [Massey 1967]). The authors have endeavoured to give full references to the original sources in all cases, where a detailed introduction of a concept seemed inappropriate.

In the survey a number of open problems have been posed, some well known, others not previously considered.

We hope that our survey will interest the reader to enter the area of current problems of combinatorial group theory and find problems to his taste.

Chapter 1
Equations in Groups and Some Related Questions

§ 1. Basic Concepts and the Theorem of Makanin

Equations in groups play an important part in many applied questions of algebra and logic. The development of this theory is also connected with intrinsic problems in the theory of groups. The greatest progress has been in the theory of equations in free groups, and in the main it is to the latter that we shall devote our attention in this chapter.

Let $\overline{X} = \{x_1, x_2, \ldots\}$ be a countable alphabet of unkowns, $\overline{C} = \{c_1, \ldots, c_r\}$ an alphabet of coefficients and $F_{\overline{C}}$ the free group on the generators $c_1, \ldots, c_r$. By a system of equations in the free group $F_{\overline{C}}$ in the unknowns $\overline{X}_0$ (where $\overline{X}_0$ is a subset of X) and coefficients $\overline{C}$ we understand a system of relations

$$\varphi_i(x_{i_1}, \ldots, x_{i_n}, c_1, \ldots, c_r) = 1 \, , \tag{1}$$

$i = 1, \ldots, m$, $x_{i_1}, \ldots, x_{i_n} \in \overline{X}_0$, where φ_i is a word in the alphabet $\overline{X}_0^{\pm 1} \cup \overline{C}^{\pm 1}$.

For brevity we shall often denote a sequence of letters of the same type by a single upper case letter of that type, e.g. $\overline{X}$, $\overline{C}$, $\overline{\Phi} = \{\Phi_1, \ldots, \Phi_m\}$, etc.

A *solution* of (1) is a set of values of the unknowns $x_i = x_i(\overline{C}) \in F_{\overline{C}}$ whose substitution in (1) transforms it into a system of relations holding in $F_{\overline{C}}$.

For each system of equations in a free group we can construct a single equation equivalent to this system [Khmelevskij 1971a]. However, it would not be expedient to limit ourselves to considering single equations.

For a system of equations there are two questions that are usually considered:

1. Determine whether (1) has at least one solution.
2. Describe the set of solutions of (1).

The second question usually requires a more precise definition, because it is not entirely clear what it means to describe this set. Of course most satisfactory would be a description by means of a finite set of parametric solutions, if such a solution exists in principle for the given equation or system.

Thus let $\overline{T} = \{t_1, \ldots, t_k\}$ be a new alphabet whose symbols will be called parameters. Each parameter $t_i \in \overline{T}$ is allowed to range over values in $F_{\overline{C}}$. The set of relations

$$x_i = X_i(\overline{T}, \overline{C}) \tag{2}$$

$i = 1, \ldots, n$ is called a *parametric solution* if the substitution of (2) for the variables x in (1) transforms (1) into a system of free equations in $F_{\overline{C}} * F_{\overline{T}}$. The concept of a parametric solution in its dependence on the situation may still be modified. Frequently one introduces parameters taking integer values. For example, Lorents [Lorents 1968] has shown that to describe the general solution of a system of equations in one unknown one can restrict the parametric words to the form $AB^{\mu}C$, where μ is an integer parameter.

Khmelevskij [Khmelevskij 1971b] constructs an algorithm to recognize solubility and to give a description of the general solution of a system, in which each equation contains at most two unknowns and has one of the following forms:

$$\varphi(x_i, c_1, \dots c_r) = \psi(x_j, c_1, \dots c_r)$$

or

$$W(x_i, x_j) = A\,,$$

where W is a word containing none of the coefficients. To this end special functions are introduced in Khmelevskij [Khmelevskij 1971b] which were later to be called Nielsen–Khmelevskij functions.

The most significant progress in the study of systems of equations in free groups was achieved by Makanin [Makanin 1982]. Thus he obtains an algorithm which allows one to answer for an arbitrary finite system of equations in a free group the following question: Does the system possess at least one solution? To construct such an algorithm, Makanin [Makanin 1982] explains his method of what are called generalized equations, used earlier by him in the solution of the analogous problem in free semigroups [Makanin 1977]. It appears that the method of generalized equations is so far the only method allowing one to find the solution of an arbitrary system of equations in a free group. Makanin [Makanin 1984] continues his research, in connexion with the attempt to give a positive solution to a problem raised by Tarski: is the elementary theory of free groups of rank $r \geq 2$ decidable? At present one has only partial results for the solution of this problem.

Makanin's algorithm is very complicated (in the sense of computing time) and its practical realization, even in quite simple situations, is very difficult. It would be useful to find new (simpler) approaches to the problem of algorithmic recognition of solubility of equations in free groups (if one exists, of course). What has been said about the problem of describing the solution sets of systems of equations in free groups has found an answer in the work of Razborov [Razborov 1987] (cf. also [Razborov 1984], where he brings some much weaker results on this topic).

The work of Razborov is based on the paper by Makanin [Makanin 1982]. The formulation of the basic results in the paper of Razborov [Razborov 1987] is preceded by a long chain of definitions, which we shall now set forth.

§ 2. Solutions of Systems and Homomorphisms

A solution of the system of equations (1) may be interpreted as a homomorphism $\Phi : F_{\overline{X},\overline{C}} \to F_{\overline{C}}$, from the free group $F_{\overline{X},\overline{C}}$ generated by the symbols from the system of unknowns and coefficients $c_1, \dots, c_r$ to the free group $F_{\overline{C}}$ such that

$$\Phi(c_i) = c_i,\ i = 1, \dots, r\,, \quad \Phi\left(\varphi_j(\overline{X}, \overline{C})\right) = 1\,, \quad j = 1, \dots, m\,. \tag{3}$$

Under the homomorphism Φ the solution of the system (1) determines relations

$$X_i(\overline{C}) = \Phi(x_i)\ , \quad i = 1, \ldots, n\ ,$$

and conversely, every solution of (1) corresponds in a natural way to a homomorphism Φ satisfying (3).

There is still another way of viewing the solution. For a given system of equations $\overline{\varphi} = 1$ of the form (1) let us denote by $H(\overline{\varphi}) = \langle \overline{X}, \overline{C} | \overline{\varphi}(\overline{X}, \overline{C}) = 1 \rangle$ the finitely presented group whose generating set comprises the unknowns and coefficients of the system $\overline{\varphi} = 1$ and whose defining relations are the words forming the left-hand sides of the equations $\overline{\varphi} = 1$. Let $F_1 = F_{\overline{C}}$ be the free group in which we are seeking a solution. Then the solutions of the system $\overline{\varphi}$ are in 1-1 correspondence with the homomorphisms $H(\overline{\varphi}) \to F_1$ which map the coefficients to themselves. From a given solution $\pi : H(\overline{\varphi}) \to F_1$ we can, by means of prefixing automorphisms of $H(\overline{\varphi})$ which preserve the coefficients, obtain a series of new solutions $\pi\sigma : H(\overline{\varphi}) \to F_1$, $\sigma \in \operatorname{Aut} H(\overline{\varphi})$.

We shall use both these points of view in the solution and adhere to the following notation, taken from [Razborov 1987]: $F_{\overline{X},\overline{C}} = F(\overline{\varphi})$, while $\hat{\hat{\pi}}_{\overline{X}} : F(\overline{\varphi}) \to F_1$ is the homomorphism corresponding to the solution $\overline{X}$ of the system (1):

$$\hat{\hat{\pi}}_{\overline{X}}(c_i) = c_i,\ i = 1, \ldots, r, \quad \hat{\hat{\pi}}_{\overline{X}}(x_j) = X_j,\ j = 1, \ldots, n\ ,$$

$\hat{\pi} : H(\overline{\varphi}) \to F_1$ is the homomorphism induced by $\hat{\hat{\pi}}_{\overline{X}}$.

For the groups $H(\overline{\varphi})$ the word problem can turn out to be insoluble and they do not satisfy the maximum condition for normal subgroups. In order to avoid some technical difficulties, a certain canonical way of passing from $H(\overline{\varphi})$ to a group $G(\overline{\varphi})$ is convenient, which does not suffer from the above two deficiencies.

A group G with distinguished elements $c_1, \ldots, c_r$ is called *residually free* if for any $g \in G$ not equal to 1 there exists a homomorphism $\pi : G \to F_1$ such that $\pi(c_i) = c_i$ and $\pi(g) \neq 1$. For a justification of the passage $H(\overline{\varphi}) \to G(\overline{\varphi})$ two subsidiary assertions are necessary which are proved in [Razborov 1987], the first one being based on a result by Guba [Guba 1986] and the second on a result by Makanin [Makanin 1984].

Lemma 1. *Let an infinite sequence*

$$G_1 \xrightarrow{\pi_1} G_2 \xrightarrow{\pi_2} \ldots \xrightarrow{\pi_{s-1}} G_s \xrightarrow{\pi_s} \ldots$$

of finitely generated residually free groups and surjective homomorphisms be given. Then almost all the homomorphisms in the sequence are isomorphisms.

□

Now let H be any group with distinguished elements $c_1, \ldots, c_r$ and $S(H)$ the intersection of the kernels of all possible homomorphisms $\hat{\pi} : H \to F_1$ such that $\hat{\pi}(c_i) = c_i$, $i = 1, \ldots, r$. The factor group $H/S(H)$ is denoted by $F_r(H)$. Clearly it is residually free.

Lemma 2. *There exists an algorithm which for any finite presentation*

$$H = \langle g_1, \dots, g_n, c_1, \dots, c_r | \varphi_j(g_1, \dots, g_n, c_1, \dots, c_r) = 1 \, , \; j = 1, \dots, m \rangle$$

and any word $\psi(g_1, \dots, g_n, c_1, \dots, c_r)$ determines whether the natural image of ψ in the group $F_r(H)$ is equal to 1. □

Since F_r is a functor from the category of groups with distinguished elements $c_1, \dots, c_r$ to the category of residually free groups with distinguished elements $c_1, \dots, c_r$, every homomorphism $\hat{\pi} : H_1 \to H_2$ induces a homomorphism $F_r(\hat{\pi}) : F_r(H_1) \to F_r(H_2)$, which is an isomorphism whenever $\hat{\pi}$ is an isomorphism.

The group $G(\overline{\varphi})$ mentioned earlier is defined by the relation $G(\overline{\varphi}) = F_r(H(\overline{\varphi}))$. Since $F_r(F_1) = F_1$, it follows that F_r associates with a homomorphism of the form $\hat{\pi}_{\overline{X}} : H(\overline{\varphi}) \to F_1$ (where $\overline{X}$ is a solution of $\overline{\varphi} = 1$) a homomorphism of the form $\pi_{\overline{X}} = F_r(\hat{\pi}_{\overline{X}}) : G(\overline{\varphi}) \to F_1$ such that $\pi_{\overline{X}}(c_i) = c_i$. Since this correspondence is 1-1, the set of distinguished homomorphisms $\pi : G(\overline{\varphi}) \to F_1$ is in 1-1 correspondence with the set of solutions of the system $\overline{\varphi} = 1$; moreover, the solution $\overline{X}$ corresponds to the homomorphism $\pi_{\overline{X}} = F_r(\hat{\pi}_{\overline{X}})$. Thus the problem of describing the general solution of the system (1) is equivalent to the problem of describing the homomorphism $G(\overline{\varphi}) \to F_1$. For the solution of this problem we shall introduce the notion of a fundamental sequence.

§ 3. Fundamental Sequences and Razborov's Theorem

A *fundamental sequence* of length n for a system of equations $\overline{\varphi} = 1$ is a triple $(\mathfrak{M}, \mathrm{Hom}, \mathrm{Aut})$ where $\mathfrak{M}$ consists of n systems of equations $\overline{\varphi}^{(1)} = 1, \dots, \overline{\varphi}^{(n)} = 1$, $\overline{\varphi}^{(1)}$ coinciding with $\overline{\varphi}$ and $\overline{\varphi}^{(n)}$ the trivial system consisting of the empty family of equations; Hom is a set of $n-1$ homomorphisms $\pi_1, \dots, \pi_{n-1}$ of the form $\pi_i : G(\overline{\varphi}^{(i)}) \to G(\overline{\varphi}^{(i+1)})$, $1 \le i \le n-1$; Aut consists of n finitely generated groups $P_1, \dots, P_n$ of automorphisms of $G(\overline{\varphi}^{(1)}), \dots, G(\overline{\varphi}^{(n)})$ respectively.

The fundamental sequence $\Phi = (\mathfrak{M}, \mathrm{Hom}, \mathrm{Aut})$ is said to be *effectively given* if the system in $\mathfrak{M}$, the homomorphisms in Hom and the finite generating systems of the groups in Aut are indicated. Moreover, a homomorphism $G(\overline{\varphi}) \to G(\overline{\psi})$ is effectively given if its action on the free generators in the homomorphism $F(\overline{\varphi}) \to F(\overline{\psi})$ inducing it are indicated.

If Φ is a fundamental sequence of length n for the system $\overline{\varphi} = 1$, $\pi : G(\overline{\varphi}^{(n)}) \to F_1$ is a fixed homomorphism to a free group and $\sigma_1, \dots, \sigma_n$ are automorphisms in $P_1, \dots, P_n$ respectively, then the composition

$$\underset{\circlearrowleft \sigma_1}{G(\overline{\varphi})} \xrightarrow{\pi_1} \underset{\circlearrowleft \sigma_2}{G\left(\overline{\varphi}^{(2)}\right)} \xrightarrow{\pi_2} \dots \xrightarrow{\pi_{n-1}} \underset{\circlearrowleft \sigma_n}{G\left(\overline{\varphi}^{(n)}\right)} \xrightarrow{\pi_n} F_1 \qquad (4)$$

equals $\pi_{\overline{X}}$ for some solution $\overline{X}$ of $\overline{\varphi} = 1$. We shall say that Φ describes the solution $\overline{X}$ of $\overline{\varphi} = 1$ if the homomorphism $\pi_{\overline{X}}$ can be expressed in the form (4) for some choice of $\sigma_1, \dots, \sigma_n$.

In fact for a description of the general solution of a system of equations by means of fundamental sequences we can limit ourselves to groups of automorphisms of a special type, which we shall now describe.

Suppose that the family $\overline{X}$ of unknowns of a system $\overline{\varphi} = 1$ splits into two parts $\overline{X}$ and $\overline{Z}$ and similarly the equations $\overline{\varphi} = 1$ split into two parts $\overline{\theta}$ and $\overline{\psi}$ for which

$$\overline{X} = \overline{Y} \cup \overline{Z}\,; \quad \overline{\varphi} = \overline{\theta} \cup \overline{\psi}\,; \quad \overline{Y} \cap \overline{Z} = \emptyset\,, \quad \overline{\theta} \cap \overline{\psi} = \emptyset\,.$$

Further, the members of $\overline{Y}$ do not occur in the equations from $\overline{\psi}$ while the equations in $\overline{\theta}$ are one of three types:

Type 1. $\overline{\theta}$ is the empty family.

Type 2. $\overline{\theta}$ consists of a single equation $q(\overline{Y}) = 1$ (with one exception which will be described below) which combines all the unknowns from $\overline{Y}$, and moreover

$$q = [y_1, y_2]\,[y_3, y_4] \dots [y_{2g-1}, y_{2g}]$$

or

$$q = y_1^2 y_2^2 \dots y_g^2\,.$$

The above-mentioned exception arises if $q = [y_1, y_2]$; then in $\overline{\theta}$, besides the equation $[y_1, y_2] = 1$ there occur one or more pairs of equations of the form

$$[y_1, U(\overline{Z}, \overline{C})] = [y_2, U(\overline{Z}, \overline{C})] = 1\,. \tag{5}$$

Type 3. The family $\overline{Y}$ of unknowns can be split into three parts $\overline{X} = \{u_1, \dots, u_k, \overline{v}, w_1, \dots, w_k\}$ such that $\tilde{\theta}$ consists with one exception of equations of the following form:

$$\begin{aligned} &u_i = U_i(\overline{Z}, \overline{C})\,, \quad 1 \le i \le k \\ &\prod_{i=1}^{k} (w_i^{-1} u_i w_i)\, q(\overline{v}) = U_0(\overline{Z}, \overline{C})\,, \end{aligned} \tag{6}$$

where the words $U_0, \dots, U_k$ are arbitrary while $q(\overline{v})$ is either empty or such that as in the preceding case, all the variables of the family $\overline{v}$ occur in q.

The exception is that when $k = q = 1$, i.e. when (6) takes the form $u = U_1(\overline{Z}, \overline{C})$, $w^{-1}uw = U_0(\overline{Z}, \overline{C})$; then $\overline{\theta}$ includes besides these equations one or more pairs of equations of the form

$$\begin{aligned} &w^{-1} U(\overline{Z}, \overline{C}) w = V(\overline{Z}, \overline{C})\,, \\ &[u, U(\overline{Z}, \overline{C})] = 1\,. \end{aligned}$$

We define a group of automorphisms $\hat{\hat{P}} \subset \operatorname{Aut}(F(\overline{\varphi}))$ as follows. If $\overline{\theta}$ is of type 1, then

$$\hat{\hat{P}} = \left\{ \hat{\hat{\pi}} \in \mathrm{Aut}\,(F(\overline{\varphi}))\,|\,\hat{\hat{\pi}}(z) = z\,,\ \hat{\hat{\pi}}(c) = c\,,\ z \in \overline{Z}\,,\ c \in \overline{C} \right\}.$$

If $\overline{\theta}$ is of type 2, then we consider first the group $\hat{\hat{P}}_1$ of automorphisms of the free group with basis $\overline{Y}$, consisting of all automorphisms which transform the cyclic word q into q or q^{-1}. Put $\hat{\hat{P}}_2 = \hat{\hat{P}}_1 * \mathrm{id}$, where id is the identity automorphism of the free group with basis $\overline{Z}, \overline{C}$.

If $q = [y_1, y_2]$ and the family $\overline{\theta}$ includes the supplementary equations (5), we put $\hat{\hat{P}} = \hat{\hat{P}}_2$. In the contrary case we take $\hat{\hat{P}}$ to be generated by $\hat{\hat{P}}_2$ and the automorphisms of the form $\overline{z} \mapsto \overline{z}$, $\overline{c} \mapsto \overline{c}$, $\overline{y} \mapsto A^{-1}\overline{y}A$ (where A is a certain element from $\overline{Z}, \overline{C}$).

If $\overline{\theta}$ is of type 3, we consider the stabilizer $\hat{\hat{P}}_1$ of the elements $u_1, \ldots, u_k$, $\prod_{i=1}^{k}(w_i^{-1}u_iw_i)q(\overline{v})$ in the group of automorphisms of the free group on $\overline{Y}$.

We put $\hat{\hat{P}} = \hat{\hat{P}}_1 * \mathrm{id}$, where id is the identity automorphism of the free group on $\overline{Z}, \overline{C}$.

The group $P = \{F_r(\hat{\hat{\pi}})|\hat{\hat{\pi}} \in \hat{\hat{P}}\}$ will be called the canonical automorphism group and is assigned the same type 1, 2 or 3 as the family $\overline{\theta}$. We remark that if $\overline{\theta}$ is of type 1 (type 3) then $\hat{\hat{P}}$ (resp. $\hat{\hat{P}}_1$) is the stabilizer of a finite set of words in a free group, while for $\overline{\theta}$ of type 2, $\hat{\hat{P}}_1$ is the stabilizer of the unordered pairs of cyclic words (q, q^{-1}) in a free group. A finite generating set for such a group is effectively found in the basic construction of McCool [McCool 1975].

The fundamental sequence $\Phi = (\mathfrak{M}, \mathrm{Hom}, \mathrm{Aut})$ is called *canonical* if all groups in Aut are generated by a finite number of canonical automorphism groups. The basic result of Razborov [Razborov 1987] may be stated as follows.

Theorem 1. *For every system of equations in a free group we can effectively construct a finite set of canonical sequences such that every solution of the system is described by one of the sequences constructed.* □

The canonical system of the form described above corresponds to a certain triangular system of equations, of which each step consists of a partition of the system of unknowns $\overline{X}$ into two parts $\overline{Y}$, $\overline{Z}$ and the system of equations into two parts $\overline{\theta}, \overline{\psi}$, of the form described above, where the equations in $\overline{\theta}$ are quadratic in the unknowns $\overline{Y}$. Thus the problem is in a certain sense reduced to a problem of describing the solution sets of quadratic equations. This class of equations allows a solution by means of a polynomial algorithm (relative to the sum of the lengths of the coefficients), which will be discussed in the next section.

§ 4. On the Structure of the Set of Solutions of Quadratic Equations in Free Groups

Usually an equation is called quadratic if each variable occurs in it not more than twice. If there is a variable which occurs only once in an equation

$$\Phi(\overline{X}, \overline{C}) = 1 , \tag{7}$$

then by expressing it in terms of the other variables and the coefficients we obtain a parametric solution, which describes all the solutions of the system. Excluding this trivial case from our considerations, we may call an equation quadratic if every unknown occurs exactly twice in it (of degree ± 1). Each such equation can by a reversible change of variables, more precisely, by an automorphism of $F_{\overline{X},\overline{C}}$ preserving the elements of $F_{\overline{C}}$ be reduced to one of the forms

$$\prod_{i=1}^{g} [x_{2i-1}, x_{2i}]\, A_0 \prod_{j=1}^{s} x_{j+2g}^{-1} A_j x_{j+2g} = 1 , \tag{8}$$

$$\prod_{i=1}^{g} x_i^2 A_0 \prod_{j=1}^{s} x_{j+g}^{-1} A_j x_{j+g} = 1 , \tag{9}$$

where the $A_j \in F_{\overline{C}}$ play the role of the coefficients ($A_j \neq 1, j = 1, \dots, s$). An equation containing no occurrence of the coefficients is called coefficient-free. Such an equation can by an automorphism of $F_{\overline{X}}$ be reduced to one of the two following forms:

$$[x_1, x_2] \dots [x_{2g-1}, x_{2g}] = 1 , \tag{10}$$

$$x_1^2 x_2^2 \dots x_g^2 = 1 . \tag{11}$$

The first investigations of quadratic equations in free groups were carried out by Lyndon [Lyndon 1959] and Mal'tsev [Mal'tsev 1962]. In these papers the basic features of a general method can be found, allowing one to solve the problem of describing the solution set of an arbitrary quadratic equation in a free group, which moreoever is fairly simple. The next results in this direction are due to Comerford and Edmunds [Comerford–Edmunds 1989] and the authors of this survey [Grigorchuk–Kurchanov, 1989a, b, c, d]. Our result is somewhat stronger, for to assert the presence of a polynomial algorithm allows us to give a description of the solution set. Also we are able to use the reduced automorphism group in the description of this set.

We return to the point of view of the solution of (7) by means of the distinguished homomorphism $\varphi : F_{\overline{X},\overline{C}} \to F_{\overline{C}}$. Let us write K_Φ for the sum of the lengths of the coefficients occurring on the left-hand side of $\Phi = 1$, i.e. $K_\Phi = \sum_{i=0}^{s} |A_i|$, if $\Phi = 1$ has one of the forms (8) or (9).

Theorem 2. *There exists an algorithm which allows us to find a (finite) set of parametric solutions $\{\varphi_i\}_{i \in I}$ of the quadratic equation $\Phi = 1$ in a number of operations which for fixed g, s is a polynomial in K_Φ, such that any solution φ*

of this equation can be written in the form $\varphi = \lambda \circ \varphi_i \circ \gamma$, *where* $\gamma \in \mathrm{Stab}_{F_{\overline{X},\overline{C}}} \Phi$, $\gamma(c_j) = c_j$, $j = 1, \dots, r$, $i \in I$ *is a suitable index and* $\lambda : F_{\overline{X},\overline{C}} \to F_{\overline{C}}$ *is any homomorphism such that* $\lambda(c_j) = c_j, j = 1, \dots, r$. □

We remark that by the theorem of McCool [McCool 1975] the group $\mathrm{Stab}_{F_{\overline{X},\overline{C}}} \Phi$ is finitely generated (and moreover finitely presented) and its finite generating system can be effectively determined.

We now slightly change our approach to the problem of describing the solution sets of a quadratic equation in a free group. For this purpose we introduce an alphabet $\overline{A} = \{A_0, \dots, A_s\}$ and consider the corresponding free group $F_{\overline{A}}$. Here we must bear in mind that the symbols A_i in (8) and (9) express the values of the coefficients $A_i = A_i(\overline{C})$, $i = 0, 1, \dots, s$. Let us denote by $V(\overline{X}, \overline{A})$ the word on the left-hand side of (8), resp. (9). A *special* parametric solution of the equation $V(\overline{X}, \overline{A}) = 1$ is a homomorphism $\varphi : F_{\overline{X},\overline{A}} \to F_{\overline{T},\overline{C}}$ such that $\varphi(A_i) = A_i(\overline{C})$, $i = 0, \dots, s$ and $\varphi(V) = 1$. Finally we write $\mathrm{Stab}_{\overline{A}} V$ for the group of automorphisms of the free group $F_{\overline{X},\overline{A}}$ preserving the word $V(\overline{X}, \overline{A})$ and the symbols $A_i (\varphi(A_i) = A_i, i = 0, 1, \dots, s)$.

Theorem 3. *There exists an algorithm which for any equation* (8) *or* (9) *allows us to find a (finite) set* $\{\varphi_i\}_{i \in I}$ *of special parametric solutions of this equation in a number of operations which for fixed* g, s *is a polynomial in* K_Φ, *such that any solution* φ *of* (8) *or* (9) *may be expressed in the form* $\varphi = \lambda \circ \varphi_i \circ \gamma$, *where* $\gamma \in \mathrm{Stab}_{\overline{A}} V$, $i \in I$ *is a suitable index and* $\lambda : F_{\overline{T},\overline{C}} \to F_{\overline{C}}$ *is a certain homomorphism such that* $\lambda(c_j) = c_j$, $j = 1, \dots, r$. □

Thus for a description of the set of solutions of a quadratic equation in a free group we can limit ourselves to the consideration of stabilizers of words in standard form, i.e. the words standing on the left of (8), (9), considered as words in $\overline{X} \cup \overline{A}$.

§ 5. Coefficient-Free Quadratic Equations

As already mentioned earlier, coefficient-free quadratic equations can by a reversible change of variables be reduced to one of the forms (10), (11). Equations of this form have been studied by Lyndon [Lyndon 1959], Zieschang [Zieschang 1964], Piollet [Piollet 1986]; in particular, Zieschang solves what is known as the rank problem, see Part I and [Lyndon–Schupp 1977]. [Piollet 1986] considers the problem of describing the set of parametric solutions of quadratic coefficient-free equations and proves that such an equation has a finite number of so-called basic parametric solutions, while every other solution can by an automorphism of $F_{\overline{X}}$ preserving the left-hand side of the equation be reduced to a specific basic one, i.e. it is obtained by the substitution for the parameters of certain C-values. Moreover, as the set of parameters in [Piollet 1986] one can take the set of unknowns $\overline{X}$ of the corresponding equation.

Let us stay with this point of view, i.e. consider a parametric solution of an equation (10) or (11) as a homomorphism $\varphi : F_{\overline{X},\overline{C}} \to F_{\overline{X},\overline{C}}$ such that $\varphi(c_j) = c_j$, $j = 1, \ldots, r$ and $\varphi(\Phi) = 1$ (where Φ is the left-hand side of the equation).

Clearly for (10) the mapping

$$\varrho_1 : \begin{cases} x_i \mapsto x_i & \text{if } i \text{ is odd, } 1 \le i \le 2g-1 \,, \\ x_j \mapsto 1 & \text{if } j \text{ is even, } 2 \le j \le 2g \,, \end{cases}$$

is a parametric solution.

For the case of (11) and even $g = 2n$ we form the parametric solutions

$$\varrho_2 : \begin{cases} x_i \mapsto x_i & \text{if } i \text{ is odd, } 1 \le i \le 2n-1 \,, \\ x_j \mapsto x_{j-1}^{-1} & \text{if } j \text{ is even, } 2 \le j \le 2n \,; \end{cases}$$

finally, in the case of (11) and odd $g = 2n+1$ we form the parametric solution

$$\varrho_3 : \begin{cases} x_i \mapsto x_i & \text{if } i \text{ is odd, } 1 \le i \le 2n-1 \,, \\ x_j \mapsto x_{j-1}^{-1} & \text{if } j \text{ is even, } 2 \le j \le 2n \,, \\ x_{2n+1} \mapsto 1 \,. \end{cases}$$

We denote by $\operatorname{Stab}\Phi$ the stabilizer of the left-hand side of the corresponding equation (10) or (11) in $\operatorname{Aut} F_{\overline{X}}$.

Theorem 4 [Grigorchuk–Kurchanov 1989a]. *Any parametric solution of an equation* (10) *or* (11) *may be written as* $\lambda \circ \varrho_i \circ \gamma$, *where* $\gamma \in \operatorname{Stab}\Phi$, $i = 1$ *if the equation being considered is* (10), $i = 2$ *if the equation is* (11) *and* g *is even and* $i = 3$ *in the remaining case, while* $\lambda : F_{\overline{X},\overline{C}} \to F_{\overline{X},\overline{C}}$ *is an endomorphism such that* $\lambda(c_j) = c_j$, $j = 1, \ldots, r$.

Any ordinary solution of such an equation may be represented in the form $\lambda \circ \varrho_i \circ \gamma$ *for suitable* $\gamma \in \operatorname{Stab}\Phi$, $i = 1, 2, 3$ *depending on the equation, where* $\lambda : F_{\overline{X},\overline{C}} \to F_{\overline{C}}$ *is a certain homomorphism such that* $\lambda(c_j) = c_j$, $j = 1, \ldots, r$. □

§ 6. The Classification of Epimorphisms from Surface Groups to Free Groups

The problem of describing solution sets of equations in groups is closely connected with the problem of classifying epimorphisms, which in turn has a link to certain questions in topology.

Let (G, K) be a pair of groups. Two epimorphisms $\varphi, \psi : G \to K$ are called *equivalent* if there exist $\alpha \in \operatorname{Aut} G$, $\beta \in \operatorname{Aut} K$ such that the diagram

$$\begin{array}{ccc} G & \xrightarrow{\varphi} & K \\ {\scriptstyle\alpha}\downarrow & & \downarrow{\scriptstyle\beta} \\ G & \xrightarrow{\psi} & K \end{array}$$

commutes.

The epimorphisms φ, ψ are are *strongly equivalent* if there exists $\gamma \in \operatorname{Aut} G$ such that the triangle

$$\begin{array}{ccc} G & \xrightarrow{\varphi} & \\ {\scriptstyle\gamma}\downarrow & & K \\ G & \xrightarrow{\psi} & \end{array}$$

commutes. The problem of classifying epimorphisms for the pair (G, K) consists in giving a description of the set of epimorphisms $G \to K$ up to one of these notions of equivalence. The same question can be put for homomorphisms between G and K. Besides, one can also consider the classification problem using twisting by elements of a certain subgroup $\mathcal{E}$ of $\operatorname{Aut} G$. These are all variants of the statement of the classification problem.

For any pair (F_m, F_r) of free groups of ranks m, r, where $m \geq r$, there is a unique epimorphism $F_m \to F_r$ up to strong equivalence.

In order to formulate the next assertion we introduce the following two series of groups

$$G_g = \left\langle x_1, \ldots, x_{2g} \middle| \prod_{i=1}^{g} [x_{2i-1}, x_{2i}] = 1 \right\rangle ,$$

$$\Gamma_g = \left\langle x_1, \ldots, x_g \middle| \prod_{i=1}^{g} x_i^2 = 1 \right\rangle ,$$

which are isomorphic to the fundamental group of a closed orientable, respectively non-orientable surface of genus g, as is well known. We shall use the following notation: $[x, y]_0 = x^{-1}y^{-1}xy$, $[x, y]_1 = x^{-1}yxy$. For even $g = 2n$ and any multi-index $\overline{\varepsilon} = \{\varepsilon_1, \ldots, \varepsilon_n\}$, $\varepsilon_i = 0, 1$, we consider the group

$$\Gamma_{\overline{\varepsilon},g} = \left\langle x_1, \ldots, x_g \middle| \prod_{i=1}^{n} [x_{2i-1}, x_{2i}]_{\varepsilon_i} = 1 \right\rangle ,$$

which is isomorphic to Γ_g in the case $\overline{\varepsilon} \neq \{0, \ldots, 0\}$. If $r \leq n$, we define the epimorphism $\beta_{\overline{\varepsilon}} : \Gamma_{\overline{\varepsilon},g} \to F_{\overline{C}}$ by the relations

$$\beta_{\overline{\varepsilon}} : \begin{cases} x_{2i-1} \mapsto c_i\,, & i = 1, \ldots, r\,, \\ x_j \mapsto 1 & \text{for the remaining indices } j,\ 1 \leq j \leq g\,. \end{cases}$$

Consider for $r < n$, the $2^r - 1$ non-zero sets $\overline{\varepsilon} = \{\varepsilon_1, \ldots, \varepsilon_r, 0, \ldots, 0\}$ as well as the set $\overline{\varepsilon} = \{0, 0, \ldots, 0, 1\}$; we thus have $q = 2^r$ epimorphisms from Γ_g to

F_r. For $r = n$ the number of non-trivial multi-indices $\bar{\varepsilon}$ is $2^r - 1$ and we obtain $q = 2^r - 1$ epimorphisms from Γ_g to F_r.

If in the case of G_g we have $r > g$ or in the case of Γ_g we have $r > [g/2]$, then by the result of Zieschang quoted above there exists no epimorphism from G_g or Γ_g to F_r. For the remaining values of the parameters g and r we have the following assertion.

Theorem 5 [Kurchanov–Grigorchuk 1989]. *The number of equivalence classes and strong equivalence classes of epimorphisms from G_g, Γ_g to F_r is finite and equal to p, resp. q, whose values follow.*

1) (Simultaneous result of Zieschang and the authors) *For G_g and $r \leq g$ we have $p = q = 1$.*

2) *For Γ_g we have*

a) *If $g = 2n + 1$, $r < n$, then $p = q = 1$.*

b) *If $g = 2n$, $r < n$, then $p = 2$, $q = 2^r$.*

c) *If $g = 2n$, $r = n$, then $p = 1$, $q = 2^r - 1$.*

In the cases 2b, c) *the q different representatives of the strong equivalence classes give the epimorphisms β_ε defined above.*

If α, β are strongly equivalent epimorphisms from G_g or Γ_g to F_r then the automorphism γ connecting them can always be chosen so as to leave the left-hand side of the corresponding equation (10) *or* (11) *fixed.* □

Below we make use of the assertion of Theorem 5 in considering the problem of the equivalence of Heegaard decompositions and we state at once a consequence of Theorem 5.

A solution $\overline{X} = (X_1(\overline{C}), \ldots, X_n(\overline{C}))$ of (10) or (11) is called *generating* if its components $X_1(\overline{C}), \ldots, X_n(\overline{C})$ generate F_r. Then it follows from Theorem 5 that the action of Stab$\,\Phi$ on the set of generating solutions of $\Phi = 1$, of the form (10) or (11), has a finite number q of orbits. The values of q are enumerated in Theorem 5.

§ 7. On the Minimal Number of Fixed Points in the Homotopy Class of Mappings and the Width of Elements in Free Groups

Let L be a compact polyhedron and $f : L \to L$ a continuous map. We denote by $MF[f]$ the minimal number of fixed points of maps in the homotopy class $[f]$ of f. Further we denote by $N(f)$ the *Nielsen number* of f (the definition of $N(f)$ can be found in [Jiang 1983]). It is known that $N(f)$ is bounded by $MF[f]$ [Jiang 1983]. An important result in fixed-point theory states that for very sparse boundaries on L the Nielsen number $N(f)$ coincides with the minimal number of fixed points in the homotopy class of f. Indeed, let us call a point $l \in L$ of a connected space L a *separating point* of L if $L \backslash l$ is not connected. The point $l \in L$ is called *locally separating* if l is a separating point for every connected open neighbourhood U of l in L.

Theorem 6 [Jiang 1983]. *Let L be a compact connected polyhedron without locally separating points. Suppose that L is not a surface of negative Euler characteristic. Then $MF(f) = N(f)$ for every continuous map $f : L \to L$.*

Thus for polyhedra satisfying the conditions of the theorem the problem of calculating $MF(f)$ is reduced to the usually much simpler problem of computing the Nielsen number $N(f)$. For compact surfaces of negative Euler characteristic the situation is more complicated. In [Jiang 1984] the first example is constructed of a surface and a map f for which $MF(f) = N(f)$ does not hold. Jiang [Jiang 1987] gives an algebraic reduction of the problem of calculating $MF(f)$ for maps of compact surfaces. Without going into the details of this reduction we only remark on the connexion of these problems with equations in free groups.

Let F_r be a free group with basis $c_1, \ldots, c_r$ (where r is finite or infinite). The *width* of an element $g \in F_r$ relative to the basis $\{c_i\}_{i=1}^r$ is the least n such that there exist $x_1, \ldots, x_n \in F_r$, indices $i(1), \ldots, i(n)$ and integers $k(1), \ldots, k(n)$ such that we have a relation

$$g = \prod_{j=1}^{n} x_j^{-1} c_{i(j)}^{k(j)} x_j \; ; \tag{12}$$

(the width of the unit element is taken to be zero).

Let us examine the word

$$R(z_1, \ldots, z_{2g+h+n}) = \prod_{i=1}^{g} [z_{2i-1}, z_{2i}] \prod_{j=2g+1}^{2g+h} z_j^2 \prod_{k=1}^{n} z_{2g+h+k}$$

(one of g, h is zero, depending on the type of orientability of the surface considered).

Jiang [Jiang 1987] defines a finitely generated free group H and a normal subgroup $K \trianglelefteq H$, generated by an element B of H which is constructed from f. Further, [Jiang 1987] uses f to construct elements σ_i, $1 \leq i \leq 2g + h + n$, in H to satisfy the relation

$$R(\sigma_1, \ldots, \sigma_{2g+h+n}) = 1 \, .$$

Then the algebraic reduction of the problem of computing $MF(f)$ is given by the following

Proposition 1 [Jiang 1987]. *The number $MF(f)$ is equal to the minimum of the widths of the elements of the form*

$$L = R\Big(u_1\sigma_1, \ldots, u_{2g+h}\sigma_{2g+h}, u_{2g+h+1}^{-1} z_{2g+h+1} u_{2g+h+1}, \ldots,$$
$$u_{2g+h+k}^{-1} z_{2g+h+k} u_{2g+h+k}\Big)$$

where the width is calculated relative to the basis $\{sBs^{-1} : s \in S\}$ *of* K, *and where* S *is a system of minimal Schreier representatives of* K *in* H *and the minimum is taken relative to all choices of* $u_1, \ldots, u_{2g+h+n}$ *from* K.

In connexion with this result, Jiang [Jiang 1987] poses the problem of finding a method of calculating the width in free groups. Since this problem is related to the task of solving quadratic equations in free groups, we naturally find that the techniques developed in this context are useful.

Proposition 2. *Let* n *be the width of an element* g *of* F_r. *Then there exists a representation of* g *in the form* (12) *with the estimate*

$$\sum_{i=1}^{n} k(i) \leq l(g) ,$$

where $l(g)$ *is the length of* g *relative to the basis of* F_r. □

Thus the calculation of the width is reduced to the known problem of solving a finite number of quadratic equations in free groups.

Corollary 1. *There exists an algorithm which permits the calculation of the width of elements in free groups.* □

This result provides an effective means of finding an upper bound for $MF[f]$. The first steps in obtaining an effective lower bound for this number was the construction of an algorithm allowing one to establish the existence of at least one fixed point in the homotopy class, i.e. whether or not the equation $MF[f] = 0$ holds. This problem was reduced to the construction of an algorithm permitting, for a quadratic equation in automorphisms

$$\Phi\left(\mu_1(x_1), \ldots, \mu_n(x_n)\right) = \mathcal{D} ,$$

$\mu_1, \ldots, \mu_n \in \operatorname{Aut} K$, $\mathcal{D} \in K$, to decide whether or not this equation has a solution in K. Here it is enough to solve this problem in the case where the μ_i are inner automorphisms.

§ 8. On Quadratic Equations in Hyperbolic Groups

The group G is called *hyperbolic* if for some finite presentation

$$G = \langle \overline{C} | r = 1 \ , \ r \in R \rangle \tag{13}$$

the word problem for this group can be solved by Dehn's algorithm (cf. [Dehn 1911], [Lyndon–Schupp 1977]). Many interesting properties of this class of groups have been established in the paper of Gromov [Gromov 1987]. It is also known that groups in the basic class of small cancellation groups are hyperbolic.

Quadratic equations in groups with small cancellation and hyperbolic groups have been studied by Lysenok [Lysenok 1989a, b]. In particular, Lysenok [Lysenok 1989b] obtains an algorithm allowing him to determine, for any equation in a hyperbolic group whether or not it has a solution. From the description of this algorithm it is not hard to deduce that this problem can actually be solved in polynomial time relative to the sum of the lengths of the coefficients in the equation, provided that the equation is given in a suitable form. This problem is also treated by Ol'shanskij [Ol'shanskij 1989].

The simplest quadratic equation is an equation of the form $x^{-1}Ax = B$. The question whether an equation of this type has a solution in a group G is equivalent to asking whether A and B are conjugate in G. For hyperbolic groups this problem was solved by Gromov [Gromov 1987]. [Lysenok 1989a, b] gave a purely combinatorial solution of this problem. An easy consequence is the solubility for hyperbolic groups of the generalized conjugacy problem, which may be formulated thus: Does there exist an algorithm which for two finite sets of elements A_i, B_i, $i = 1, \ldots, k$ in a group G allows one to establish whether $x \in G$ exists such that

$$x^{-1}A_i x = B_i \quad i = 1, \ldots, k\,?$$

The presence of such an algorithm for the class of hyperbolic groups allows one to determine, for any $\varphi \in \operatorname{Aut} G$, whether it is inner. We shall use this remark below in the proof that the word problem is solvable in groups of mapping classes of compact surfaces.

[Lysenok 1989b] also considers the problem of describing the set of all solutions of quadratic equations in hyperbolic groups, which is rather harder than the mere existence problem. For this purpose he constructs, for any quadratic equation in a hyperbolic group G, a finite family of quadratic equations of the form

$$\Phi(x_1, \ldots, x_n, c_1, \ldots, c_r) = \prod_{i=1}^{k} y_i^{-1} r_i y_i\,, \quad r_i^{\pm 1} \in R\,, \quad 1 \le i \le k\,, \tag{14}$$

in the free group $F_{\overline{C}}$ generated by the same symbols as G, and any solution of the equation $\Phi = 1$ in G is the canonical projection on G of the solutions of one of the equations of the family (14). We note that for the solution of each of the equations (14) one can use the description of the solution sets of quadratic equations in free groups formulated in Theorems 2, 3.

Unfortunately this approach does not give a good bound for the number of equations in (14) or for k, and it applies only to hyperbolic groups all of whose abelian subgroups are cyclic. We shall now describe another approach to this problem, for which we shall need the following result.

Proposition 3 [Lysenok]. *The centralizer $C(g)$ of an element g in a hyperbolic group G is finitely generated. If the hyperbolic group is given by the representation* (13), *for which Dehn's algorithm applies, then the finite generating set of $C(g)$ can be effectively found.*

We can now extend the notion of a parametric solution of an equation (8) or (9). For this purpose we provide the parameter $t^{(i)}_{g_i}$ with two indices, a natural number i and an element g_i of G. We fix a family $g_1, \ldots, g_s$ where g_i may equal g_j even when i and j differ. Then the family of words

$$x_i = X_i\left(c_1, \ldots, c_r, t^{(1)}_{g_1}, \ldots t^{(s)}_{g_s}\right), \quad i = 1, \ldots, n, \tag{15}$$

is called a parametric solution of the equation $\Phi = 1$ if we obtain a solution of this equation by substituting for $t^{(i)}_{g_i}$ any element of the centralizer $C(g_i)$. We shall call the result of such a substitution of values for the parameters in (15) a *specialization* of parameters in (15).

Theorem 7 [Grigorchuk, Lysenok]. *Let G be a hyperbolic group given by a finite presentation* (13) *for which Dehn's algorithm applies. Then there exists a number Q which can be effectively calculated from* (13) *such that for any quadratic equation $\Phi = 1$ of the form* (8) *or* (9) *a finite family of parametric solutions of these equations in G can effectively be found to satisfy the following conditions.*

(1) *The lengths of all elements g_i defining the domain of values of the parameters are bounded above by Q.*

(2) *Any solution of the equation $\Phi = 1$ in G may be transformed by a certain automorphism of $F_{\overline{X},\overline{C}}$ stabilizing the cyclic word Φ and the generating symbols c_i, $i = 1, \ldots, r$, to a specified one of the parametric solutions found.*

If the group (13) *is fixed and the numbers g, s in* (8), (9) *are bounded, then the number of operations of the algorithm needed to write out the family of parametric solutions of the equation being considered is bounded above by a polynomial function in the overall length K_Φ of the coefficients occurring in the equation.* □

The proof of this theorem is based on a combination of the techniques of solving quadratic equations in free groups developed by the authors and the technique of analysing quadratic equations in hyperbolic groups due to Lysenok [Lysenok 1989a, b].

Chapter 2
Splitting Homomorphisms and Some Problems in Topology

§ 1. Heegaard Decompositions of 3-Manifolds and their Equivalence

A 3-manifold M containing a family $\overline{\mathcal{D}} = \{\mathcal{D}_1, \dots, \mathcal{D}_n\}$ of pairwise distinct proper images of embedded 2-cells (disks) such that the result of the section of M along $\cup_i \mathcal{D}_i$ is a 3-cell, is called a *cube with n handles.* By van Kampen's theorem the fundamental group $\pi_1(M)$ is a free group of rank n. Two cubes M_1, M_2 with n_1, n_2 handles respectively are homeomorphic if and only if $n_1 = n_2$ and M_1, M_2 are both orientable or both non-orientable.

A *Heegaard decomposition* of genus g of a closed connected 3-manifold M is a pair (V_1, V_2), where V_i, $i = 1, 2$ is a cube with g handles, $M = V_1 \cup V_2$ and $V_1 \cap V_2 = \partial V_1 = \partial V_2 = T$ is a closed surface (orientable or not, depending on the orientation types of V_1 and V_2). The genus of T is g in the orientable case and $2g$ in the non-orientable case.

It is well known that every closed connected 3-manifold possesses a Heegaard decomposition [Hempel 1976]. If (W_1, W_2) is a Heegaard decomposition of a manifold N which is homeomorphic to M, then it is said to be equivalent to (V_1, V_2) if there exists a homeomorphism h from M to N such that $h(V_i) = W_i$, $i = 1, 2$.

To a Heegaard decomposition (V_1, V_2) of a manifold M a commutative diagram

$$\begin{array}{ccccc} & & \pi_1(V_1) & & \\ & \phi_1 \nearrow & & \searrow i_1 & \\ \pi_1(T) & & \xrightarrow{\ \phi_1 \times \phi_2\ } & & \pi_1(V_1) \times \pi_1(V_2)\,, \\ & \phi_2 \searrow & & \nearrow i_2 & \\ & & \pi_2(V_2) & & \end{array} \qquad (1)$$

may be associated, where Φ_i is the homomorphism induced by the embedding $V_i \hookrightarrow M$ and i_1, i_2 the coordinatewise embeddings (in (1) we have deliberately omitted mention of the base point $x_0 \in T$ relative to which the fundamental groups are defined).

Using van Kampen's theorem we may show that the fundamental group of M is isomorphic to the factor group $\pi_1(T)/(\operatorname{Ker}\Phi_1)(\operatorname{Ker}\Phi_2)$ [Stallings 1966]. Thus the manifold M is simply connected, i.e. $\pi_1(M)$ is trivial, precisely when the relation

$$(\operatorname{Ker}\Phi_1) \cdot (\operatorname{Ker}\Phi_2) = \pi_1(T) \qquad (2)$$

holds.

We now come to the notion of a splitting homomorphism. Let G be a group isomorphic to $\pi_1(T)$ and K, L groups isomorphic to the free group F_g of rank

g. A *splitting homomorphism* is any homomorphism $\varphi \times \psi : G \to K \times L$, for which the coordinatewise homomorphisms $\varphi : G \to K$, $\psi : G \to L$ are epimorphisms.

Two splitting homomorphisms

$$\varphi_1 \times \psi_1 : G_1 \to K_1 \times L_1$$
$$\varphi_2 \times \psi_2 : G_2 \to K_2 \times L_2$$

($G_i \cong \pi_1(T)$, $K_i \cong L_i \cong F_i$, $i = 1, 2$) are called *π-equivalent* if there exist isomorphisms $\alpha : G_1 \to G_2$, $\beta_1 : K_1 \to K_2$, $\beta_2 : L_1 \to L_2$ such that the diagram

$$\begin{array}{ccc} G_1 & \xrightarrow{\varphi_1 \times \psi_1} & K_1 \times L_1 \\ {\scriptstyle\alpha}\downarrow & & \downarrow{\scriptstyle\beta_1 \times \beta_2} \\ G_2 & \xrightarrow{\varphi_2 \times \psi_2} & K_2 \times L_2 \end{array}$$

commutes.

Theorem 1. *Let T be a closed surface of genus g in the orientable case and $2g$ in the non-orientable case. Then any splitting homomorphism*

$$\varphi \times \psi : \pi_1(T) \to F_g \times F_g$$

is π-equivalent to a splitting homomorphism associated with the Heegaard decomposition of genus g of some closed 3-manifold M. $\square$

In the orientable case this theorem was proved by Jaco [Jaco 1970]. The proof was fairly difficult, but has since been simplified by Waldhausen [Waldhausen 1978]. The non-orientable case has apparently not previously been established. We remark that Theorem 1 follows easily from the uniqueness up to equivalence of epimorphisms from G_g (resp. Γ_{2g}) to the free group F_g which was part of the assertion of Theorem 5 of Chap. 1.

Since for any epimorphism $\xi : G \to L$, $\operatorname{Ker} \xi = H$, there exists an isomorphism $\beta : L \to G/H$ rendering the diagram

$$\begin{array}{ccc} G & \xrightarrow{\xi} & L \\ & {\scriptstyle\lambda}\searrow & \downarrow{\scriptstyle\beta} \\ & & G/H \end{array} \tag{3}$$

commutative (where $\lambda : G \to G/H$ is the canonical homomorphism), it follows that every splitting homomorphism is π-equivalent to one of the form

$$\pi_1(T) \xrightarrow{\varphi \times \psi} \pi_1(T)/H' \times \pi_1(T)/H'' , \tag{4}$$

where

$$\varphi : \pi_1(T) \to \pi_1(T)/H' , \ \psi : \pi_1(T) \to \pi_1(T)/H''$$

are the canonical epimorphisms and H', H'' are normal subgroups of $\pi_1(T)$ with factor groups isomorphic to F_g.

Let us fix such a subgroup $H_0 \trianglelefteq \pi_1(T)$, $\pi_1(T)/H_0 \cong F_g$. By Theorem 5 of Chap. 1 it follows that there are automorphisms $\xi, \eta \in \operatorname{Aut}\pi_1(T)$ such that $\xi(H_0) = H'$, $\eta(H_0) = H''$.

Lemma 1. *Let $\lambda : G \to L$ be an epimorphism, $H = \operatorname{Ker}\lambda$ and $\xi \in \operatorname{Aut} G$. Then there exists an isomorphism β_ξ for which the diagram*

$$\begin{array}{ccc} G & \xrightarrow{\lambda} & L \\ {\scriptstyle\xi}\downarrow & & \downarrow{\scriptstyle\beta_\xi} \\ G & \xrightarrow{\mu} & G/\xi(H) \end{array}$$

commutes, where $\mu : G \to G/\xi(H)$ is the canonical homomorphism.

By the commutativity of the diagram (3) the homomorphism λ may be taken to be canonical $\lambda : G \to G/H$. In this case we obtain $\beta_\xi(gH) = \xi(g)\xi(H)$ for each $g \in G$ and the assertion of the lemma follows. □

Using Lemma 1 on the splitting homomorphism (4) with $H' = \xi(H_0)$, $H'' = \eta(H_0)$, we reach the commutative diagram

$$\begin{array}{ccc} \pi_1(T) & \xrightarrow{\varphi\times\psi} & \pi_1(T)/\xi(H_0) \times \pi_1(T)/\eta(H_0) \\ {\scriptstyle\xi^{-1}}\downarrow & & \downarrow{\scriptstyle\beta_1\times\beta_2} \\ \pi_1(T) & \xrightarrow{\lambda\times\mu} & \pi_1(T)/H_0 \times \pi_1(T)/\xi^{-1}\eta(H_0) \end{array},$$

where λ, μ are the canonical homomorphisms and β_1, β_2 the appropriate homomorphisms whose existence follows from Lemma 1.

Thus we have shown that every splitting homomorphism is π-equivalent to one of the form

$$\pi_1(T) \xrightarrow{\lambda\times\mu} \pi_1(T)/H_0 \times \pi_1(T)/\xi(H_0)\,, \tag{5}$$

where $H_0 \trianglelefteq \pi_1(T)$ is any normal subgroup such that $\pi_1(T)/H_0 \cong F_g$ and $\xi \in \operatorname{Aut}\pi_1(T)$.

We now realize the surface T as the boundary of a cube U with g handles; V is another copy of a cube with g handles (of the same orientation type as U), $h : U \to V$ is any homeomorphism and $\varphi : \pi_1(T) \to \pi_1(U)$ the epimorphism induced by $\Phi : T \hookrightarrow U$, $H_0 = \operatorname{Ker}\varphi$.

Let us construct the closed manifold $M = U \cup_\psi V$, where ψ is a homeomorphism of T and glue U and V along their boundaries by means of ψ according to the rule: $h\psi(z) = z$, $z \in \partial U$. Then the splitting homomorphism associated with this decomposition of M will be equivalent to

$$\lambda \times \mu : \pi_1(T) \to \pi_1(T)/H_0 \times \pi_1(T)/\psi^{-1}(H_0)\,,$$

where λ, μ are the canonical homomorphisms.

By a theorem of Nielsen [Nielsen 1927] every automorphism of $\pi_1(T)$ can be written in the form $\psi_{\#}^{-1}$ for a suitable homeomorphism ψ of T. Thus, if $\xi = \psi_{\#}^{-1}$, then the splitting homomorphism (5) is realized on the example of the manifold M and the corresponding Heegaard decomposition. □

Theorem 2. *Let (V_1, V_2), (W_1, W_2) be Heegaard decompositions of 3-manifolds M and N. Then the associated splitting homomorphisms are π-equivalent if and only if the decompositions (V_1, V_2), (W_1, W_2) themselves are equivalent.*

Let (V_1, V_2) and (W_1, W_2) be equivalent decompositions and $h : M \to N$ the corresponding homeomorphism: $h(V_1) = W_1$, $h(V_2) = W_2$, $h(x_0) = y_0$, $x_0 \in \partial V_1$. Then we have the commutative diagram

$$\begin{array}{ccc} \pi_1(\partial V_1, x_0) & \xrightarrow{\xi \times \mu} & \pi_1(V_1, x_0) \times \pi_1(V_2, x_0) \\ \alpha \big\downarrow & & \big\downarrow \beta_1 \times \beta_2 \\ \pi_1(\partial W_1, y_0) & \xrightarrow{\nu \times \eta} & \pi_1(W_1, y_0) \times \pi_1(W_2, y_0) \end{array} , \tag{6}$$

where $\alpha = \overline{h}_{\#}$, $\overline{h}$ being the restriction of h to ∂V_1 and $\beta_i = h_{i\#}$, h_i being the restriction of h to V_i, $i = 1, 2$ while ξ, μ, ν, η are the epimorphisms of the fundamental group induced by the embeddings of ∂V_1, ∂W_1 in V_1, V_2, W_1, W_2, respectively.

Conversely, for any splitting homomorphisms $\xi \times \mu$ and $\nu \times \eta$ there exist isomorphisms α, β_1, β_2 such that the diagram (6) commutes. There is a homeomorphism h of the pointed space $(\partial V_1, x_0)$ onto $(\partial W_1, y_0)$ such that $h_{\#} = \alpha$ [Nielsen 1927]. This homeomorphism h may be extended to homeomorphisms from V_1 to W_1 and V_2 to W_2 precisely when $h_{\#}(\operatorname{Ker} \xi) \leq \operatorname{Ker} \nu$ and $h_{\#}(\operatorname{Ker} \mu) \leq \operatorname{Ker} \eta$ ([Griffiths 1964], [Grigorchuk–Kurchanov–Zieschang 1989]). But these inclusions follow by the commutativity of (6) and the fact that $h_{\#} = \alpha$. This completes the proof. □

§ 2. The Poincaré Conjecture and Three Algorithmic Problems Connected with 3-Manifolds

A closed 3-manifold is called a *standard sphere* if it is homeomorphic to the sphere S^3, given in $\mathbb{R}^4$ by the equation $x^2 + y^2 + z^2 + u^2 = 1$.

A closed 3-manifold M is called a *homology 3-sphere* if M has the same homology groups as S^3 or equivalently , $\pi_1(M)$ has a trivial commutator factor group:

$$\pi_1(M) / [\pi_1(M), \pi_1(M)] = \{1\} .$$

Finally, a closed 3-manifold M is called a *homotopy 3-sphere* if M has the same homotopy groups as S^3, or equivalently, a trivial fundamental group $\pi_1(M)$.

Poincaré was the first to construct an example of a closed manifold which is a homology sphere but not a homotopy sphere (this example is discussed in detail in [Fomenko 1983]).

A fundamental conjecture of Poincaré, stated in 1904, asserts that every closed connected simply connected 3-sphere is homeomorphic to a standard 3-sphere. In other words, every homotopy sphere is a standard sphere. This conjecture has not been proved (nor disproved) up to now.

On the other hand, at about the same time Poincaré raised a problem usually known as the *algorithmic* Poincaré problem. It consists of the following: Let closed 3-dimensional manifolds be given in the form of a list (with an effective enumeration). Does there exist an algorithm which by means of a code defined for manifolds will determine whether a manifold is a homology sphere or not? Although Poincaré himself did not have a natural method of coding 3-dimensional manifolds, such a generally recognized method of coding now exists, see for example [Volodin–Kuznetsov–Fomenko 1974], [Fomenko 1983, 1984]. We shall now present an algebraic version of this method of coding.

As already noted earlier, for every closed 3-manifold there exists a Heegaard decomposition. The Heegaard decomposition corresponds to a splitting homomorphism and this homomorphism is equivalent to a splitting homomorphism of the form (5), which for a given group H_0 is defined by an automorphism $\xi \in \operatorname{Aut} \pi_1(T)$. This suggests the following method of coding 3-manifolds: Fix the cubes U, V with g handles, a homeomorphism $h : U \to V$ and name an automorphism $\xi \in \operatorname{Aut} \pi_1(\partial U)$ the code of the manifold $M = U \cup_\psi V$, where ψ is a homeomorphism of T such that $\xi = \psi_\#^{-1}$ (we recall that the gluing of the boundaries of U and V is realized by the rule $h\psi(z) = z$, $z \in \partial U$). Thus as the code here there appears in fact a splitting homomorphism of the form (5), where $H_0 = \operatorname{Ker} \varphi$ and $\varphi : \pi_1(\partial U) \to \pi_1(U)$ is the epimorphism induced by the embedding $\partial U \hookrightarrow U$.

We now go over to the algebraic point of view, where as the group H_0 we take the normal subgroup generated by the elements $a_1, \dots, a_g$ in the presentation (7) of the fundamental group $\pi_1(T)$ in the orientable case, and generated by $b_1, \dots, b_g$ in the presentation (8) of $\pi_1(T)$ in the non-orientable case:

$$G_g = \left\langle a_1, \dots, a_g, b_1, \dots, b_g \mid \prod_{i=1}^{g} [a_i, b_i] = 1 \right\rangle, \tag{7}$$

$$\Gamma_{2g} = \left\langle a_1, \dots, a_g, b_1, \dots, b_g \mid \prod_{i=1}^{g} \left(a_i^{-1} b_i a_i b_i\right) = 1 \right\rangle. \tag{8}$$

The manifold corresponding to $\xi \in \operatorname{Aut} \pi_1(T)$ is denoted by $M(\xi)$, and the corresponding splitting homomorphism (5) by $\Phi(\xi)$. We remark that the map $\xi \mapsto M(\xi)$ is not bijective and the difficulty of the classification problem for

3-manifolds is due to this fact. Besides, it is not excluded that one and the same manifold may have Heegaard decompositions (V_1, V_2), (W_1, W_2) of the same genus such that the associated splitting homomorphisms are inequivalent. Therefore the problem of classifying splitting homomorphisms is much coarser than the problem of classifying 3-manifolds up to homeomorphism. Nevertheless it is of independent interest and should be treated as a separate problem.

Problem 1. *Construct an algorithm which for any two automorphisms $\varphi, \psi \in \operatorname{Aut} \pi_1(T)$ determines whether or not the splitting homomorphisms $\Phi(\varphi)$, $\Phi(\psi)$ are equivalent.*

The algorithmic Poincaré problem may be formulated in purely algebraic terms. Indeed, as Waldhausen [Waldhausen 1968] has proved, any two Heegaard decompositions of the same genus of the standard sphere S^3 are equivalent. For every integer $g \geq 1$ we construct a Heegaard decomposition of genus g, whose associated splitting homomorphism is equivalent to

$$\pi_1(T) \xrightarrow{\lambda \times \mu} \pi_1(T)/H_0 \times \pi_1(T)/H_1 \,, \tag{9}$$

where $\pi_1(T)$ is the fundamental group of the oriented surface T, H_0 is the normal subgroup generated by $a_1, \ldots, a_g$ and H_1 the normal subgroup generated by $b_1, \ldots, b_g$ (λ, μ are the canonical epimorphisms; that $\lambda \times \mu$ is epimorphic follows from the relation $H_0 H_1 = G_g$). We shall call (9) the *standard splitting homomorphism.* Then the algorithmic Poincaré problem is equivalent to the following question:

Problem 2. *Does there exist an algorithm which for any automorphism $\varphi \in \operatorname{Aut} \pi_1(T)$ determines whether or not the splitting homomorphism $\Phi(\varphi)$ is equivalent to the standard homomorphism* (9)*?* □

The algorithmic Poincaré problem has so far only been solved for Heegaard decompositions of genus 2, by Birman and Hilden [Birman–Hilden 1973]. Other solutions were obtained independently by Volodin and Fomenko. A history of the question and the corresponding algorithm may be found in [Fomenko 1984].

Another algorithmic question which deserves detailed attention is

Problem 3. *Does there exist an algorithm which for an automorphism $\varphi \in \operatorname{Aut} \pi_1(T)$ decides whether or not the group $\pi_1(M(\varphi))$ is trivial?* □

From (2) it follows that $\pi_1(M(\varphi))$ is trivial if and only if

$$H_0 \cdot \varphi(H_0) = \pi_1(T) \,. \tag{10}$$

Let us put $\varphi(a_i) = V_i(\bar{a}, \bar{b})$, $i = 1, \ldots, g$. The factorization of the relation (10) by the normal subgroup H_0 leads to the conclusion that $\pi_1(M(\varphi))$ is trivial if and only if the group given by the presentation

$$\langle b_1, \ldots, b_g | W_i(\bar{b}) = 1 \, , \; i = 1, \ldots, g \rangle \, , \tag{11}$$

is trivial, where $W_i(\bar{b})$ is obtained from $V_i(\bar{a}, \bar{b})$ by replacing each a_i by the empty word.

A presentation of a group by generators and defining relations is called *balanced* if the number of generators is equal to the number of relations. Although, by a theorem of Adyan [Adyan 1955], there exists no algorithm which for every finite presentation of a group by generators and relations gives a response to the question whether or not the group is trivial, the possibility is not excluded that such an algorithm exists for balanced presentations. This question is closely connected with the problem of [Andrews–Curtis 1965] (cf. Sect. 8). On the other hand, the above-mentioned problem is related to the outcome of the problem for $F_g \times F_g$. Indeed, the subgroup $\pi_1(M(\varphi))$ is trivial if and only if the splitting homomorphism $\Phi(\varphi)$ is an epimorphism. Let us denote by $V_i(\bar{c})$, $W_i(\bar{c})$ the images of the generators a_i, b_i, $i = 1, \ldots, g$ under the homomorphism $\Phi(\varphi)$. Problem 3 will admit a positive solution if we have an algorithm which for any $2g$ elements of $F_g \times F_g$ will determine whether or not they form a generating set of this group. Consequently we shall begin by establishing whether or not such a set can be reduced to the standard generating set $\{(c_i, 1), (1, c_i), i = 1, \ldots, g\}$ by elementary Nielsen transformations.

§ 3. Information on Aut $\pi_1(T)$ and Some of its Subgroups and Factor Groups

To begin with, let T be an orientable surface of genus g. Then $\pi_1(T)$ may be expressed as a factor group of the free group F_{2g} with generators $a_1, \ldots, a_g$, $b_1, \ldots, b_g$ by the normal closure R of the word $V = [a_1, b_1] \ldots [a_g, b_g]$. Nielsen [Nielsen 1927] showed that any automorphism of $\pi_1(T)$ is induced by an automorphism of the free group F_{2g} which preserves R. By Magnus's theorem [Lyndon–Schupp 1977] such an automorphism of F_{2g} transforms V to $K^{-1}V^{\varepsilon}K$, where $\varepsilon = \pm 1$ and K is some element of F_{2g}. This remark is very useful in showing that for a certain set of permutations of $a_1, \ldots, a_g, b_1, \ldots, b_g$ the action defines an automorphism of $\pi_1(T)$.

An analogous statement can be made in the case of a non-orientable surface T; we need only replace the word V by $W = a_1^2 a_2^2 \ldots a_g^2$.

In this circle of problems (as well as in some other geometrical problems) an important role is played by the group $MC(T)$ of *mapping classes*, defined as the factor group of the group of homeomorphisms of T by the normal subgroup consisting of all homeomorphisms isotopic to the identity. As an abstract group $MC(T)$ is isomorphic to $\operatorname{Out} \pi_1(T) \cong \operatorname{Aut} \pi_1(T)/\operatorname{Inn} \pi_1(T)$, the outer automorphism group of $\pi_1(T)$. The groups $\operatorname{Aut} \pi_1(T)$ and $MC(T)$ are finitely presented. One approach to finding a finite presentation for these groups is based on ideas of [McCool 1975]. On the other hand, [Hatcher and

Thurston 1980] and [Wajnryb 1983] use topological arguments to find a set of defining relations for $MC(T)$ in the orientable case. The presentation for the group of mapping classes for orientable surfaces of genus 2 found by Birman [Birman 1974a] looks particularly simple. It has this form: generators $\sigma_1, \ldots, \sigma_5$; defining relations

$$\begin{aligned}
&\sigma_i\sigma_j = \sigma_j\sigma_i \quad \text{if} \quad |i-j| \geq 2\,,\ 1 \leq i\,,\ j \leq 5\,,\\
&\sigma_i\sigma_{i+1}\sigma_i = \sigma_{i+1}\sigma_i\sigma_{i+1}\,,\ 1 \leq i \leq 4\,,\\
&(\sigma_1\sigma_2 \ldots \sigma_5)^6 = 1\,,\\
&(\sigma_1\sigma_2\sigma_3\sigma_4\sigma_5^2\sigma_4\sigma_3\sigma_2\sigma_1)^2 = 1\,,\\
&\sigma_1\sigma_2\sigma_3\sigma_4\sigma_5^2\sigma_4\sigma_3\sigma_2\sigma_1 \rightleftarrows \sigma_i\,,\ 1 \leq i \leq 5
\end{aligned}$$

(where the expression $x \rightleftarrows y$ means that x and y commute). From this presentation it is clear that $MC(T)$ in the case of genus 2 is obtained from the braid group $B(6)$ with six threads by adjoining some simple relations.

As already noted earlier, a presentation for the groups $MC(T)$ in the case of orientable surfaces of genus $g \geq 3$ is constructed by Hatcher and Thurston [Hatcher–Thurston 1980] and Wajnryb [Wajnryb 1983]; however, they are fairly unwieldy and it is not clear how they can be used. A more favourable situation prevails for the generating set of $MC(T)$. The first to find a finite generating set of $MC(T)$ was Dehn [Dehn 1938] who singled out a certain class of homeomorphisms of T, now called Dehn twists. Dehn showed that the group $MC(T)$ is generated by $3g-1$ twists around simple closed curves of a special form on T. This result was made more precise in the papers of Lickorish [Lickorish 1963, 1964, 1965], where, besides, a generating set in the non-orientable case is given. In these papers the number of generators increases with g. However, Suzuki [Suzuki 1977] showed that in the orientable case one can manage with four generators. We shall give the automorphisms in $\operatorname{Aut}\pi_1(T)$ whose images in $MC(T) = \operatorname{Aut}\pi_1(T)/\operatorname{Inn}\pi_1(T)$ generate this group.

$$\begin{aligned}
&\alpha_0 : \begin{cases} a_1 \to b_1^{-1}\,,\ a_j \to a_j (1 \neq 1)\,,\\ b_1 \to b_1^{-1}a_1b_1\,,\ b_j \to b_j (j \neq 1)\,; \end{cases}\\
&\alpha_1 : \begin{cases} a_i \to a_{i+1}\,,\\ b_i \to b_{i+1}\,,\ 1 \leq i \leq g\,,\ i+1 = 1\,,\ \text{if } i = g\,; \end{cases}\\
&\alpha_2 : \begin{cases} a_i \to a_i\,,\ 1 \leq i \leq g\,,\\ b_1 \to b_1a_1^{-1}\,,\ b_j \to b_j\,,\ 2 \leq j \leq g\,; \end{cases}\\
&\alpha_3 : \begin{cases} a_2 \to b_2a_2(b_1^{-1}a_1b_1)(a_2^{-1}b_2^{-1}a_2)\,,\\ a_j \to a_j (j \neq 2)\,,\\ b_1 \to b_1(a_2^{-1}b_2^{-1}a_2)\,,\\ b_j \to b_j (2 \leq j \leq g)\,. \end{cases}
\end{aligned} \tag{12}$$

We shall now introduce certain subgroups of $\operatorname{Aut}\pi_1(T)$. Let T at first be orientable. For the presentation (7) we denote, as before, by H_0, H_1 the normal subgroups generated by $a_1, \dots, a_g$ and $b_1, \dots, b_g$ respectively and put L^-, $L^+ \leq \operatorname{Aut} \pi_1(T)$ for the subgroups consisting of all automorphisms stabilizing H_0, H_1:

$$L^- = \{\varphi \in \operatorname{Aut} \pi_1(T) : \varphi(H_0) = H_0\}$$
$$L^+ = \{\varphi \in \operatorname{Aut} \pi_1(T) : \varphi(H_1) = H_1\} \ .$$

(We remark that L^- and L^+ are conjugate in $\operatorname{Aut} \pi_1(T)$.) If T is realized as the boundary of a ball V with handles and H_0 is the kernel of the homomorphism induced by the inclusion $T \hookrightarrow V$, then the geometric sense of L^- is that its elements are homeomorphisms of T which extend to homeomorphisms of V. This circumstance is the reason for the importance of L^- and its conjugate L^+ in the study of 3-dimensional manifolds.

If T is non-orientable, then for the presentation (8) of the fundamental group $\pi_1(T)$ we denote by H_* the normal subgroup generated by $b_1, \dots, b_g$ and by $L^* \leq \operatorname{Aut} \pi_1(T)$ the subgroup

$$L^* = \{\varphi \in \operatorname{Aut} \pi_1(T) : \varphi(H_*) = H_*\} \ .$$

The group L^* has the same geometric sense as the group L^- in the oriented case.

We define the groups $MC^-(T)$, $MC^+(T)$, $MC^*(T)$ as the images of L^-, L^+, L^* under the canonical homomorphism $\operatorname{Aut} \pi_1(T) \to MC(T)$. The generators of the group $M^-(T)$ have been calculated by Suzuki [Suzuki 1977]. They are the images of the automorphisms α_1, α_2, α_3 defined by (12) and the images of the automorphisms

$$\alpha_4 : \begin{cases} a_1 \to b_1^{-1} a_1^{-1} b_1 \ , \ a_j \to a_j (2 \leq j \leq g) \ , \\ b_1 \to b_1^{-1} s_1^{-1} \ , \ b_j \to b_j (2 \leq j \leq g) \ ; \end{cases}$$

(where s_i by definition is the word $b_i^{-1} a_i^{-1} b_i a_i$, $i = 1, \dots, g$),

$$\alpha_5 : \begin{cases} a_1 \to s_1^{-1} a_2 s_1 \ , \ a_2 \to a_1 \ , \\ a_j \to a_j (3 \leq j \leq g) \ , \\ b_1 \to s_1^{-1} b_2 s_1 \ , \ b_2 \to b_1 \ , \\ b_j \to b_j (3 \leq j \leq g) \ ; \end{cases}$$

$$\alpha_6 : \begin{cases} a_i \to a_i (1 \leq i \leq g) \ , \\ b_1 \to a_1 b_1 a_2^{-1} s_2 (b_1^{-1} a_1^{-1} b_1) \ , \\ b_2 \to b_2 a_2 (b_1^{-1} a_1^{-1} b_1) a_2^{-1} \ , \\ b_j \to b_j (j \neq 1, 2) \ . \end{cases}$$

We do not know whether such a compact generating set has been found for $MC^*(T)$ in the non-orientable case.

We shall now consider some simple algorithmic questions related to the presentation of the above groups. The word problem (cf. [Lyndon–Schupp 1977], [Magnus–Karrass–Solitar 1966]) for the group $\pi_1(T)$ in the orientable case was solved positively by Dehn, who for this purpose devised an algorithm which now bears his name. Consequently the word problem for $\operatorname{Aut}\pi_1(T)$ can also be solved, for to show that two automorphisms $\varphi, \psi \in \operatorname{Aut}\pi_1(T)$ coincide it is enough to show that $\varphi(g) = \psi(g)$ in $\pi_1(T)$ when g runs over a generating set of $\pi_1(T)$. Clearly the word problem for $\pi_1(T)$ is also solvable when T is a non-orientable manifold, because $\pi_1(T)$ contains a subgroup of index two isomorphic to G_g.

The membership problem in $H_0 \leq \pi_1(T)$ for the orientable case (in $H^* \leq \pi_1(T)$ for the non-orientable case) is solvable. Indeed, the factor groups $\pi_1(T)/H_0$ and $\pi_1(T)/H_*$ are isomorphic to free groups, hence the word problem is solvable for them. Therefore to prove the inclusion $W(\overline{a}, \overline{b}) \in H_0$ it is enough to replace in W the a_i by the empty symbol to obtain a word $V(\overline{b})$ and then to check whether $V(\overline{b})$ reduces to the unit element in $F_g = \langle b_1, \ldots, b_g \rangle$. In the non-orientable case the question is solved similarly.

From these results it follows that the membership problem for the subgroup $L^- \leq \operatorname{Aut}\pi_1(T)$ (orientable case), $L^* \leq \operatorname{Aut}\pi_1(T)$ (non-orientable case) is solvable, for to check whether $\varphi \in L^-$ it is enough to check the $2g$ conditions

$$\varphi(a_i) \in H_0 \,, \quad \varphi(b_i) \in H_0 \,, \quad i = 1, \ldots, g \,.$$

Hence the cosets of $\operatorname{Aut}\pi_1(T)/L^-(\operatorname{Aut}\pi_1(T)/L^*)$ form a recursive set.

The word problem for $MC(T)$ and the membership problem for the subgroups $MC^-(T)$, $MC^*(T)$ is also solvable, for to check that $\varphi \in \operatorname{Inn}\pi_1(T)$ we need to verify the existence of a solution for the system of equations

$$\begin{cases} \varphi(a_i) = x^{-1} a_i x \,, \\ \varphi(b_i) = x^{-1} b_i x \,, \quad i = 1, \ldots, g \end{cases}$$

in $\pi_1(T)$ for the orientable case and an analogous system for the non-orientable case. This problem is solvable for any hyperbolic group (cf. Chap. 1, Sect. 8), that is for $\pi(T)$ where $g \geq 2$ (the case of genus 1 is trivial).

As mentioned earlier, the subgroups $L^- \leq \operatorname{Aut}\pi_1(T)$ (T orientable) and $L^* \leq \operatorname{Aut}\pi(T)$ (T non-orientable of even genus) are characterized by the property that the corresponding homeomorphisms of T extend to the interior of the handle-body bounded by this surface (by the condition that H_0 resp. H_* is the kernel of the homomorphism $\pi_1(T) \to \pi_1(T)$ induced by the inclusion $T = \partial V \to V$). The group Homeom V of homeomorphisms of the solid handle-body acts by automorphisms on the fundamental group $\pi_1(V)$. Let us see which automorphisms of $\pi_1(V)$ are realized in this way. In the translation to the algebraic language we consider a homomorphism $L^- \to \operatorname{Aut} F_g$ or $L^* \to \operatorname{Aut} F_g$ (depending on the orientation type) which is defined as follows. Let $\varphi \in L^-$, $\varphi(b_i) = W_i(\overline{a}, \overline{b})$; if we replace a_j in W_i by the empty symbol we obtain an automorphism $\hat{\varphi} : F_g \to F_g$, $\hat{\varphi}(\overline{b}_i) = W_i(\overline{1}, \overline{b})$. The map $\varphi \mapsto \hat{\varphi}$ is the homomorphism referred to earlier.

We have shown that the image of L^- is the whole of $\operatorname{Aut} F_g$, in other words, every automorphism of the fundamental group $\pi_1(V)$ is geometric, i.e.

induced by a certain homeomorphism of V. It is simpler still to prove this as follows. Let us compute the images in $\operatorname{Aut} F_g$ of the generating automorphisms $\alpha_1, \alpha_3, \alpha_4, \alpha_5$ of L^-. We obtain the automorphisms of the free group

$$\hat{\alpha}_1 : \{b_i \to b_{i+1}\,,\ 1 \le i \le g\,,\ i+1 = 1\,,\ \text{if } i = g\,;$$

$$\hat{\alpha}_3 : \begin{cases} b_1 \to b_1 b_2^{-1}\,, \\ b_j \to b_j,\ (j \neq 1)\,; \end{cases}$$

$$\hat{\alpha}_4 : \begin{cases} b_1 \to b_1^{-1}\,, \\ b_j \to b_j,\ (j \neq 1)\,; \end{cases}$$

$$\hat{\alpha}_5 : \begin{cases} b_1 \to b_2,\ b_2 \to b_1\,, \\ b_j \to b_j\ (2 \le j \le g)\,, \end{cases}$$

which generate $\operatorname{Aut} F_g$, where $F_g = \langle b_1, \dots, b_g \rangle$ [Magnus–Karrass–Solitar 1966].

Concerning the non-orientable case, i.e. the homomorphism $L^* \to \operatorname{Aut} \pi_1(T)$, we have a more complicated situation, because it is no longer true that every automorphism of $\pi_1(V)$ is geometric. Thus in [Grigorchuk–Kurchanov–Zieschang [1989] it is shown, based on results in [Grigorchuk–Kurchanov 1989d] that

$$[\operatorname{Aut} \pi_1(V) : \operatorname{Top} \pi_1(V)] = 2^g - 1\,,$$

where $\operatorname{Top} \pi_1$ is the group of topological automorphisms of the solid non-orientable handle-body V. In the same paper the topological nature of this phenomenon is revealed.

§ 4. On the Problem of the Equivalence of Splitting Homomorphisms

Let G, K be a pair of groups for which there exists at least one epimorphism $\varphi : G \to K$ which is unique up to equivalence (condition (i)). In this case $\operatorname{Aut} G$ acts transitively on the set of kernels $H \trianglelefteq G$, such that $G/H \cong K$. Let us fix some kernel $H_0 = \operatorname{Ker} \varphi_0$ with the property $G/H_0 \cong K$.

By a splitting homomorphism for the pair (G, K) we shall understand any homomorphism $\varphi_1 \times \varphi_2 : G \to K \times K$ such that the components $\varphi_1, \varphi_2 : G \to K$ are surjective. Two splitting homomorphisms $\varphi_1 \times \varphi_2$ and $\psi_1 \times \psi_2$ for the pair (G, K) are called π-equivalent if there exist automorphisms $\alpha \in \operatorname{Aut} G$, $\beta_1, \beta_2 \in \operatorname{Aut} K$ such that the diagram

$$\begin{array}{ccc} G & \xrightarrow{\varphi_1 \times \varphi_2} & K \times K \\ {\scriptstyle\alpha}\downarrow & & \downarrow{\scriptstyle\beta_1 \times \beta_2} \\ G & \xrightarrow{\psi_1 \times \psi_2} & K \times K \end{array}$$

commutes. Every splitting homomorphism for the pair (G, K) satisfying condition (i) is π-equivalent to the splitting homomorphism

$$G \xrightarrow{\xi \times \eta} G/H_0 \times G/\varphi(H_0)\,, \tag{13}$$

where $\varphi \in \operatorname{Aut} G$ and ξ, η are the canonical homomorphisms.

We now extend the concept of equivalence for splitting homomorphisms. Let $\mathcal{J}$ be a subgroup of $\operatorname{Aut} G$. Two splitting homomorphisms $\varphi_1 \times \varphi_2$, $\psi_1 \times \psi_2$ are called π-equivalent relative to $\mathcal{J}$ if there exist isomorphisms $\alpha \in \mathcal{J}$, $\beta_1, \beta_2 \in \operatorname{Aut} K$ such that the above diagram commutes.

Denote by $\mathcal{E}$ the subgroup

$$\mathcal{E} = \{\varphi \in \mathcal{J} : \varphi(H_0) = H_0\} \tag{14}$$

and by Φ_φ, $\varphi \in \operatorname{Aut} G$, the splitting homomorphism of the form (13) defined by $\varphi, \varphi \in \operatorname{Aut} G$.

Theorem 3. *Two splitting homomorphisms Φ_φ and Φ_ψ, $\varphi, \psi \in \operatorname{Aut} G$, are π-equivalent relative to $\mathcal{J}$ if and only if the double cosets $\mathcal{E}\varphi\mathcal{E}$, $\mathcal{E}\psi\mathcal{E}$ of $\mathcal{E}$ in $\mathcal{J}$ coincide.*

Suppose that $\psi = \varphi_1\varphi\varphi_2$, where $\varphi_1, \varphi_2 \in \mathcal{E}$. Then $\varphi_1\varphi\varphi_2(H_0) = \varphi_1\varphi(H_0)$ and the following diagram is commutative:

$$\begin{array}{ccc} G & \xrightarrow{\xi \times \eta} & G/H_0 \times G/\varphi_1\varphi(H_0) \\ {\scriptstyle \varphi_1^{-1}}\downarrow & & \downarrow{\scriptstyle \beta_1 \times \beta_2} \\ G & \xrightarrow{\lambda \times \mu} & G/\varphi_1^{-1}(H_0) \times G/\varphi(H_0) \end{array}\,,$$

where β_1, β_2 are the canonical isomorphisms determined by Lemma 1 and ξ, η, λ, μ are canonical homomorphisms. Since $\varphi_1^{-1}(H_0) = H_0$, one half of the assertion is proved.

Let us now show that if a splitting homomorphism Φ_φ is equivalent to Φ_ψ relative to $\mathcal{J}$, then $\mathcal{E}\varphi\mathcal{E} = \mathcal{E}\psi\mathcal{E}$. Suppose that the isomorphisms $\alpha \in \mathcal{J}$, $\beta_1, \beta_2 \in \operatorname{Aut} K$ render the diagram

$$\begin{array}{ccc} G & \longrightarrow & G/H_0 \times G(\varphi(H_0) \\ {\scriptstyle \alpha}\downarrow & & \downarrow{\scriptstyle \beta_1 \times \beta_2} \\ G & \longrightarrow & G/H_0 \times G/\psi(H_0) \end{array}$$

commutative. From this diagram it follows that the isomorphism α transforms $(H_0, \varphi(H_0))$ to $(H_0, \psi(H_0))$, i.e. $\alpha(H_0) = H_0$, $\alpha\varphi(H_0) = \psi(H_0)$. Therefore $\alpha \in \mathcal{E}$, $\gamma = \psi^{-1}\alpha\varphi \in \mathcal{E}$ and $\psi\gamma = \alpha\varphi$, which means that the double cosets of φ and ψ coincide.

Corollary. *Two splitting homomorphisms Φ_φ and Φ_ψ are π-equivalent relative to $\mathcal{J}$ if and only if the left coset $\mathcal{E}\varphi$ has a non-empty intersection with $\psi\mathcal{E}$.* □

Another interesting question connected with splitting homomorphisms is the question of the uniqueness up to π-equivalence of a surjective splitting homomorphism (if it exists). Thus, suppose that for the pair of groups (G, K) there exists an epimorphism $\varphi_0 \times \varphi_1 : G \to K \times K$ which also satisfies condition (i). Let us write $H_0 = \operatorname{Ker} \varphi_0$, $H_1 = \operatorname{Ker} \varphi_1$; then $H_0 H_1 = G$. We are interested in conditions under which the splitting homomorphism is unique up to π-equivalence. We remark that as a rule the question of the uniqueness up to π-equivalence of a splitting homomorphism $G \to K \times K$ is equivalent to the question of uniqueness up to ordinary equivalence. The condition on the pair (G, K) ensuring this equivalence consists in the following: The group $\operatorname{Aut} K \times \operatorname{Aut} K$ is embedded in a natural way in $\operatorname{Aut}(K \times K)$ and moreover there is an automorphism $\xi \in \operatorname{Aut}(K \times K)$ of order 2 permuting the coordinates: $\xi(K_1, K_2) = (K_2, K_1)$. We require K to fulfil the following conditions:

(ii) $|\operatorname{Aut}(K \times K) : (\operatorname{Aut} K \times \operatorname{Aut} K)| = 2$.

Moreover, we impose on G the condition

(iii) There exists an automorphism $\varrho \in \operatorname{Aut} G$ such that $\varrho(H_0) = H_1$, $\varrho(H_1) = H_0$.

Proposition 1. *Assume that the pair (G, K) satisfies the conditions* (i)–(iii). *Then the uniqueness of a splitting epimorphism $G \to K \times K$ up to equivalence holds if and only if it holds up to π-equivalence.* □

We remark that the conditions of Proposition 1 are fulfilled by the pair $(\pi_1(T_g), T_g)$ in the case of interest to us.

Consider the following group

$$\mathcal{F} = \{\alpha \in \operatorname{Aut} G | \alpha(H_1) = H_1\}. \tag{15}$$

Since (i) holds, every splitting homomorphism is π-equivalent to one of the form

$$G \xrightarrow{\lambda \times \mu} G/H_0 \times G/\varphi(H_1), \tag{16}$$

where $\varphi \in \operatorname{Aut} G$ and μ, λ are the canonical homomorphisms. We shall call $\varphi \in \operatorname{Aut} G$ a generating automorphism if it satisfies the relation

$$H_0 \cdot \varphi(H_1) = G. \tag{17}$$

The set of generating automorphisms is denoted by $\mathcal{P}$. Clearly (16) is surjective precisely when φ is a generating automorphism.

Theorem 4. *A splitting homomorphism $G \to K \times K$ satisfying (i) is unique up to π-equivalence if and only if $\mathcal{P} = \mathcal{E}\mathcal{F}$.*

Let $\varphi \in \mathcal{E}\mathcal{F}$, $\varphi = \alpha_0 \alpha_1$, $\alpha_0 \in \mathcal{E}$, $\alpha_1 \in \mathcal{F}$. Then there exist isomorphisms β_1, β_2 such that the diagram

$$\begin{array}{ccc} G & \xrightarrow{\xi\times\eta} & G/H_0 \times G/\varphi(H_1) \\ {\scriptstyle\alpha_0^{-1}}\downarrow & & \downarrow{\scriptstyle\beta_1\times\beta_2} \\ G & \xrightarrow{\lambda_0\times\mu_0} & G/\alpha_0^{-1}(H_0) \times G/\alpha_1(H_1) \end{array}$$

commutes (where $\xi, \eta, \lambda_0, \mu_0$ are canonical homomorphisms). Since $\alpha_0^{-1}(H_0) = H_0$, $\alpha_1(H_1) = H_1$, it follows that $\lambda_0 \times \mu_0$ and hence $\xi \times \eta$ are surjective. Thus we have shown that the epimorphism Φ_φ defined by (16) is equivalent to the standard epimorphism

$$G \xrightarrow{\lambda_0\times\mu_0} G/H_0 \times G/H_1 \,. \tag{18}$$

Conversely, assume that every epimorphism of the form (16) is equivalent to the standard epimorphism $\lambda_0 \times \mu_0$. Suppose that for some isomorphisms α, β_1, β_2 the diagram

$$\begin{array}{ccc} G & \xrightarrow{\lambda\times\mu} & G/H_0 \times G/\varphi(H_1) \\ {\scriptstyle\alpha}\downarrow & & \downarrow{\scriptstyle\beta_1\times\beta_2} \\ G & \xrightarrow{\lambda_0\times\mu_0} & G/H_0 \times G/H_1 \end{array}$$

commutes. From this diagram it follows that $\alpha(H_0) = H_0$ and $\alpha\varphi(H_1) = H_1$. Put $\gamma = \alpha\varphi \in \mathcal{F}$; then $\varphi = \alpha^{-1}\gamma \in \mathcal{E}\mathcal{F}$ and the theorem is proved. □

§ 5. On an Algebraic Reduction of the Poincaré Conjecture and the Algorithmic Poincaré Problem

In the study of topology at the beginning of the 1960's it was shown that the Poincaré conjecture is closely linked to certain questions in the theory of groups and may be formulated in purely algebraic terms. Since then the concept of a splitting homomorphism was introduced by Stallings [Stallings 1966] and Jaco [Jaco 1970] proved (in the orientable case) that every splitting homomorphism is associated to a Heegaard decomposition, leading to the following variant of an algebraic reformulation of the Poincaré conjecture:

Theorem 5. *The Poincaré conjecture holds precisely if for each $g \geq 2$, every epimorphism $\Phi : \pi_1(T_g) \to F_g \times F_g$ (where T_g is a closed orientable surface of genus g) factors through a free product.* □

We say that Φ factors through a free product if there exists a non-trivial free product $A * B (A, B \neq \{1\})$ and homomorphisms ψ, α such that $\psi : \pi_1(T_g) \to A * B$ is an epimorphism and the diagram

$$\begin{array}{ccccc} & & \pi_1(T_g) & & \\ & \swarrow^{\psi} & & \searrow^{\phi} & \\ A * B & & \xrightarrow{\alpha} & & F_g \times F_g \end{array} \tag{19}$$

commutes.

After Waldhausen [Waldhausen 1968], who showed that any two Heegaard decompositions of the standard sphere of the same genus are equivalent, the question whether the Poincaré conjecture holds can be reformulated in the following attractive form:

Theorem 6 [Hempel 1976]. *The Poincaré conjecture holds if and only if, for each $g \geq 2$ there exists a unique epimorphism $\pi_1(T_g) \to F_g \times F_g$ up to π-equivalence.* □

Since the pair $(\pi_1(T_g), T_g)$ satisfies (i)–(iii), we can in the formulation of Theorem 6 replace π-equivalence by ordinary equivalence.

From Theorem 6 it follows that in the formulation of Theorem 5 the abstract groups A, B can be replaced by the concrete groups $\pi_1(T_1)$, $\pi_1(T_{g-1})$.

The subgroups $\mathcal{E}, \mathcal{F}$ defined by (14) and (15), where H_0 is the normal subgroup generated by $a_1, \ldots, a_g$ and H_1 the normal subgroup generated by $b_1, \ldots, b_g$ from a presentation (7), will be denoted by L^-, L^+, respectively. The set of generating automorphisms will be denoted by $\mathcal{P}$, as before. A consequence of Theorem 3 and Theorem 6 may be formulated as the assertion below:

Theorem 7. *The Poincaré conjecture holds if and only if for each $g \geq 2$ the set $\mathcal{P}$ of generating automorphisms of $\pi_1(T_g)$ coincides with L^-L^+.* □

We remark that L^- is conjugate to L^+ by any automorphism interchanging H_0 and H_1, for example by η:

$$\eta(a_i) = b_{g-i+1}\ , \quad \eta(b_i) = a_{g-i+1}\ , \quad i = 1, \ldots, g\ . \tag{20}$$

Since the normal subgroup $\operatorname{Inn}\pi_1(T_g) \leq \operatorname{Aut}\pi_1(T_g)$ is contained in the intersection $L^- \cap L^+$, we can formulate Theorem 7 in terms of the group $MC(T_g) = \operatorname{Aut}\pi_1(T_g)/\operatorname{Inn}\pi_1(T_g)$ and its subgroups $MC^-(T_g)$, $MC^+(T_g)$, which are the images of L^-, L^+ by the canonical homomorphism. Information on $MC^-(T_g)$ can be found in Sect. 4 above. The reformulation of the Poincaré conjecture in terms of the group of mapping classes analogous to that of Theorem 7 can be extracted immediately from Waldhausen's theorem. This was practically done in [Hempel 1976, Theorem 14.10], where however the hypothesis is incorrect, because the homeomorphism h which occurs in the formulation of this theorem does not exist. A similar assertion occurs also in [Birman 1974b].

From Theorem 3 it follows that the question of algorithmic recognizability of the equivalence of two intertwining homomorphisms Φ_φ and Φ_ψ is equivalent to the question whether the set of double cosets $L^-\backslash\operatorname{Aut}\pi_1(T_g)/L^-$ of L^- in $\operatorname{Aut}\pi_1(T_g)$ is recursive. The solution of this problem for all $g \geq 2$ allows one to solve the problem of the equivalence of Heegaard decompositions. Also of interest is the question whether the double coset $L^- \varphi L^-$ defined by the automorphism (20) is recursive, because this is equivalent to the algorithmic Poincaré problem. (The set $L^- \eta L^-$ consists of the automorphisms used as codes for the 3-manifold S^3.)

§ 6. Some Analogues with the Group of Symplectic Matrices and the Torelli Group

We shall now describe some parallels between the group $\operatorname{Aut} \pi_1(T_g)$ and its subgroups L^-, L^+ and the group $Sp_{2g}(\mathbb{Z})$ of symplectic matrices and its subgroups of lower resp. upper triangular matrices. We recall that a matrix S of order $2g$ is called *symplectic* if it satisfies the relation $SJS' = \pm J$, where

$$J = \begin{pmatrix} 0 & I_g \\ -I_g & 0 \end{pmatrix},$$

the dash denotes transposition and $0, I_g$ are the zero resp. unit matrix of order g.

Every automorphism $\varphi \in \operatorname{Aut} \pi_1(T_g)$ corresponds to an automorphism of the abelian group $\pi_1(T_g)/[\pi_1(T_g), \pi_1(T_g)]$, whose matrix we denote by S_0. We thus obtain an antirepresentation

$$S_p : \operatorname{Aut} G_g \to Sp_{2g}(\mathbb{Z}) .$$

In this mapping the elements of L^- correspond to the lower triangular matrices

$$\begin{pmatrix} a & 0 \\ b & c \end{pmatrix},$$

while those of L^+ correspond to the upper triangular matrices

$$\begin{pmatrix} a & b \\ 0 & c \end{pmatrix}.$$

For this reason the elements of L^- are called lower triangular and those of L^+ upper triangular automorphisms.

The 3-manifold $M(\varphi)$ corresponding to $\varphi \in \operatorname{Aut} \pi_1(T)$ by the second coding procedure given in our paper is a homology sphere if and only if the abelianization

$$\pi_1(M(\varphi))/\left[\pi_1(M(\varphi)), \pi_1(M(\varphi))\right]$$

of its fundamental group is trivial, which is clearly so precisely when the matrix D in the representation

$$S_\varphi = \begin{pmatrix} A & B \\ C & D \end{pmatrix} \in Sp_{2g}(\mathbb{Z}) \qquad (21)$$

belongs to $GL_g(\mathbf{Z})$.

Lemma 2. *Every symplectic matrix* (21) *such that* $D \in GL_g(\mathbb{Z})$ *can be written in the form* Q^+Q^-, *where* Q^+, $Q^- \in Sp_{2g}(\mathbb{Z})$ *are upper resp. lower triangular matrices.* □

Using a generating set of L^- we may show that the group of lower triangular automorphisms covers the group of lower triangular matrices and similarly for L^+. Therefore we have the relation $Sp(\mathcal{P}) = Sp(L^+)Sp(L^-)$, i.e. for any $g \geq 2$ the sets $\mathcal{P}$ and L^+L^- do not differ in their projection on the symplectic group.

In this connexion the kernel of the antihomomorphism Sp, called the Torelli group, is of particular interest. Johnson [Johnson 1983] shows that the Torelli group is finitely generated if $g \geq 3$. On the other hand, McCullough and Miller [McCullough–Miller 1986] show that in the case of genus 2 this group does not have a finite generating set. In the same paper a homomorphism from the Torelli group to the group of matrices over Laurent polynomials in non-commuting indeterminates[1] is constructed. Unfortunately this provides no information on the kernel of this map.

The group $\operatorname{Aut}\pi_1(T_g)$ is generated by its subgroups L^-, L^+. It is not known whether there exists an integer n such that we have

$$\operatorname{Aut}\pi_1(T_g) = (L^-L^+)^n = L^-L^+L^-L^+ \ldots L^-L^+$$

($2n$ factors). This is an argument showing that in the search scheme for a proof of the Poincaré conjecture the most progress will emerge from a proof that every generating automorphism in L^+L^- also lies in L^-L^+.

§ 7. Algebraic Reduction of the Problem of the Equivalence of Links

Two links Q_1, Q_2 embedded in S^3 are said to be equivalent if there exists a homeomorphism $h : S^3 \to S^3$ such that $h(Q_1) = Q_2$. One of the hard problems in topology is that of the classification of knots and links up to equivalence. An important step in this direction was the finding of an algorithm allowing one to decide for two combinatorially defined links whether or not they are equivalent. In [Birman 1974a] an algebraic form of the problem is given, which we shall present to the reader.

Let $M(0, 2m)$ be the group of mapping classes of the sphere fixing $2m$ points and E the subgroup of $M(0, 2m)$ consisting of those mapping classes which can be extended to a body with m handles $\mathcal{D}^3 \setminus \bigcup_{i=1}^m l_i$, where $\mathcal{D}^3$ is a sphere and l_i, $i = 1, \ldots, m$ form a family of unlinked, unknotted arcs embedded in $\mathcal{D}^3$ as shown in Fig. 1.

The group $M(0, 2m)$ admits the presentation

[1] In other words, the group ring of a free group (Transl. note).

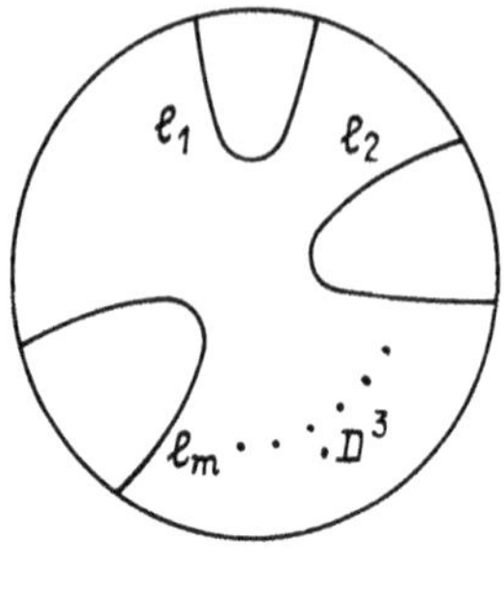

Fig. 1

$$\langle \omega_1, \ldots, \omega_{2m-1} | \omega_i\omega_j = \omega_j\omega_i,\ |i-j| \geq 2\,,$$
$$\omega_i\omega_{i+1}\omega_i = \omega_{i+1}\omega_i\omega_{i+1},\ i = 1, \ldots, 2m-2\,,$$
$$\omega_1 \ldots \omega_{2m-2}\omega_{2m-1}^2\omega_{2m-2} \ldots \omega_1 = 1,\ (\omega_1\omega_2 \ldots \omega_{2m-1})^{2m} = 1\rangle$$

and it is a homomorphic image of the braid group

$$B(2m) = \langle \sigma_1, \ldots, \sigma_{2m-1} | \sigma_i\sigma_j = \sigma_j\sigma_i,\ |i-j| \geq 2\,,$$
$$\sigma_i\sigma_{i+1}\sigma_i = \sigma_{i+1}\sigma_i\sigma_{i+1},\ i = 1, \ldots, 2m-2\rangle$$

under the map $\sigma_i \mapsto \omega_i$, $i = 1, \ldots, 2m-1$.

In [Birman 1974a] the corresponding link is constructed for every element $\varphi \in M(0, 2m)$. Let F_{2m} be the free group with basis $x_1, \ldots, x_{2m}$. The braid group $B(2m) \leq \operatorname{Aut} F_{2m}$ acts by automorphisms of F_{2m} [Birman 1974a].

Denote by $H_0 \trianglelefteq F_{2m}$ the normal subgroup generated by the elements $x_{2i-1}x_{2i}^{-1}$, $i = 1, \ldots, m$ (clearly we have $F_{2m}/H_0 \cong F_m$) and by $\tilde{E} \leq B(2m)$ the subgroup of all automorphisms mapping H_0 into itself. We shall see that E is the projection of $\tilde{E}$ under the natural homomorphism $B(2m) \to M(0, 2m)$. In [Birman 1974a] the following result is proved.

Theorem 8. *Let $\varphi_1, \varphi_2 \in M(0, 2m)$ and let V_1, V_2 be the links associated with φ_1, φ_2 respectively. Then V_1 and V_2 are equivalent if φ_2 lies in the double coset relative to E of one of the elements*

$$\varphi_1,\ \varphi_1^{-1},\ \operatorname{Rev}\varphi_1,\ \operatorname{Rev}\varphi_1^{-1}\,,$$

where if $\varphi = \omega_{\mu_1}^{\varepsilon_1} \ldots \omega_{\mu_\tau}^{\varepsilon_\tau}$, then $\operatorname{Rev}\varphi = \omega_{\mu_\tau}^{\varepsilon_\tau} \ldots \omega_{\mu_2}^{\varepsilon_2}\omega_{\mu_1}^{\varepsilon_1}$. □

It can be shown that if two elements $\beta_1, \beta_2 \in B(2m)$ map to the same element $\beta \in B(0, 2m)$, then either both or neither belong to $\tilde{E}$. Thus we obtain the following consequence of Theorem 8.

Corollary 2. *If the set of double cosets $\tilde{E}\backslash B(2m)/\tilde{E}$ is recursive, then there exists an algorithm allowing us to decide for any two links Q_1, Q_2 given in combinatorial terms whether or not they are equivalent.* □

The group $M(0,2m)$ gives rise to another simpler group $B(2m)$ which in any case is easier to study. Much progress in the search for a proof that the set $\tilde{E}\backslash B(2m)/\tilde{E}$ is recursive will be made by finding a generating set for $\tilde{E}$ (as well as clearing up the question whether or not it is finite).

The action of $B(2m)$ on the set of kernels $H \trianglelefteq F_{2m}$ such that $F_{2m}/H \cong F_m$ is not transitive, therefore not every splitting homomorphism $F_{2m} \to F_m \times F_m$ is π-equivalent relative to B_{2m} to a splitting homomorphism of the form

$$F_{2m} \to F_{2m}/H_0 \times F_{2m}/\varphi(H_0)\,, \tag{22}$$

where $\varphi \in B(2m)$. If we confine our attention to splitting homomorphisms of the form (22) and look for an algorithm establishing π-equivalence relative to $B(2m)$ of any pair of splitting homomorphisms of the form (22), then this question is equivalent to the question whether the set $\tilde{E}\backslash B/\tilde{E}$ is recursive.

§ 8. On the Andrews–Curtis Conjecture

For a discussion of the algorithmic problem of the recognition of a homotopy sphere it should be noted that the answer to this question will be positive if there exists an algorithm recognizing the triviality of a group given by a balanced presentation. For this question as for the question of the truth of the Poincaré conjecture there is a conjecture announced by Andrews and Curtis [Andrew–Curtis 1965]:

AC-conjecture Suppose that the group

$$\langle a_1, \ldots, a_n | R_1, \ldots, R_n \rangle$$

is trivial. Then the set of words $R_1, \ldots, R_n$ may be reduced to the set $A = \{a_1, \ldots, a_n\}$ by the transformations

(1) $R_i \to R_i^{-1}$, R_j unchanged for $j \neq i$,

(2) $(R_i, R_j) \to (R_iR_j, R_j)$, R_k unchanged for $k \neq i$,

(3) $R_i \to W^{-1}R_iW$, R_j unchanged for $j \neq i$, $W \in A \cup A^{-1}$. □

We now formulate a stronger conjecture. Let $R = \{R_1, \ldots, R_n\}$, $Q = \{Q_1, \ldots, Q_n\}$ be two sets of words in the alphabet $A \cup A^{-1}$. By the normal closure of R with respect to Q we understand the group generated by $T^{-1}R_iT$, where $i = 1, \ldots, n$ and T runs over the subgroup $F_Q = \langle Q_1, \ldots, Q_n \rangle$ generated by Q. This group will be denoted by R^Q. Clearly R^A is simply the normal closure of R in the free group.

GK-conjecture If the normal closure of $\overline{R} = \{R_1, \ldots, R_n\}$ with respect to $\overline{Q} = \{Q_1, \ldots, Q_n\}$ coincides with F_A then the set of words $\{\overline{R}, \overline{Q}\}$ may be reduced to $\{\overline{A}, \overline{1}_n\}$, $\overline{1}_n = \{1, 1, \ldots, 1\}$ n units, by transformations (1), (2), (3′), (4) and (5), where

(3′) $R_i \to W^{-1}R_iW$, R_j unchanged for $j \neq i$, $W \in Q \cup Q^{-1}$,

(4) $Q_i \to Q_iR_j$, Q_k unchanged for $k \neq i$,

(5) the Nielsen automorphisms on the set $\overline{Q}$.

Clearly the truth of the GK-conjecture implies that of the Andrews–Curtis conjecture.

Theorem 9. *The GK-conjecture holds if and only if, for any $g \geq 2$, there exists an epimorphism*

$$F_{2g} \to F_g \times F_g \ , \tag{23}$$

unique up to equivalence. □

Proof. Since for any pair of natural numbers n, $N > n$ there is a unique epimorphism $F_N \to F_n$ up to strong equivalence, it follows that any splitting homomorphism of the form (23) is equivalent to one of the form

$$F_{2g} \to F_{2g}/H_0 \times F_{2g}/\varphi(H_1) \ , \tag{24}$$

where $F_{2g} = \langle a_1, \ldots, a_g, b_1, \ldots, b_g \rangle$, $H_0 = \langle \overline{a_1, \ldots, a_g} \rangle$ is the normal closure of the a's and $H_1 = \langle \overline{b_1, \ldots, b_g} \rangle$ is the normal closure of the b's, while $\varphi \in \operatorname{Aut} F_{2g}$. Conversely, the homomorphism (24) is equivalent to $\mu \times \psi = \Phi : F_{2g} \to F_g \times F_g$, where $F_g = \langle c_1, \ldots, c_g \rangle$ and

$$\begin{aligned} &\mu(a_i) = 1 \ , \quad \mu(b_i) = c_i \ , \quad i = 1, \ldots, g \ ; \\ &\psi(a_i) = R_i(\overline{C}) \ , \quad \psi(b_i) = Q_i(\overline{C}) \ , \quad i = 1, \ldots, g \ ; \end{aligned} \tag{25}$$

where the words $R_i(\overline{C})$, $Q_i(\overline{C})$ are defined as follows. Let $a'_i = \varphi(a_i)$, $b'_i = \varphi(b_i)$; then $a_i = \overline{R}_i(\overline{a}', \overline{b}')$, $b_i = \overline{Q}_i(\overline{a}', \overline{b}')$, $i = 1, \ldots, g$ for some words $\overline{R}_i(\overline{a}', \overline{b}')$, $\overline{Q}_i(\overline{a}', \overline{b}')$ (inverse substitutions). Let us replace in $R_i(\overline{a}', \overline{b}')$, $Q_i(\overline{a}', \overline{b}')$ the b'_i by the empty symbol and a'_i by c_i. This yields words $R_i(\overline{C})$, $Q_i(\overline{C})$. Now the homomorphism $\mu \times \psi : F_{2g} \to F_g \times F_g$ is defined by the relations (25), which may be written as a table

$$\left(\begin{array}{c|c} 1 \ldots 1 & c_1 \ldots c_g \\ R_1 \ldots R_g & Q_1 \ldots Q_g \end{array} \right) \tag{26}$$

(in the first row are the values of μ and in the second row those of ψ). We have the following assertion.

Lemma 3. *The splitting homomorphism $\mu \times \psi$ defined by (26) is an epimorphism if and only if $F_g = R^Q$.* □

If the GK-conjecture is true for some value of the parameter g, then it follows by Lemma 3 that the splitting homomorphism $F_{2g} \to F_g \times F_g$ is unique up to equivalence. For a proof of the opposite implication we need the following assertion:

Lemma 4 (Grigorchuk, Kurchanov, Lysenok). *Let $F_N = \langle a_1, \ldots, a_m, b_1, \ldots, b_n \rangle$ be a free group of rank $N = m + n$, $H \trianglelefteq F_N$ the normal subgroup generated by $a_1, \ldots, a_m$ and $\mathcal{J} \leq \operatorname{Aut} F_N$ the subgroup consisting of all* *automorphisms mapping H into itself. Then $\mathcal{J}$ is generated by the automorphisms*

(1) *the Nielsen automorphisms on the set* $a_1, \ldots, a_n$;
(2) $a_i \to b_j^{-1} a_i b_j$, $a_k \to a_k$, $k \neq i$, $b_l \to b_l$;
(3) *the Nielsen automorphisms on the set* $b_1, \ldots, b_n$;
(4) $b_i \to b_i a_j$, $a_k \to a_k$, $b_l \to b_l$, $l \neq i$. □

Let us apply Lemma 4 for a group F_{2g}, $m = n = g$. For this purpose we introduce the groups

$$L_g^- = \{\varphi \in \operatorname{Aut} F_{2g} | \varphi(H_0) = H_0\} ,$$

$$L_g^+ = \{\varphi \in \operatorname{Aut} F_{2g} | \varphi(H_1) = H_1\} .$$

According to Theorem 4 we have $\mathcal{P} = L^- L^+$, where $\mathcal{P}$ is the set of generating automorphisms. Hence the homomorphism defined by (26) is equivalent to a standard homomorphism if and only if a certain automorphism of L_g^- transforms it to the standard homomorphism given by the table

$$\left(\begin{array}{c|c} 1 \ldots 1 & c_1 \ldots c_g \\ c_1 \ldots c_g & 1 \ldots 1 \end{array} \right) .$$

Taking account of the form of the generators of L_g^- described in Lemma 4, we see that the set of words $\{\overline{R}, \overline{Q}\}$ is reduced by (1), (2), (3′), (4), (5) to $\{\overline{C}, \overline{1}\}$ as was to be shown. □

Coming back to our initial consideration, the Andrews–Curtis conjecture, we see that its correctness is equivalent to the possibility of reducing any splitting homomorphism to standard form, given by

$$\left(\begin{array}{c|c} 1 \ldots 1 & c_1 \ldots c_g \\ R_1 \ldots R_g & c_1 \ldots c_g \end{array} \right) ,$$

where the normal closure of the words $R_1, \ldots, R_g$ coincides with $F_g = \langle c_1, \ldots, c_g \rangle$.

The group $\operatorname{Aut} F_{2g}$ is generated by its subgroups L_g^-, L_g^+. The standard image defines an antihomomorphism $\varrho : \operatorname{Aut} F_{2g} \to GL_{2g}(\mathbb{Z})$, in which the groups L_g^-, L_g^+ correspond to the groups of lower resp. upper triangular matrices, i.e. matrices of the form

$$\begin{pmatrix} A & 0 \\ B & C \end{pmatrix} , \quad \begin{pmatrix} A & B \\ 0 & C \end{pmatrix} .$$

If we replace the given generating system of F_{2g} by $x_1, \ldots, x_{2g}$, then the kernel $IA(F_{2g})$ of the antihomomorphism ϱ is generated by the automorphism

$$x_i \to x_i [x_j, x_k] , \quad j \neq k , \quad x_s \to x_s , \quad s \neq i .$$

Clearly each of these automorphisms belongs either to L_g^- or to L_g^+. To prove that every generating automorphism $\varphi \in \operatorname{Aut} F_{2g}$ lies in $L_g^- L_g^+$ we can restrict ourselves to elements from the kernel $(IA(F_{2g})$. It is not known whether n exists such that

$$\operatorname{Aut} F_{2g} = (L_g^- L_g^+)^n = L_g^- L_g^+ \dots L_g^- L_g^+$$

($2n$ factors).

The weight of progress in the direction of a proof of the Andrews–Curtis conjecture would emerge from a proof of the fact that every generating homomorphism in $L_g^+ L_g^-$ is contained in $L_g^- L_g^+$.

Great interest attaches also to the question whether the set $L_g^- L_g^+ \subset \operatorname{Aut} F_{2g}$ is recursive, because this is connected with the question of the existence of an algorithm allowing us to answer the following question for a balanced presentation: does this presentation define the trivial group or not?

Concerning Andrews–Curtis conjecture see also [Hog–Angeloni–Metzler 1991] and the bibliography of this paper.

Chapter 3
On the Rate of Growth of Groups and Amenable Groups

§ 1. On the Growth of Graphs and of Riemannian Manifolds

The notion of growth is widely used in mathematics. In this chapter we consider this concept as applied to groups. However, at first we shall consider a notion of geometric growth, whose combinatorial form leads to the notion of growth of groups.

So to begin with, let Γ be a connected infinite graph, in each of whose vertices not more than L edges meet, where L is some constant. Let $e \in V(\Gamma)$ be a fixed vertex. We introduce a metric on Γ by taking the distance $d(x, y)$ between two vertices $x, y, \in V(\Gamma)$ as the length of the shortest path joining x to y (the length of a path is the number of its edges).

We denote by $B_e(r)$ the sphere of radius r with centre at the point e:

$$B_e(r) = \{x \in V(\Gamma) : d(e, x) \le r\} .$$

The function

$$\gamma(r) = \gamma_{\Gamma,e}(r) = \#(B_e(r)) ,$$

where $\#(B)$ denotes the number of elements in B, is called the *growth function of the graph* Γ. For example, for the graphs depicted in Fig. 2(a), (b) this

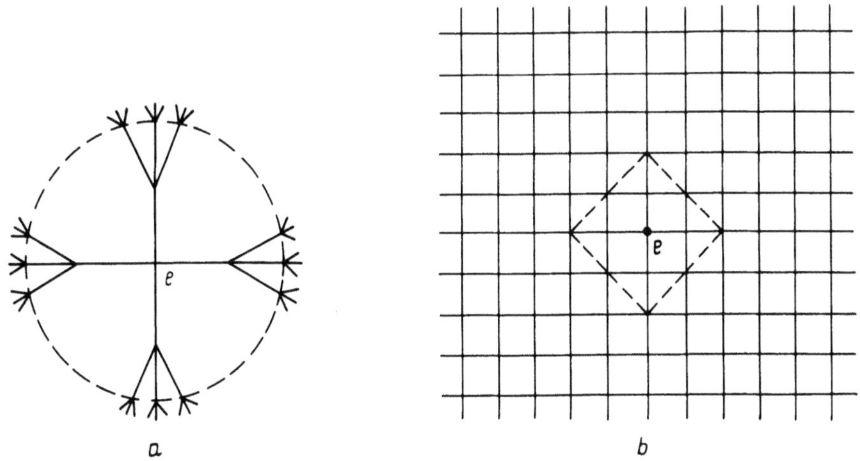

Fig. 2

function equals in case (a), $\gamma(r) = 2.3^r - 1$ and in case (b), $\gamma(r) = 2r^2 + r + 1$ (the dotted line in Fig. 2 defines the boundary of the sphere of radius 2).

The order of growth of the function $\gamma(r)$ as $r \to +\infty$ characterizes the order of growth of Γ at infinity. Naturally it has to be made precise what we understand by order of growth, and its dependence on the point e will be analysed below. We shall consider these questions in detail for Cayley graphs of finitely generated groups. For now we only fix attention on the fact that for the graph illustrated in Fig. 2(a) the function $\gamma(r)$ grows exponentially while for the graph 2(b) it grows as a power, in other words polynomially. While the growth is quadratic for the graph 2(b), a k-dimensional integer lattice is an example of polynomial growth of degree k.

Now let M be a complete unbounded Riemannian manifold, $e \in M$ a fixed point, $B_e(r)$ a geodesic ball of radius r with centre at e:

$$B_e(r) = \{x \in M : d(e, x) \leq r\} ,$$

where $d(x, y)$ is the Riemannian metric on M. The function

$$v(r) = v_{M,e}(r) = \mathrm{Vol}\,(B_e(r)) ,$$

where Vol$\,(B)$ is the volume of B is naturally called the *growth function of the Riemannian manifold* at infinity. The order of its growth characterizes the growth of M at ∞. If instead of M we have a Lobachevskij plane we obtain an example of an exponential growth function $v(r)$, and if we take a k-dimensional Euclidean space, we have an example of polynomial growth of degree $v(r)$ (Fig. 3(a), (b)).

It is not difficult to construct examples of graphs and Riemannian manifolds of many different growth orders. However, if the construction of the manifold

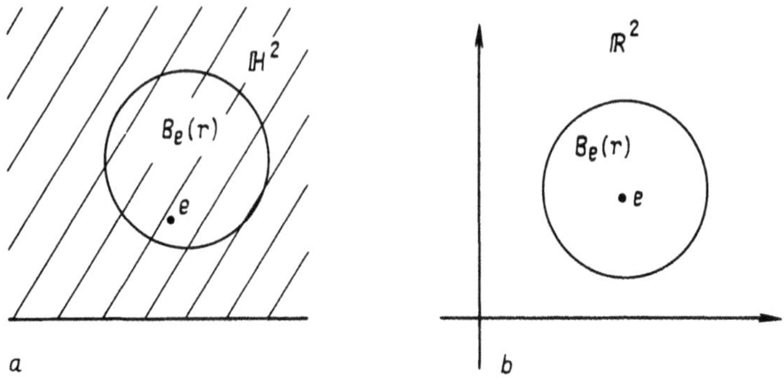

Fig. 3

or graph has a large symmetry group, the growth will most likely be either polynomial or exponential. The origins of this prognosis will become clear from the account that follows. We now go on to consider the combinatorial form of the notion of growth for a Riemannian manifold with a large symmetry group.

§ 2. On the Notion of Growth of a Finitely Generated Group

Let G be a finitely generated group and $A = \{a_1, \dots, a_n\}$ a generating set. We consider the Cayley graph $\Gamma = \Gamma(G, A)$ of G constructed from this generating set. As distinguished vertex $e \in V(\Gamma)$ we choose the unit element. Then the *growth function* $\gamma(r)$ *of the group* G relative to the generating set A is the growth function $\gamma_{\Gamma,e}(r)$ of the graph Γ.

For example, for the free group F_2 on two generators the frame of its Cayley graph is illustrated in Fig. 2(a), therefore $\gamma(r) = 2 \cdot 3^r - 1$, while for the free abelian group $\mathbb{Z}^2$ of rank 2, the frame of its Cayley graph relative to a basis is given by Fig. 2 (b), so $\gamma(r) = 2r^2 + 2r + 1$.

More formally, $\gamma(r)$ is defined in the following manner. Let $\partial(g)$ be the length of the element g relative to the generating set A, i.e. the length of the shortest representation

$$g = a_{i_1}^{\varepsilon_1} \dots a_{i_n}^{\varepsilon_n}, \quad \varepsilon_j = \pm 1, \quad j = 1, \dots, n,$$

of g as a product of the generators and their inverses. We define on G a left-invariant metric $d(g, h)$, $g, h \in G$, by the formula $d(g, h) = \partial(g^{-1}h)$. Let $B_1(r)$ be the sphere of radius 1 with the unit element as centre, constructed in the metric space (G, d). Then the growth function of G relative to the generating set A is defined by the relation

$$\gamma(r) = \gamma_{G,A}(r) = \#(B_1(r)) ,$$

and this is easily seen to agree with the earlier definition.

To eliminate the dependence of the growth function on the generating set we introduce the following equivalence relation on the set of positive functions of integer arguments: $f_1(r) \sim f_2(r)$ if and only if there exists a constant $C \in \mathbb{N}$ such that

$$f_1(r) \leq f_2(Cr) , \quad f_2(r) \leq f_1(Cr) , \quad r = 1, 2, \dots .$$

The equivalence class $[\gamma(r)]$ of $\gamma(r)$ is called the *growth rate* of the group G and does not depend on the generating system. It characterizes the growth of G.

In what follows it will be convenient also to introduce a preordering on the set of growth functions by writing $\gamma_1(r) \preceq \gamma_2(r)$ if there exists $C \in \mathbb{N}$ such that $\gamma_1(r) \leq \gamma_2(Cr)$ for $r = 1, 2, \dots$. In this way the set of all growth rates of finitely generated groups becomes a partially ordered set.

The examples considered above show that if $G = F_m$ is a free group of rank $m \geq 2$, then $\gamma(r) \sim e^r$, while if $G = \mathbb{Z}^d$, the free abelian group of rank d, then $\gamma(r) \sim r^d$.

Clearly if H is a subgroup of finite index in G, then the growth rates of G and H are the same. In what follows we shall say that G is *virtually nilpotent*, *virtually soluble* etc. if G has a nilpotent (soluble) subgroup of finite index.

In the geometric and algebraic examples considered above, analogies were traced, leading to the impression that the notion discussed here are one and the same. The first to establish a precise link between geometric growth and the growth of groups were Efremovich [Efremovich 1953] and Shvarts [Shvarts 1955]. This connexion is for example expressed in the following statement:

Theorem 1 [Grigorchuk 1989b]. *Let $\tilde{M} \to M$ be a regular covering of a compact Riemannian manifold M corresponding to the normal subgroup $H \trianglelefteq \pi_1(M)$. Then we have an equivalence*

$$v_{\tilde{M}}(r) \sim \gamma_{\pi_1(M)/H}(r) , \tag{1}$$

where $v_{\tilde{M}}(r)$ is the growth function of $\tilde{M}$ and $\gamma_{\pi_1(M)/H}(r)$ is the growth rate of the group $\pi_1(M)/H$. □

The particular case of this assertion relating to the universal covering (i.e. when H is the trivial group) was proved by Shvarts [Shvarts 1955]. Figure 4 illustrates two important cases where Theorem 1 can be applied.

We remark that the factor group $G = \pi_1(M)/H$ is isomorphic to the group of covering transformations of the covering $\tilde{M} \to M$ and the action of G on $\tilde{M}$ has a compact fundamental domain, which confirms the correctness of (1).

In 1968 Milnor [Milnor 1968a] independently arrived at the notion of growth rate and related it to conditions on the curvature of a Riemannian manifold M. With this paper there began interest in the study of growth of groups.

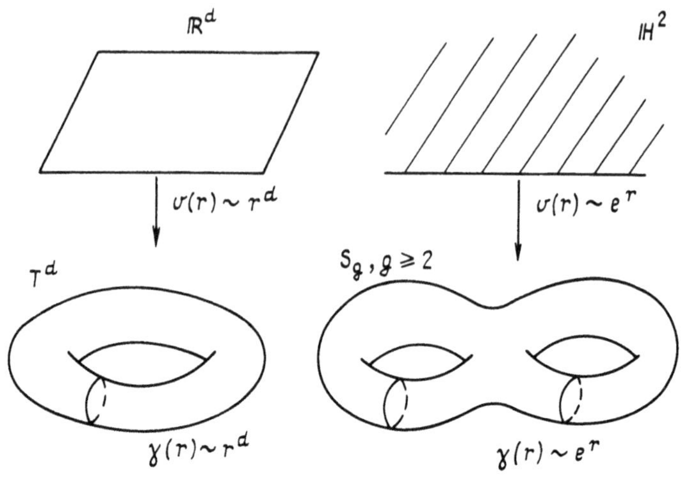

Fig. 4

Already the first examples of nilpotent groups considered by [Milnor 1968a] showed that some nilpotent groups had polynomial growth (in [Wolf 1968] this was shown for all nilpotent groups). Bass [Bass 1972] found a formula for the exponent of the growth rate of a nilpotent group. He showed that if G is a nilpotent group and G_k, $k = 1, 2, \ldots$ its lower central series, then $\gamma_G(r) \sim r^d$, where

$$d = \sum_k k.\dim_{\mathbb{Z}} (G_k/G_{k+1}),$$

and $\dim_{\mathbb{Z}}(G_k/G_{k+1})$ denotes the torsion free rank of the abelian group G_k/G_{k+1}.

An extension of the class of groups considered to the class of soluble groups does not add new information on the growth of groups, because it can be shown that soluble groups either have exponential growth or they contain a nilpotent subgroup of finite index. No new information on the growth of groups is added by linear groups, because according to the Tits alternative [Tits 1972] a finitely generated group of matrices over a field either contains a free subgroup on two generators or is virtually soluble.

The following examples give some idea of the nilpotent and soluble cases.

The nilpotent group

$$\langle a, b, c | [a, b] = c, \ [a, c] = [b, c] = 1 \rangle$$

of class two has a growth rate of degree 4 [Milnor 1968a]. Indeed, if $a^i b^j c^k$ lies in the spheres $B_1(r)$, then $|i| \leq r$, $|j| \leq r$, and $|k| \leq r^2/4$, from which the estimate $[\gamma(r)] \preceq r^4$ follows. It is not hard to show that $[\gamma(r)] \succeq r^4$.

The group

$$\Gamma = \langle a, b, c | cac^{-1} = a^2 b,\ cbc^{-1} = ab,\ [a, b] = 1 \rangle$$

is two-step soluble since it is isomorphic to an extension of the form

$$1 \to \mathbb{Z} \times \mathbb{Z} \to \Gamma \to \mathbb{Z} \to 1 .$$

It has exponentional growth, because the matrix

$$\begin{pmatrix} 2 & 1 \\ 1 & 1 \end{pmatrix}$$

describing the action of c on $\mathbb{Z} \times \mathbb{Z}$ has the eigenvalue $\lambda = (3 + \sqrt{5})/2 > 1$. In fact it is not hard to prove that 2^r words $ca^{\varepsilon_1} ca^{\varepsilon_2} \dots ca^{\varepsilon_n}$, $\varepsilon_j = 0, 1$, $j = 1, \dots, n$, represent distinct elements, therefore $\gamma(r) \geq 2^r$.

Milnor [Milnor 1968b] was the first to realize the difficulty of the problem of constructing new examples of group growths. In [Milnor 1968b] he raises the following question: "Is it true that the growth of any finitely generated group is either exponential or of the form of a power r^d?" In this or another form the question was raised in [Milnor 1968b], [Wolf 1968], [Bass 1972], [Adyan 1975] and other papers. It is now usually known as Milnor's problem.

In favour of a positive solution of Milnor's problem there is the result of Adyan [Adyan 1975] which shows that the free periodic Burnside group

$$B(m, n) = \langle a_1, \dots, a_m | X^n = 1 \rangle$$

has exponential growth for $m \geq 2$ and odd $n \geq 665$.

We shall now undertake an attempt to describe groups of polynomial growth. A complete answer to this question was given by Gromov [Gromov 1981] who showed that this class of groups exhausts the virtually nilpotent groups. Combined with Bass's formula this shows that if a group has polynomial growth, then the exponent of this growth is a natural number.

Gromov's proof is very geometrical and uses the concept of a limit of a sequence of metric spaces. As v.d. Dries and Wilkie [v.d. Dries–Wilkie 1984] have shown, this proof gains by being expressed in the language of non-standard analysis. Moreover, this provides a strengthening of Gromov's theorem. Thus if $\gamma(r) \leq cr^d$ holds for infinitely many integers r then it follows that G is virtually nilpotent. [v.d. Dries–Wilkie 1984] is one of the rare examples so far of an application of non-standard analysis to algebra.

§ 3. On the Proof of Gromov's Theorem and Some Related Results

Let us once more state Gromov's theorem:

Theorem 2. *A finitely generated group has polynomial growth if and only if it contains a nilpotent subgroup of finite index.* □

The hard part of the proof, accomplished by M. Gromov, is the implication: polynomial growth $\Rightarrow$ virtually nilpotent. The proof is an induction on the exponent, for which the following two simple assertions can be used:

Lemma 1. *If Γ has polynomial growth of exponent $\leq d$ and there is an exact sequence*

$$1 \to N \to \Gamma \to \mathbb{Z} \to 1 ,$$

then N is finitely generated of polynomial growth with exponent $\leq d-1$. □

Lemma 2. *If we have an exact sequence*

$$1 \to N \to \Gamma \to \Delta \to 1 ,$$

and both N and Δ contain polycyclic subgroups of finite index, then Γ contains a polycyclic subgroup of finite index. □

The strategy of the proof consists in this, that in Γ of polynomial growth we look for a subgroup of finite index having a homomorphism on an infinite cyclic group. The existence of such a subgroup follows at once from Tits' theorem if Γ can be homomorphically mapped on an infinite subgroup of a connected Lie group.

Let us make a weaker assertion: Γ may be mapped on a finite group of sufficiently large order, contained in a connected Lie group G. By the classical theorem of Jordan there exists $n = n(G)$ such that this finite subgroup contains an abelian subgroup of index $\leq n$. Let Γ_0 be the intersection of all subgroups of index $\leq n$ in Γ. Then the factor group $\Gamma_0/[\Gamma_0, \Gamma_0]$ is infinite and admits a homomorphism onto $\mathbb{Z}$, so the induction hypothesis applies.

Gromov's basic idea consists in constructing a Lie group G with the help of a limit of a sequence of metric spaces formed from a finite generating set of Γ. More precisely, let $d(g,h)$ be the left-invariant word metric on $\Gamma : d(g,h) = \partial(g^{-1}h)$ defined above. We consider the sequence of metric spaces (Γ, d_k), $k = 1, 2, \ldots$, where d_k is given by $d_k = 2^{-k}d$. If Γ has polynomial growth of degree $\leq d$, then Gromov shows that some subsequence (Γ, d_{k_i}) converges in a precisely defined sense to a locally compact, locally connected homogeneous metric space Y of Hausdorff dimension $\leq d$. Roughly speaking, for any $r > 0$ the spheres of radius r in (Γ, d_k) and in Y can be mapped isometrically into a common metric space such that each point on the sphere of unit radius in Y turns out to be close to some point on the unit sphere in (Γ, d_k), for large k. Applying the theorem of Montgomery–Zippin (containing the solution of Hilbert's 5th problem, [Gromov 1981]), Gromov shows that the group of isometries of Y is a Lie group with a finite number of connected components, after which he completes the proof of his theorem by constructing a suitable homomorphism from a subgroup Γ of finite index in G.

Gromov's theorem allows a generalization of growth to semigroups with cancellation. The definition of growth rate, exponent and other related notions is analogous to the group case. It is known that the growth of semigroups may be very whimsical. For example, in [Trofimov 1982] a continuum of non-isomorphic semigroups of quadratic growth $\sim r^2$ is constructed, it is shown that a semigroup may have polynomial growth of fractional degree (for example $\sim r^{5/2}$) and so on.

If we narrow the class of semigroups considered by requiring left and right cancellation, the situation becomes more favourable.

Although not every cancellation semigroup can be embedded in a group (relevant examples were constructed by Mal'tsev), cancellation semigroups of subexponential growth, i.e. of slower than exponential growth, possess groups of left and right fractions [Grigorchuk 1988]. Thus the theory of cancellation semigroups of polynomial growth is essentially the theory of semigroups of polynomial growth, embeddable in groups.

The concept of a nilpotent semigroup was first introduced by Mal'tsev [Mal'tsev 1953]. Let $x, y, \xi_1, \ldots, \xi_n$ be variable symbols taking values in a semigroup S. We define sequences of words $X_n, Y_n, n = 0, 1, \ldots$, on these symbols as follows: $X_0 = x$, $Y_0 = y$ and recursively

$$X_{n+1} = X_n \xi_{n+1} Y_n \,, \quad Y_{n+1} = Y_n \xi_{n+1} X_n \,.$$

The semigroup S whose elements satisfy the identity $X_n = Y_n$ (but not $X_{n-1} = Y_{n-1}$) is called *nilpotent of class* n.

The reason for this definition is the following theorem of Mal'tsev: A group G is nilpotent of class n if and only if G satisfies the identity $X_n = Y_n$ (but not $X_{n-1} = Y_{n-1}$). He has also shown that every nilpotent cancellation semigroup of class n can be embedded in a nilpotent group of class n.

Let S be a semigroup and S_0 a subsemigroup. We shall say that S_0 is of finite index in S if there is a finite subset K of S such that for each $s \in S$ there exists $k \in K$ such that $sk \in S_0$. In [Grigorchuk 1988] the following statement is proved.

Theorem 3. *A finitely generated cancellation semigroup has polynomial growth if and only if it contains a nilpotent subsemigroup of finite index.* □

The essence of the proof consists in estimating the growth of the group of fractions $G = S^{-1}S$ in terms of the growth of S. If S has polynomial growth, then G also has polynomial growth, and Gromov's theorem can be used. What is of interest is a construction of a cancellation semigroup of subexponential growth such that its group of fractions $G = S^{-1}S$ has a higher growth rate than S. In this way one may expect to find new examples of groups of intermediate growth and thus find a negative solution of Problem No. 12 in [Wagon 1985].

§ 4. Example of a Group of Intermediate Growth and the Construction Scheme of such a Group

Milnor's problem was solved negatively by Grigorchuk [Grigorchuk 1983] (cf. also [Grigorchuk 1984a, b, 1985a]), where not only a continuous series of groups of intermediate growth between polynomial and exponential was constructed, but the following result was also proved:

Theorem 4. *There exists a continuum of finitely generated groups of polynomial growth. There exist groups of incommensurable growth rates.* □

Despite the negative character of the answer to Milnor's question, the examples of groups of intermediate growth enrich the theory of groups. For certain applications an acquaintance with groups of intermediate growth is needed, cf. [Grigorchuk 1984a, 1985b, c, 1989a, b].

We shall now construct one concrete example of a group of intermediate growth. This group G is taken from [Grigorchuk 1980a], where it is constructed as example of an infinite finitely generated group, all of whose elements have finite order. Moreover, G is a 2-group, i.e. for each element g there exists an integer k such that $g^{2^k} = 1$.

The group G is constructed as a group of transformations of the interval $[0, 1]$ with all dyadic rational points omitted. Its generators are a, b, c, d whose action on $[0, 1]$ is according to the rules

$$a : \frac{\quad P \quad}{0 \qquad 1} \qquad c : \frac{P \;\; I \;\; PP \ldots}{0 \;\; \frac{1}{2} \;\; \frac{3}{4} \;\; \ldots 1}$$

$$b : \frac{P \;\; P \;\; I \ldots}{0 \;\; \frac{1}{2} \;\; \frac{3}{4} \ldots 1} \qquad d : \frac{I \;\; P \;\; PI \ldots}{0 \;\; \frac{1}{2} \;\; \frac{3}{4} \;\; \ldots 1}$$

(the letter P above an interval Δ means interchanging the two halves of Δ, the letter I denotes the identity permutation). We note that on the 2nd, 3rd and 4th copy of $[0, 1]$ the following infinite periodic sequences of the alphabet $\{P, I\}$ are written:

$$\begin{matrix} P & P & I & P & P & I & \ldots \\ P & I & P & P & I & P & \ldots \\ I & P & P & I & P & P & \ldots \end{matrix}$$

The generators a, b, c, d satisfy the relations

$$a^2 = b^2 = c^2 = d^2 = 1 \,, \quad bc = cb = d \,, \quad bd = db = c \,, \quad cd = dc = b \,,$$

which follow immediately from the definitions of the corresponding transformations. However, this is not a complete system of defining relations for G and moreover, G cannot be described by a finite number of defining relations. An economic system of defining relations for G was found by Lysenok [Lysenok 1985].

Theorem 5 [Grigorchuk 1984a]. *The growth function of G satisfies the estimate*

$$e^{\sqrt{r}} \preceq \gamma(n) \preceq e^{r^\alpha} ,$$

where $\alpha = \log_{32} 31 < 1$. □

We shall now explain a scheme, according to which all examples of groups of intermediate growth known at present can be practically arranged.

Suppose that G is a group containing a normal subgroup H of finite index which admits an embedding in a direct product of $m \geq 2$ copies of G:

$$H \overset{\psi}{\cong} \tilde{H} \leq \underbrace{G \times G \times \ldots \times G}_{m} \tag{2}$$

(ψ is an isomorphism). Suppose further that G has a finite generating set A, for which the length function $\partial(g)$ satisfies the following condition: if $g \in H$,

$$\psi(g) = (g_1, g_2, \ldots, g_m) \in G \times G \times \ldots \times G ,$$

then

$$\sum_{i=1}^{n} \partial(g_i) \leq \lambda(\partial g) , \tag{3}$$

where λ is a fixed number less than one.

Theorem 6. *The growth function of a group G satisfying* (2) *and* (3) *has the upper bound*

$$\gamma_G(r) \preceq e^{r^\alpha} ,$$

where $\alpha < 1$ is some constant (depending only on m and λ). □

The existence of this upper bound is the deciding moment for the proof that certain groups have intermediate growth. Indeed, to show that the growth of G is higher than polynomial, it is sufficient, by Gromov's theorem, to show that G is not virtually nilpotent and this usually presents no difficulty. For example, such properties of groups as being periodic, infinite, finitely generated exclude the possibility of G having a nilpotent subgroup of finite index.

For all examples of groups of intermediate growth that are known at present, we have the lower bound $e^{\sqrt{r}}$ for the growth rate. The reason for this universal property of the function $e^{\sqrt{r}}$ is revealed in [Grigorchuk 1989a], where a link is found between this question and the question of the growth of the graded algebra associated with the group. We shall have more to say of this link below, but for now we indicate one method of obtaining a lower bound which applies to all groups in [Grigorchuk 1984a, b], and in particular, to our example here.

Suppose that the group G, as above, satisfies the condition (2), but in addition require that $\tilde{H}$ is of finite index in $G \times \ldots \times G$.

Theorem 7. *Under the above conditions there exists a positive constant* β *such that* $\gamma_G(r) \succeq e^{r^\beta}$. □

The proof of this result is not hard. Indeed, using the fact that the growth of a group and its subgroup of finite index coincide, and further, that

$$\gamma_{G^{(m)}}(r) \sim \gamma_G^m(r) ,$$

where $G^{(m)}$ is the direct product of m copies of G, we find that the above condition on G yields the relation

$$\gamma_G(r) \sim \gamma_G^m(r) , \tag{4}$$

where $m \geq 2$. Condition (4) represents the so-called Δ^2-condition, which is widely used in the theory of Orlicz-spaces and from which the lower bound $\gamma_G(r) \succeq e^{r^\beta}$ for some β easily follows [Krasnosel'skij–Rutitskij 1958].

It is not known whether (2) implies the residual finiteness of G; however, all known examples of groups of intermediate growth belong to this class of groups. Moreover, the groups constructed by [Grigorchuk 1984a, b] are residually-p groups, where p is a prime number. In this connexion the following questions arise:

Question 1. *Is it true that every group of intermediate growth is residually finite?*

Question 2. *Is it true that every infinite, finitely generated simple group has exponential growth?*

§ 5. On the Structure of the Set of Growth Degrees of Groups that are Residually-p Groups

In this section we shall be concerned with questions of the structure of growth functions of groups that are residually-p groups, where p is a prime number. For this class of groups the information on the structure of the set of growth functions is very complete. As shown in [Grigorchuk 1984a, 1985c], we have the following assertion.

Theorem 8. *For any prime number p the partially ordered set of growth functions of groups that are residually-p groups contains a chain of the power of the continuum and an antichain of the power of the continuum.* □

Here as usual we understand by a chain a lineary ordered set and by an antichain a set whose elements are pairwise incomparable.

Theorem 9. *For any function* $f(r)$ *growing more slowly than* e^r *but faster than any polynomial function there exists a finitely generated group which is residually finite and such that* $\gamma_G(r)$ *and* $f(r)$ *are incomparable in the sense of the preorder* $\preceq$. □

Theorem 10. *If a finitely generated group G is residually-p and not virtually nilpotent, then* $\gamma_G(r) \succeq e^{\sqrt{r}}$. □

Thus $[e^{\sqrt{r}}]$ is a universal lower bound for the growth rates of groups that are residually-p but not virtually nilpotent.

Theorem 11. *For any prime p there is a finitely generated p-group G and a constant $\alpha < 1$ such that*

$$\gamma_G(r) \preceq e^{r^\alpha} . \quad \square$$

In this connexion we can pose the following open questions:

Question 3. *Does there exist a finitely generated group G, residually finite and such that*

$$\gamma_G(r) \sim e^{\sqrt{r}} \text{ ?} \quad \square$$

Question 4. *If the response to Question 3 is negative, how closely can one approach the function $e^{\sqrt{r}}$ in the scale of growth functions of groups?* □

A proof of Theorem 10 is contained in [Grigorchuk 1989a] and the connexions between the growth of groups and the growth of certain graded algebras are described there. Let G be a group, p a prime number, $F_p[G]$ the group algebra over the Galois field F_p, $\Delta \leq F_p[G]$ its *augmentation ideal*, i.e. the ideal generated by all elements of the form $g - 1$, $g \in G$.

By the *lower central series* (Zassenhaus series) of the group G we understand the sequence of characteristic subgroups

$$G_n, \; n = 1, 2, \ldots, \quad \text{where} \quad G_n = \{g \in G : g - 1 \in \Delta^n\} .$$

The same series may be defined by the recurrence relation

$$G_n = [G, G_{n-1}] \, G^p_{\langle \frac{n}{p} \rangle} , \quad G_1 = G ,$$

where $\langle \frac{n}{p} \rangle$ is the least integer $\geq n/p$, and G_i^p is the group generated by the pth powers of the elements of G_i.

From the descending sequence of ideals Δ^n, $n = 0, 1, \ldots$ we can construct the associated graded algebra

$$A_G = \bigoplus_{n=0}^{\infty} A_n = \bigoplus_{n=0}^{\infty} (\Delta^n / \Delta^{n+1}) ,$$

in which multiplication is defined in standard fashion by means of representatives of the residue classes. The power series

$$f_G(t) = \sum_{n=0}^{\infty} a_n(G) t^n ,$$

where $a_n(G) = \dim_{F_p}(A_n)$, is called the *Hilbert–Poincaré* series of A_G. If $d = a_1(G) < \infty$, then this series defines a function $f_G(t)$ which is analytic in a neighbourhood of zero of radius $r > 1/d$.

The study of the asymptotic growth of the coefficients $a_n(G)$ as $n \to \infty$ is of interest from the point of view of the general theory of Hilbert–Poincaré series of graded algebras and also in connexion with the following assertion.

Proposition 1. *For any generating set $A = \{a_1, \ldots, a_m\}$ of a group G we have the estimate*

$$a_r(G) \leq \gamma_G(r) ,$$

where $\gamma_G(r) = \gamma_{G,A}(r)$ is the growth function of G relative to the generating set A. □

The results of Lazard on analytic pro-p-groups entail the following assertion

Theorem 12. *Let G be a group which is residually-p, such that $a_1(G) < \infty$. Then the following conditions are equivalent:*

(1) *The coefficients $a_n(G)$ of the Hilbert–Poincaré series of A_G grow polynomially;*

(2) *the pro-p-completion of G is an analytic pro-p-group;*

(3) *the series $f_G(t)$ represents a rational function of the form*

$$\prod_{i=1}^{l}(1-t^{m_i})^{-r_i} \prod_{j=1}^{s} \frac{1-t^{pc_j}}{1-t^{c_j}} ,$$

where l, s, m_i, r_i, c_j are natural numbers.

If none of the equivalent conditions (1)–(3) hold, then $a_n \succeq e^{\sqrt{n}}$. □

From this theorem and Proposition 1, Theorem 10 immediately follows because the estimate $\gamma(r) \prec e^{\sqrt{r}}$ entails the estimate $a_r \prec e^{\sqrt{r}}$, from which it follows that G is virtually soluble, and hence virtually nilpotent, thanks to the Milnor–Wolf theorem.

The sequence $a_n(G)$ is connected with other important characteristics of G. With the help of the lower p-central series defined by (5) we can construct a graded Lie p-algebra (restricted Lie algebra)

$$\mathcal{L}_G = \bigoplus_{n=1}^{\infty} L_n = \bigoplus_{n=1}^{\infty} G_n/G_{n+1} ,$$

in which the bracket operation is induced by commutation in G and the unary operation $a \mapsto a^{[p]}$ is induced by the pth power operation on the elements of G. By Quillen's theorem [Zalesskij–Mikhalev 1973] the algebra A_G is isomorphic to the universal p-envelope of $\mathcal{L}_G$. Let us put $b_n(G) = \dim_{F_p} L_n$. By Quillen's theorem we have

$$\sum_{n=0}^{\infty} a_n(G)t^n = \prod_{n=1}^{\infty} \left(\frac{1-t^{p^n}}{1-t^n}\right)^{b_n(G)} ,$$

showing the relation between the sequences $\{a_n(G)\}$ and $\{b_n(G)\}$.

The growth of the sequence $b_n(G)$ coincides with that of the sequence of indices $[G : G_n]$ of the terms of the p-central series. In some cases this allows us to find $b_n(G)$ asymptotically and hence $a_n(G)$. In this way it is shown in [Grigorchuk 1989a] that for the group Γ of intermediate growth constructed above (cf. Sect. 3.4) we have $a_n(G) \sim e^{\sqrt{n}}$. This shows that among the infinite finitely generated periodic groups there exist groups with minimum possible growth $\sim e^{\sqrt{n}}$ of the coefficients $a_n(G)$. On the other hand, the p-groups of Golod (1964) constructed by the theorem of Golod and Shafarevich (1964) have maximum possible, viz. exponential growth of the $a_n(G)$ [Grigorchuk 1989a].

§ 6. On an Application of the Theory of Groups of Polynomial Growth to Geometry

Let L be a non-compact orientable surface on which a group G of homeomorphisms acts properly discontinuously, and moreover such that the factor space L/G is compact. Such a surface is given, for example, by a regular covering $L \to S$ of a closed orientable surface S of genus $g \geq 1$ constructed from a normal subgroup $H \trianglelefteq \pi_1(S)$. A surface of the form just described will be called a Γ-surface. The first natural question arising in the study of Γ-surfaces is the question of its topological type.

We recall that a topological description of non-compact surfaces (of which there are uncountably many non-homeomorphic ones) is given by v. Kerékjártó [Massey 1967]. A complete system of inveriants distinguishing non-homeomorphic surfaces is given by the triple $(\beta(L), \beta'(L), \beta''(L))$, where $\beta(L)$ is the ideal boundary (space of ends of L), $\beta'(L)$ the subset of non-orientable ends and $\beta''(L)$ the subset of non-planar ends, as well as the genus and orientation-type of the surface. The topological classification of Γ-surfaces gives

Theorem 13. *Suppose that a non-compact orientable surface L admits a properly discontinuous action by a group Γ of homeomorphisms such that the factor space L/Γ is compact. Then L is of one of the following six types:*

T_1 *the punctured sphere;*
T_2 *the twice punctured sphere;*
T_3 *the sphere from which a perfect Cantor set has been removed;*
T_4 *an orientable surface with one non-planar end;*
T_5 *an orientable surface with two non-planar ends;*
T_6 *an orientable surface with a perfect Cantor set of non-planar ends.* □

Figure 5 illustrates some geometric realizations of T_3–T_6.

For the proof of this theorem with the help of v. Kerékjártó's theorem we can use the theorem of Stallings [Stallings 1966] giving a description of a finitely generated group with more than one end.

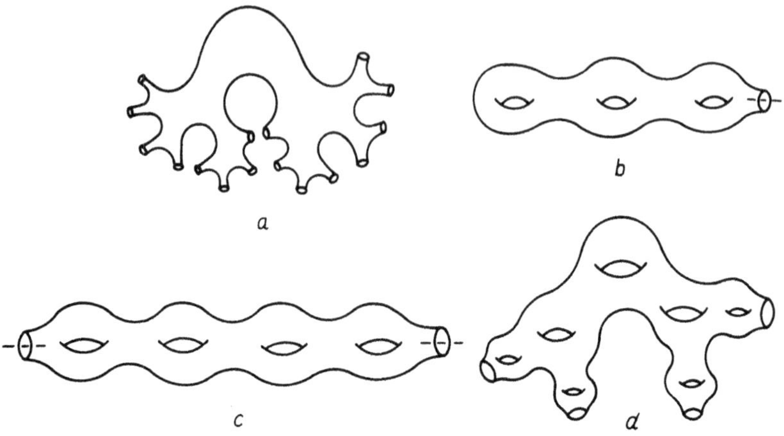

Fig. 5

In the theory of Riemann surfaces or in the theory of Riemannian 2-manifolds, equimorphisms and quasi-conformal mappings play an important role; so the question of classification up to equimorphism or quasi-conformal equivalence will also be considered.

An *equimorphism* of two metric spaces is an invertible transformation which is uniformly continuous in both directions. The equimorphism classification problem was considered for infinitesimal spaces by Efremovich [Efremovich 1953].

Let L be a Γ-surface and G a properly discontinuous group of homeomorphisms of L such that L/G is compact. Suppose that L is equipped with a Riemannian G-invariant metric. Such a 2-manifold will be called a Γ-surface. We denote by $\kappa(i)$, $i = 1, \dots, 6$ the number of equimorphism types of a hyperbolic 2-dimensional Γ-surface and of topological type T_i, $i = 1, \dots, 6$.

Theorem 14 [Grigorchuk 1989b]. *We have the relations*

$$
\begin{aligned}
&(1) \quad \kappa(1) = 2 \quad (\beta(1) = 1)\,; \\
&(2) \quad \kappa(2) = 1 \quad (\beta(2) = 2)\,; \\
&(3) \quad \kappa(3) = 4 \quad (\beta(3) = 2^{\omega})\,; \\
&(4) \quad \kappa(4) = 2^{\omega} \quad (\beta(4) = 1)\,; \\
&(5) \quad \kappa(5) = 1 \quad (\beta(5) = 2)\,; \\
&(6) \quad \kappa(6) = 2^{\omega} \quad (\beta(6) = 2^{\omega})
\end{aligned}
$$

(in the brackets the number $\beta(i)$ of ends corresponding to a surface of topological type T_i is given for comparison). □

For a proof of this theorem one uses essentially the volume invariant of the covering and a certain modification, as well as the result on the growth of groups formulated above. For example, the relation (4) may be proved as follows. Let S be a closed orientable surface of genus 2, equipped with a metric of constant negative curvature. Its fundamental group

$$\pi_1(S) = \langle a_1, b_1, a_2, b_2 | [a_1, b_1][a_2, b_2] = 1 \rangle$$

admits a homomorphism onto the free group F_2 of rank 2, and hence to any group with 2 generators. In [Grigorchuk 1985a] a continuum of periodic 2-generator groups G_ω, $\omega \in \Omega$, is constructed having distinct growth rates. For each point ω we construct an epimorphism

$$\lambda_\omega : \pi_1(S) \to G_\omega ,$$

and denote its kernel by K_ω. Let $S_\omega \to S$ be the regular covering of the Riemannian 2-manifold S constructed from the subgroup $K_\omega \trianglelefteq \pi_1(S)$. Then the growth of S_ω, according to Theorem 1, is equal to that of the group of covering transformations $\pi_1(S)/K_\omega \cong G_\omega$. This provides us with a continuum of Riemannian Γ-manifolds S_ω of different growth rates. In virtue of its periodicity the group G_ω has one end, so the surface S_ω, $\omega \in \Omega$, belongs to the topological type T_4.

§ 7. Regularly Filtered Surfaces and Amenable Groups

Let L be a 2-dimensional Riemannian manifold; L is said to be *regularly filtered* if on L there is a sequence of polyhedral domains $\{W_i\}_{i=1}^{\infty}$ with the following properties:

$$(1) \quad \bigcup_{i=1}^{\infty} W_i = L ;$$

$$(2) \quad W_i \subset W_{i+1} ;$$

$$(3) \quad \lim_{i \to \infty} \frac{l(\partial W_i)}{S(W_i)} = 0 ,$$

where $l(\partial W_i)$ is the length of the boundary ∂W_i and $S(W_i)$ is the area of W_i.

For example, the Euclidean plane $\mathbb{R}^2$ is regularly filtered. The sequence W_i may be taken to be the squares $\{(x, y) \in \mathbb{R}^2 : |x| + |y| \leq i\}$, $i = 1, 2, \ldots$. On the other hand, the Lobachevskij plane is not regularly filtered.

The concept of a regular filtration was introduced by Ahlfors for Riemann surfaces in his theory of covering surfaces.

Another domain of application of these concepts is the theory of Laplace operators on Riemannian manifolds. It turns out that the hypothesis of a regular filtration implies the presence of the point 0 in the spectrum of such an operator [Brooks 1989].

We now pass to the combinatorial form of the notion of regular filtration, the concept of an amenable group. It was introduced by von Neumann [von

Neumann 1929] in connexion with the Banach–Tarski paradox. The original definition was based on the concept of an invariant mean and in the language of measure theory it runs as follows.

A group G is called *amenable* if there exists on it a finitely additive measure μ, defined on the σ-algebra of all subsets of G and such that

$$\text{(i)} \quad \mu(G) = 1 \,,$$
$$\text{(ii)} \quad \mu(gA) = \mu(A) \quad \text{for all} \quad g \in G, A \subset G \,.$$

The second condition expresses the left invariance of μ.

The class of amenable groups is closed under the operations of taking 1) subgroups, 2) factor groups, 3) direct limits and 4) extensions [v. Neumann 1929]. There follows, for example, the amenability of all virtually soluble and locally finite groups. On the other hand, the free group of rank 2 is not amenable and hence no group containing a 2-generator free subgroup can be amenable.

The original criterion for the amenability of groups was found by Følner [Følner 1955] who showed that amenability is equivalent to the existence of a sequence F_n, $n = 1, 2, \ldots$ of finite sets satisfying the conditions

$$(1') \quad \bigcup_{n=1}^{\infty} F_n = G \,;$$
$$(2') \quad F_n \subset F_{n+1} \,;$$
$$(3') \quad \lim_{n \to \infty} |gF_n \Delta F_n| \,/\, |F_n| = 0 \quad \text{for all} \quad g \in G \,.$$

($E \Delta F$ denotes the symmetric difference of E and F.) Conditions (1′) and (2′) copy (1) and (2) given above, and it is not hard to see that (3′) is also related to (3) because it means that for large n the number of elements on the "boundary" of the set F_n is small compared with the number of elements in this set. A sequence $\{F_n\}_{n=1}^{\infty}$ satisfying (1′), (2′), (3′) is called a *Følner* sequence.

Besides the outward similarity of (1′)–(3′) and (1)–(3) we have direct links between the notions of a completely regular filtration of a Riemannian manifold and the amenability of a group, as is shown in the following assertion:

Theorem 15. *Let M be a compact Riemannian manifold and $\tilde{M} \to M$ a regular covering by a Riemannian manifold corresponding to the normal subgroup $H \trianglelefteq \pi_1(M)$. Then the manifold $\tilde{M}$ has a completely regular filtration if and only if the group of covering transformations $G \cong \pi_1(M)/H$ is amenable.* □

Various attempts have been made to describe the class of amenable groups. We shall call finite abelian groups as well as groups obtained from them by a transfinite application of the operations (1)–(4) *elementary* amenable groups. This class of groups was the only source of examples starting at the time of von Neumann, which led Day in 1957 to raise the question whether the class

EG of elementary groups coincides with the class AG of amenable groups [Day 1957]. A negative answer to this question was given by Grigorchuk [Grigorchuk 1985a], thanks to the negative solution of Milnor's problem. The point is that groups of subexponential growth are amenable (as Følner sequence for it we can take some subsequence of the sequence of spheres

$$B_1(r) = \{r \in G : \partial(g) \leq r\}$$

with centre at the unit element). The fact that the amenable group G_0 in [Grigorchuk 1985a] is not elementary follows, for example, from its periodicity.

A second question raised by Day [Day 1957] implies the following: "Is it true that the class AG coincides with the class NF of groups containing no 2-generator free subgroups?"

Some mathematicians erroneously ascribe this question to von Neumann with references to von Neumann [von Neumann 1929], but in this paper and in his other publications there is no allusion to the question whether $AG = NF$.

For a solution of Day's second problem it is necessary to introduce a new combinatorial criterion for amenability.

Let G be a finitely generated group with generating set $A = \{a_1, \dots, a_m, a_1^{-1}, \dots, a_m^{-1}\}$. Then G is isomorphic to F_m/H, where F_m is the free group on $\bar{a}_1, \dots, \bar{a}_m$ and H is a normal subgroup. Let us denote by H_n the set of elements of length n in H. The quantity

$$\alpha_H = \overline{\lim}_{n\to\infty} \sqrt[n]{|H_n|}$$

introduced in [Grigorchuk 1978] is called the *growth exponent* of H. It has the bounds

$$\sqrt{2m-1} < \alpha_H \leq 2m-1$$

[Grigorchuk 1978, 1980b]. The significance of this notion lies in the following assertion.

Theorem 16. *The group G is amenable if and only if $\alpha_H = 2m - 1$.* □

Thus the amenability of G is connected with the size of the kernel H in the representation of G as a factor group F_m/H. This criterion holds also for homogeneous spaces of finitely generated groups, and this circumstance is exploited in [Grigorchuk 1979] to disprove a natural extension of the conjecture $AG = NF$.

After the papers [Kesten 1959] and [Novikov–Adyan 1968] it seemed likely that counter-examples to the conjecture $AG = NF$ should be sought in the class of periodic groups. Moreover, in 1976 Adyan conjectured that the free periodic Burnside group $B(m,n)$ for $m \geq 2$ and odd $n \geq 665$ is not amenable. This conjecture was proved by him [Adyan 1982], where the amenability criterion stated in Theorem 16 was used.

The combinatorial amenability criterion can usually be successfully applied only in situations when the set of defining relations of the group satisfies a small cancellation condition. This leitmotif runs through [Grigorchuk 1980b], which extends a variant taken from [Grigorchuk 1979]. One of the results from [Grigorchuk 1980b] may in rough terms be stated thus: if the defining relations of G have small overlap, the number of relators is not too large, while their length is large enough, then $\alpha_H < \sqrt{2m-1} + \varepsilon$, where $\varepsilon > 0$ is a small constant. In [Adyan 1982] a much stronger result in this direction is obtained. He considers a class of groups for which the word problem can be solved by Dehn's algorithm (finitely presented groups with this property are usually called hyperbolic) and he introduces two numerical invariants: the convergence coefficient λ of the Dehn algorithm and the quantity:

$$\beta_r = \overline{\lim} \sqrt[n]{R_n} \,,$$

(R_n is the number of defining relations of length n), characterizing the growth of the set of defining relations, and giving a bound for the growth exponent α_H in terms of these two quantities. This allows him, using results from [Novikov–Adyan 1968], and [Adyan 1975], as well as the amenability criterion, to prove the non-amenability of $B(m,n)$ for $m \geq 2$ and odd $n > 665$.

The first counter-example to the conjecture $AG = NF$ was constructed by Ol'shanskij [Ol'shanskij 1980] using the criterion of Theorem 16. The starting point for this work was [Ol'shanskij 1979], in which an example was given of a finitely generated non-cyclic infinite group, all of whose proper subgroups were cyclic, thus solving a well known problem of the theory of groups. Grigorchuk assumed that for fast growth of the parameters involved in the construction, one has the bound $\alpha_H < 2m-1$ and so obtained an example of a non-amenable group without free subgroups. He proposed to Ol'shanskij to try to apply this construction for forming a corresponding example; this was successfully realized in [Ol'shanskij 1980].

One example of a non-amenable group without free subgroups was recently constructed by Gromov [Gromov 1987], where the non-amenability is controlled by means of the so-called T-property of Kazhdan. So far the question on the existence of finitely presented non-amenable groups without free subgroups remains open.

Bibliography*

Adyan, S.I. (1955): Algorithmic unsolvability of problems of recognition of certain properties of groups. Dokl. Akad. Nauk SSSR *103*, 533–535 (Russian). Zbl. 65, 9

Adyan, S.I. (1975): The Burnside Problem and Identities in Groups. Moscow, Nauka. Zbl. 306.20045. English transl.: Berlin, Springer 1979

*For convenience of the reader, references to reviews in Zentralblatt für Mathematik (Zbl.), compiled using the MATH database, and Jahrbuch über die Fortschritte der Mathematik (Jbuch) have, as far as possible, been included in this bibliography.

Adyan, S.I. (1982): Random walks on free periodic groups. Izv. Akad. Nauk SSSR, Ser. Mat. *46*, No. 6, 1139–1149. Zbl. 512.60012. English transl.: Math. USSR, Izv. *21*, 425–434

Andrews, J.J., Curtis, M.L. (1965): Free groups and handlebodies. Proc. Am. Math. Soc. *16*, 192–195. Zbl. 131, 383

Bass, H. (1972): The degree of polynomial growth of finitely generated nilpotent groups. Proc. Lond. Math. Soc., III. Ser. *25*, 603–614. Zbl. 259.20045

Birman, J.S. (1974a): Braids, links and mapping class groups. Ann. Math. Stud. No. *82*. Zbl. 305.57013

Birman, J.S. (1974b): Poincaré's conjecture and the homeotopy group of a closed orientable 2-manifold. J. Aust. Math. Soc. *17*, No. 2, 214–221. Zbl. 282.55003

Birman, J.S., Hilden, H.M. (1973): The homeomorphism problem for S^3. Bull. Am. Math. Soc. *79*, No. 5, 1006–1010. Zbl. 272.57001

Brooks, R. (1986): Combinatorial Problems in Spectral Geometry. Lect. Notes Math. *1201*, 14–32. Zbl. 616.53037

Collins, D.J., Zieschang, H. (1990): Combinatorial group theory and fundamental groups. In: Encycl. Math. Sci. *58*. Berlin, Springer (Part I of this vol.)

Comerford, I.P., Edmunds, C.C. (1989): Solutions of equations in free groups. Proc. Conf. group theory, Singapore 1987, 347–355. Zbl. 663.20023

Day, M.M. (1957): Amenable semigroups. Ill. J. Math. *1*, 509–544. Zbl. 78, 294

Dehn, M. (1911): Über unendliche diskontinuierliche Gruppen. Math. Ann. *71*, 116–144

Dehn, M. (1938): Die Gruppen der Abbildungsklassen. Acta Math. *69*, 135–206. Zbl. 19, 253

v.d. Dries, L., Wilkie, A.J. (1984): Gromov's theorem on groups of polynomial growth and elementary logic. J. Algebra *89*, No. 2, 349–374. Zbl. 552.20017

Efremovich, V.A. (1953): Geometry of proximate Riemannian manifolds. Usp. Mat. Nauk *8*, No. 5, 189–191 (Russian). Zbl. 53, 115

Følner, E. (1955): On groups with full Banach mean value. Math. Scand. *3*, 243–254. Zbl. 67, 12

Fomenko, A.T. (1983): Differential Geometry and Topology. Supplementary Chapters. Moscow, Izd. MGU, 216 pp. Zbl. 517.53001. English transl.: New York, Plenum 1987

Fomenko, A.T. (1984): Topological variational problems. Moscow, Izd. MGU. Zbl. 679.49001. English transl.: New York, Gordon and Breach 1990

Golod, E.S. (1964): On nil algebras and residually-p groups. Izv. Akad. Nauk SSSR, Ser. Mat. *28*, 273–276. Zbl. 215, 392. English transl.: Transl., II. Ser., Am. Math. Soc. *48*, 103–106 (1965)

Golod, E.S., Shafarevich, I.R. (1964): On class field towers. Izv. Akad. Nauk SSSR, Ser. Mat. *28*, 261–272. Zbl. 136, 26. English transl.: Transl., II. Ser., Am. Math. Soc. *48*, 91–102 (1965)

Griffiths, H.B. (1964): Automorphisms of a 3-dimensional handlebody. Abh. Math. Semin. Univ. Hamb. *26*, 191–210. Zbl. 229.57005

Grigorchuk, R.I. (1978): Symmetric random walks on discrete groups. In: Multicomponent random systems. Ed. R.L. Dobrushin and Ya.G. Sinai. Moscow, Nauka, 132–152. English transl.: Adv. Probab. Rel. Top., Vol. 6, M. Dekker 1980, 285–325. Zbl. 475.60007

Grigorchuk, R.I. (1979): Invariant measures on homogeneous spaces. Ukr. Math. Zh. *31*, No. 5, 490–497. Zbl. 434.28009. English transl.: Ukr. Math. J. *31*, 388–393

Grigorchuk, R.I. (1980a): On the Burnside problem for periodic groups. Funkts. Anal. Prilozh. *14*, No. 1, 53–54. English transl.: Funct. Anal. Appl. *14*, 41–43. Zbl. 595.20029

Grigorchuk, R.I. (1980b): Symmetric random walks on discrete groups. In: Adv. Probab. Rel. Top. *6*, 285–385, Marcel Dekker. Zbl. 475.60007

Grigorchuk, R.I. (1983): Milnor's problem on the growth of groups. Dokl. Akad. Nauk SSSR *271*, No. 1, 30–33. English transl.: Sov. Math., Dokl. *28*, 23–26. Zbl. 547.20025

Grigorchuk, R.I. (1984a): The growth degrees of finitely generated groups and the theory of invariant means. Izv. Akad. Nauk SSSR, Ser. Mat. *48*, No. 5, 939–985. English transl.: Math. USSR, Izv. *25*, 259–300. Zbl. 583.20023

Grigorchuk, R.I. (1984b): Construction of p-groups of intermediate growth that have a continuum of factor groups. Algebra Logika *23*, No. 4, 383–394. English transl.: Algebra Logic *23*, 265–273. Zbl. 573.20037

Grigorchuk, R.I. (1985a): Growth degrees of p-groups and torsion-free groups. Mat. Sb., Nov. Ser. *126*, No. 2, 194–214. Zbl. 568.20033.English transl.: Math. USSR, Sb. *54*, 185–205

Grigorchuk, R.I. (1985b): A relationship between algorithmic problems and entropy characteristics of groups. Dokl. Akad. Nauk SSSR *284*, No. 1, 24–29. English transl.: Sov. Math., Dokl. *32*, 355–360. Zbl. 596.20022

Grigorchuk, R.I. (1985c): Groups with intermediate growth function and their applications. Summary of Doctoral thesis, Moscow (Russian)

Grigorchuk, R.I. (1988): Cancellative semigroups of polynomial growth. Mat. Zametki *43*, No. 3, 305–319. Zbl. 643.20036. English transl.: Math. Notes *43*, No. 3, 175–183

Grigorchuk, R.I. (1989a): On the Hilbert–Poincaré series of graded algebras associated with groups. Mat. Sb., Nov. Ser. *180*, No. 2, 207–225. English transl.: Math. USSR, Sb. *66*, No. 1, 211–229. Zbl. 695.16009

Grigorchuk, R.I. (1989b): Topological and metric types of surfaces that regularly cover a closed surface. Izv. Akad. Nauk SSSR, Ser. Mat. *53*, No. 3, 498–536. Zbl. 686.57001. English transl.: Math. USSR, Izv. *34*, No. 3, 517–553

Grigorchuk, R.I., Kurchanov, P.F. (1989a): Description of the solution set of strictly quadratic coefficient-free equations in free groups. 11th All-Union Symp. on the theory of groups, Sverdlovsk 1979, 38–39 (Russian). Zbl. 691.20002

Grigorchuk, R.I., Kurchanov, P.F. (1989b): On the description of the solution set of quadratic equations in free groups. Internat. Algebra Conf. in memory of A.I. Mal'tsev, Novosibirsk, p. 36 (Russian). Zbl. 697.00008

Grigorchuk, R.I., Kurchanov, P.F. (1989c): On the complexity of the description of the solution set of quadratic equations in free groups. Sov. School on Found. of Math. and Function theory (Suslin Lecture). Saratov, p. 73 (Russian). Zbl. 687.00001

Grigorchuk, R.I., Kurchanov, P.F. (1989d): On quadratic equations in free groups. Int. Conf. in memory of A.I. Mal'tsev, Novosibirsk. p. 36 (Russian). Zbl. 725.20001

Grigorchuk, R.I., Kurchanov, P.F., Zieschang, H. (1989): Equivalence of homomorphisms of surface groups to free groups and some properties of 3-dimensional handle-bodies. Int. Conf. in memory of A.I. Mal'tsev, Novosibirsk, 1989 (Russian). Zbl. 697.00008

Gromov, M. (1981): Groups of polynomial growth and expanding maps. Publ. Math., Inst. Hautes Etud. Sci. *53*, 53–73. Zbl. 474.20018

Gromov, M. (1987): Hyperbolic groups. In: Essays in Group Theory. S.M. Gersten (ed.), Berlin-Heidelberg-New York, Springer, 75–263. Zbl. 634.20015

Guba, V.S. (1986): Equivalence of infinite systems of equations in free groups and semigroups to finite subsystems. Mat. Zametki *40*, No. 3, 321–324. Zbl. 611.20020. English transl.: Math. Notes *40*, 688–690

Hatcher, A., Thurston, W. (1980): A presentation for the mapping class group of a closed orientable surface. Topology *19*, 221–237. Zbl. 447.57005

Hempel, J. (1976): 3-manifolds. Ann. Math. Stud. *86*. Zbl. 345.57001

Hog–Angeloni, C., Metzler, W. (1991): Andrews–Curtis–Operationen und höhere Kommutatoren der Relatoren-Gruppe. J. Pure Appl. Algebra *75*, 37–45
Jaco, W. (1969): Heegaard splittings and splitting homomorphisms. Trans. Am. Math. Soc. *144*, 365–379. Zbl. 199, 586
Jiang, B. (1983): Lectures on Nielsen fixed point theory. Contemp. Math. *14*, Zbl. 512.55003
Jiang, B. (1984): Fixed points and braids. Invent. Math. *75*, 69–74. Zbl. 565.55005
Jiang, B. (1987): Surface maps and braid equations I. Peking Univ. 1987 Preprint. Appeared in: Lect. Notes Math. *1369*, 125–141 (1989). Zbl. 673.55003
Johnson, D. (1983): The structure of the Torelli group I. A finite set of generators for I. Ann. Math., II. Ser. *118*, No. 3, 423–442. Zbl. 549.57006
Kesten, H. (1959): Symmetric random walks on groups. Trans. Am. Math. Soc. *92*, No. 2, 336–354. Zbl. 92, 335
Khmelevskij, Yu.I. (1971a): Equations in free semigroups. Tr. Mat. Inst. Steklova *107*. Zbl. 224.02037. English transl.: Proc. Steklov Inst. Math. *107*, 1–270 (1976)
Khmelevskij, Yu.I. (1971b): Systems of equations in free groups I, II. Izv. Akad. Nauk SSSR, Ser. Mat. *35*, No. 6, 1237–1268; *36*, No. 1, 110–179. English transl.: Math. USSR, Izv. *5*, 1245–1276; *6*, 109–180. Zbl. 299.20019; Zbl. 299.20020
Krasnosel'skij, M.A., Rutitskij, Ya.B. (1958): Convex Functions and Orlicz Spaces. Moscow, Gos. Izd. Fiz.-Mat. Lit. Zbl. 84, 101. English transl.: Noordhoff, Groningen (1961)
Kurchanov, P.F., Grigorchuk, R.I. (1989): Classification of epimorphisms from the fundamental groups of a surface to free groups. Int. Conf. in memory of A.I. Mal'tsev, Novosibirsk 1989, p. 72 (Russian). Zbl. 725.20001
Lickorish, W.B.R. (1963): Homeomorphisms of non-orientable 2-manifolds. Proc. Camb. Philos. Soc. *59*, 307–317. Zbl. 115, 408
Lickorish, W.B.R. (1964): A finite set of generators for the homeotopy group of a 2-manifold. Proc. Camb. Philos. Soc. *60*, 769–778. Zbl. 131, 208
Lickorish, W.B.R. (1965): On the homeomorphisms of a non-orientable surface. Proc. Camb. Philos. Soc. *61*, 61–64. Zbl. 131, 208
Lorents, A.A. (1968): Representations of sets of solutions of systems of equations in one unknown in free groups. Dokl. Akad. Nauk SSSR *178*, 290–292. Zbl. 175, 295. English transl.: Sov. Math., Dokl. *9*, 81–84
Lyndon, R.C. (1959): The equation $a^2b^2 = c^2$ in free groups. Mich. Math. J. *6*, 89–95. Zbl. 84, 28
Lyndon, R.C., Schupp, P.E. (1977): Combinatorial Group Theory. Berlin-Heidelberg-New York, Springer. (Ergebn. Math. Grenzgeb. *89*). Zbl. 368.20023
Lysenok, I.G. (1985): A system of defining relations for the Grigorchuk group. Mat. Zametki *38*, No. 4, 503–516. English transl.: Math. Notes *38*, 784–792. Zbl. 595.20030
Lysenok, I.G. (1989a): On certain algorithmic properties of hyperbolic groups. Izv. Akad. Nauk SSSR, Ser. Mat. *53*, No. 4, 814–832. Zbl. 692.20022. English transl.: Math. USSR, Izv. *35*, No. 1, 145–163
Lysenok, I.G. (1989b): Algorithmic problems and quadratic equations in groups. Summary of Cand. thesis, Moscow (Russian)
Magnus, W., Karrass, A., Solitar, D. (1966): Combinatorial Group Theory: Presentations of Groups in Terms of Generators and Relations. New York-London-Sydney, Interscience Publishers, John Wiley and Sons Inc. Zbl. 138, 256
Makanin, G.S. (1977): The problem of solvability of equations in free semigroups. Mat. Sb., Nov. Ser. *103* (145), No. 2, 147–236. Zbl. 371.20047. English transl.: Math. USSR, Sb. *32*, 129–198
Makanin, G.S. (1982): Equations in free groups. Izv. Akad. Nauk SSSR, Ser. Mat. *46*, No. 6, 1199–1273. Zbl. 511.20019. English transl.: Math. USSR, Izv. *21*, 483–546

Makanin, G.S. (1984): Decidability of the universal theory and positive theory of free groups. Izv. Akad. Nauk SSSR, Ser. Mat. *48*, No. 4, 735–749. English transl.: Math. USSR, Izv. *25*, 75–88. Zbl. 578.20001

Mal'tsev, A.I. (1953): Nilpotent semigroups. Uch. Zap. Ivan. Ped. In-ta. *4*, 107–111 (Russian). Zbl. 87, 255

Mal'tsev, A.I. (1962): On the equation $zxyx^{-1}y^{-1}z^{-1} = aba^{-1}b^{-1}$ in free groups. Algebra Logika *1*, No. 5, 45–50 (Russian). Zbl. 144, 14

Massey, W.S. (1967): Algebraic Topology: An Introduction. New York, Harcourt, Brace and World. Zbl. 153, 249

McCool, J. (1975): Some finitely presented subgroups of the automorphism group of a free group. J. Algebra *35*, 205–213. Zbl. 325.20025

McCullough, D., Miller, A. (1986): The genus 2 Torelli group is not finitely generated. Topology Appl. *22*, 43–49. Zbl. 579.57007

Milnor, J.W. (1968a): A note on curvature and fundamental groups. J. Differ. Geom. *2*, 1–7. Zbl. 162, 254

Milnor, J.W. (1968b): Problem 5603. Am. Math. Mon. *75*, No. 6, 685–686

Milnor, J.W. (1968c): Growth in finitely generated solvable groups. J. Differ. Geom. *2*, 447–449. Zbl. 176, 298

Neumann, J. von (1929): Zur allgemeinen Theorie des Maßes. Fundam. Math. *13*, 73–116. Jbuch 55, 151

Nielsen, J. (1927): Om flytningsgruppen i den hyperbolske plan. Mat. Tidsskrift B, 65–75. Jbuch 53, 543

Novikov, P.S., Adyan, S.I. (1968): Defining relations and the word problem for free periodic groups of odd order. Izv. Akad. Nauk SSSR, Ser. Mat. *32*, No. 4, 971–979. Zbl. 194, 33. English transl.: Math. USSR, Izv. *2*, 935–942 (1968)

Ol'shanskij, A.Yu. (1979): An infinite simple torsion-free Noetherian group. Izv. Akad. Nauk SSSR, Ser. Mat. *43*, No. 6, 1328–1393. Zbl. 431.20027. English transl.: Math. USSR, Izv. *15*, 531–588

Ol'shanskij, A.Yu. (1980): The problem of the existence of invariant means on groups. Usp. Mat. Nauk *35*, No. 4, 199–200. Zbl. 452.20032. English transl.: Russ. Math. Surv. *35*, No. 4, 180–181

Ol'shanskij, A.Yu. (1989): Diagrams of homomorphisms of surface groups. Sib. Mat. Zh. *30*, No. 6, 150–171. English transl.: Sib. Math. J. *30*, No. 6, 961–979

Piollet, D. (1986): Solutions d'une équation quadratique dans le groupe libre. Discrete Math. *59*, 115–123. Zbl. 599.20035

Razborov, A.A. (1984): On systems of equations in free groups. Izv. Akad. Nauk SSSR, Ser. Mat. *48*, No. 4, 779–832. English transl.: Math. USSR, Izv. *25*, 115–162. Zbl. 579.20019

Razborov, A.A. (1987): On systems of equations in free groups. Summary of Cand. thesis, Moscow (Russian)

Shvarts, A.S. (1955): A volume invariant of coverings. Dokl. Akad. Nauk SSSR *105*, No. 1, 32–34 (Russian). Zbl. 66, 159

Stallings, J.R. (1966): How not to prove the Poincaré conjecture. Ann. Math. Stud. *60*, 83–88. Zbl. 152, 226

Stoilov, S. (1964): Theory of Functions of a Complex Variable (Russian). Vol. 2, Moscow, Nauka. Romanian original: Bucureşti (1958). Zbl. 102, 291

Suzuki, S. (1977): On homeomorphisms of a 3-dimensional handlebody. Can. J. Math. *29*, No. 1, 111–124. Zbl. 339.57001

Tits, J. (1972): Free subgroups of linear groups. J. Algebra *20*, No. 2, 250–270. Zbl. 236.20032

Trofimov, V.I. (1982): Growth functions of algebraic systems. Summary of Cand. thesis, Sverdlovsk (Russian)

Volodin, I.A., Kuznetsov, V.E., Fomenko, A.T. (1974): On the algorithmic recognition problem of the standard three-dimensional sphere. Usp. Mat. Nauk *29*, No. 5, 71–168. Zbl. 303.57002. English transl.: Russ. Math. Surv. *29*, No. 5, 71–172
Wagon, S. (1985): The Banach–Tarski paradox. Encyclopedia of Mathematics and its Applications, *24*. Zbl. 569.43001
Wajnryb, B. (1983): A simple presentation for the mapping class group of an orientable surface. Isr. J. Math. *45*, No. 2, 3, 157–174. Zbl. 533.57002
Waldhausen, F. (1968): Heegaard Zerlegungen der 3-Sphäre. Topology *7*, 195–203. Zbl. 157, 545
Waldhausen, F. (1978): Some problems on 3-manifolds. Proc. Symp. Pure Math. *32*, Part 2, 313–322. Zbl. 397.57007
Wolf, J: (1968): Growth of finitely generated solvable groups and curvature of Riemannian manifolds. J. Differ. Geom. *2*, 421–446. Zbl. 207, 518
Zalesskij, A.E., Mikhalev, A.V. (1973): Group rings. Itogi Nauk. Tekh., Ser. Sovrem. Probl. Mat. *2*, 5–118. Zbl. 293.16013. English transl.: J. Sov. Math. *4*, 1–78
Zieschang, H: (1964): Alternierende Produkte in freien Gruppen. Abh. Math. Semin. Univ. Hamb. *27*, 13–31. Zbl. 135, 418

Index of Notation

T_g	oriented surface of genus g 170
π_1	fundamental group 170
F_m, F_C	free group of rank m, on generating set C 170, 172
$\langle X \mid R \rangle$	group on the generating set X with defining relations R 174
$\mathrm{Stab}_G X$	stabilizer of X in G 179
G_g	fundamental group of orientable surface of genus g 181
Γ_g	fundamental group of non-orientable surface of genus g 181
$MF[f]$	minimal number of fixed points of maps in $[f]$ 182
$N(f)$	Nielsen number of f 182
$M(\xi)$	manifold corresponding to automorphism ξ 191
$\Phi(\xi)$	splitting homomorphism corresponding to ξ 191
$MC(T)$	group of mapping classes 193
L^-, L^+, L^*	subgroups of $\mathrm{Aut}\,\pi_1(T)$ 195
$\mathrm{Top}\,\pi_1(V)$	group of topological automorphisms of V 197
Φ_φ	splitting homomorphism 200
$MC^-(T_g), MC^+(T_g)$	images of L^-, L^+ in $MC(T_g)$ 201
$Sp_{2g}(\mathbb{Z})$	symplectic group 202
J	standard skew-symmetric matrix 202
$M(0,2m)$	group of mapping classes fixing $2m$ points 203
$B(2m)$	braid group 204
Rev	reverse of a word 204
$GL_n(\mathbb{Z})$	general linear group 207

$V(\Gamma)$	vertex set of graph Γ 208
$B_e(r)$	sphere of radius r, centre e 208
$\gamma_{\Gamma,e}(r)$	number of elements of Γ in $B_e(r)$ 208
$v_{M,e}(r) = \mathrm{Vol}(B_e(r))$	volume of $B_e(r)$ in a manifold M 209
$\Gamma(G, A)$	Cayley graph of a group G generated by a set A 210
$\partial(g)$	length of a word g 210
$\gamma_G(r)$	growth function of a group G 210
$F[G]$	group algebra of G over F 219
Δ	augmentation ideal 219
G_n	lower central series 219
$f_G(t)$	Hilbert-Poincaré series for G 220
$a_n(t)$	coefficient of f_G 220
L_G	restricted Lie algebra associated with G 220
$b_n(G)$	dimension of the completion of L_G 220
$T_n(n = 1, \ldots, 6)$	surface of the Kerékjártó classification 221
$B(m, n)$	Burnside group 225

Author Index

References are to citations of works by the author concerned or to a place in the text where the author's name is mentioned. However, uses of a proper name in a context in which the name has become standard, such as the Poincaré conjecture, are listed in the subject index rather than here.

Adyan, S.I. 98, 149f., 193, 213, 225f.
Ahlfors, L.V. 223
Alexander, J.W. 121
Alperin, R.C. 46
Andrews, J.J. 193, 205
Anisimov, A.V. 151 ff.
Arnol'd, V.I. 127

Baer, R. 40, 83f.
Bass, H. 40, 212f.
Baumslag, G. 58, 60, 126
Bestvina, M. 54
Birman, J.S. 83, 85, 192, 194, 201, 203f.
Boileau, M. 97, 114, 122
Boone, W.W. 148, 150
Borel, A. 134
Borisov, V.V. 146
Brodskij, S.D. 58, 100, 102
Brooks, R. 223
Brown, E.M. 119
Brunner, A.M. 45, 59
Burde, G. 48, 120f., 126
Burnside, W. 98
Bungaard, S. 76

Cannon, J.J. 156
Cartan, H. 129
Cayley, A. 22
Chiswell, I.M. 46, 61
Church, A. 143
Cohen, D.E. 31, 61, 134, 136, 142
Coldewey, H.-D. 4f., 9, 13, 27, 29, 40, 63, 65ff., 71, 75f., 82ff., 85 ff., 89f., 129
Collins, D.J. 45, 55, 59, 61, 97, 150
Comerford, L.P. 169, 178
Coxeter, H.S.M. 4, 29
Crowell, R.H. 119
Culler, M. 60
Curtis, M.L. 193, 205

Day, M.M. 171, 224f.
Dehn, M. 3, 11, 22, 55, 81, 83, 90, 94, 143, 169, 184, 194, 196
Dicks, W. 140
Dries, L. van den 213
Dunwoody, M.J. 139, 141, 156
Dyck, W. von 3, 8
Dyer, J.L. 54

Eckmann, B. 86
Edmunds, C.C. 169, 178
Efremovich, V.A. 170, 211, 222
Eilenberg, S. 129
Epstein, D.B.A. 107, 156

Fine, B. 60
Følner, E. 224
Fomenko, A.T. 191f.
Fox, R.H. 76
Freudenthal, H. 136, 141
Fuks, D.B. 9
Funcke, K. 119f.

Garside, F.A. 86, 126
Gersten, S.M. 53f.
Gerstenhaber, M. 101
Gilman, J. 85
Goldstein, R.Z. 49, 54
Golod, E.S. 221
Greendlinger, M.D. 94
Griffiths, H.B. 190
Grigorchuk, R.I. 169ff., 178, 180, 182, 186, 190, 197, 207, 211, 215ff., 218f., 221ff., 224f.
Gromov, M. 99f., 169f., 184, 213f., 226
Gruenberg, K.W. 133, 135
Grushko, I.A. 44
Guba, V.S. 174
Gurevich, G.A. 57

Haken, W. 150
Hall, M. Jr. 98, 131
Handel, M. 54, 85
Harer, J.L. 134
Harrison, N. 46
Hatcher, A. 85, 193f.
Hempel, J. 111, 187, 201
Higgins, P.J. 49, 51
Higman, G. 38, 100, 148
Hilden, H.M. 192
Hilton, P.J. 133, 135
Hoare, A.H.M. 75
Hog-Angeloni, C. 208
Holt, D.F. 141, 156
Hopcroft, J.E. 151, 154
Hopf, H. 107, 136, 139
Houghton, C.H. 142
Howie, J. 60, 102f.
Huebschmann, J. 61, 135
Hurwitz, A. 77f.

Imrich, W. 46

Jaco, W. 54, 188, 200
Jiang, B. 182ff.
Johnson, D. 203

Kampen, E.R. van 20, 91
Kaplansky, I. 141
Karrass, A. 4, 40, 57f., 140, 195, 197
Kazhdan, D.A. 226
Kerckhoff, S.P. 86, 90
Kerékjártó, B. von 221
Kesten, H. 225
Khmelevskij, Yu.I. 172f.
Klein, F. 3
Kneser, H.A. 141

Krasnosel'skij, M.A. 218
Kurchanov, P.F. 169, 178, 180, 182, 190, 197, 207
Kurosh, A.G. 3, 12, 40, 131
Kuznetsov, V.E. 191

Lazard, M. 220
Lehner, J. 88
Levi, F.W. 40
Lewin, J. 100
Lewin, T. 58, 100, 126
Lickorish, W.B.R. 194
Linnell, P.A. 86, 141
Lorents, A.A. 172
Lyndon, R.C. 4, 34f., 45, 49, 51, 55, 60f., 91, 104, 169, 178f., 184, 193, 195
Lysenok, I.G. 169, 185f., 207, 216

Macbeath, A.M. 75ff.
Maclachlan, C. 85
Magnus, W. 4, 9, 48, 55, 58, 61, 85, 102, 195, 197
Makanin, G.S. 86, 126, 169, 172ff.
Mal'tsev, A.I. 169, 178, 215
Markov, A.A. 143, 148, 150
Massey, W.S. 4, 13, 135, 171, 221
McCool, J. 52, 59, 177, 179, 193
McCullough, D. 203
Metzler, W. 208
Mikhalev, A.V. 220
Miller, A. 203
Miller, R.T. 85
Milnor, J.W. 170, 211ff., 216
Moïse, E.E. 106
Moldavanskij, D.I. 55, 57
Morgan, J.W. 60
Moser, W.O.J. 4
Moss, K.N. 46
Muller, D.E. 156
Müller, H. 86
Murasugi, K. 57

Neumann, B.H. 38
Neumann, H. 38, 40
Neumann, J. von 171, 223ff.
Newman, B.B. 57
Nielsen, J. 3, 32, 45, 47f., 51f., 76, 80f., 85f., 190, 193
Novikov, P.S. 98, 148, 225f.
Novikov, S.P. 13, 61, 105, 109, 111, 114, 139

Ol'shanskij, A.Yu. 98, 169, 185, 226
Orlik, P. 113

Papakyriakopoulos, C.D. 102, 107
Paterson, M.S. 156
Peczynski, N. 45, 74
Pietrowski, A. 59, 140
Piollet, D. 179f.
Poenaru, V. 150
Poincaré, H. 3, 108, 191
Post, E.L. 148
Pride, S.J. 59

Rabin, M.O. 149f.
Rado, T. 61, 87
Razborov, A.A. 169, 173f., 177
Reidemeister, K. 3f., 26, 28f., 67, 107, 109
Robinson, D.J.S. 138
Rosenberger, G. 45, 60, 74
Rost, M. 122
Rothaus, O.S. 101
Rotman, J.J. 146
Rutitskij, Ya.B. 218

Sanov, I.M. 98
Schiek, H. 94
Schönflies, A. 66
Schreier, O. 28, 52, 129ff.
Schubert, H. 119
Schupp, P.E. 4, 34f., 38, 55, 104, 156, 179, 184, 193, 195
Schur, I. 138
Scott, G.P. 54, 108, 136, 138, 141f.
Seifert, H. 5, 9, 20, 111, 119
Selberg, A. 76f.
Series, C.M. 83
Serre, J.-P. 40, 134f.
Shafarevich, I.R. 8f., 221
Shalen, P.B. 46, 54
Shvarts, A.S. 170, 221
Siegel, C.L. 82
Solitar, D. 4, 40, 57f., 140, 195, 197
Spanier, E.H. 5, 87, 109
Stallings, J.R. 4, 111, 138f., 187, 200, 221
Stammbach, U. 133, 135
Stillwell, J. 90
Suzuki, S. 194f.
Swan, R.G. 141f.

Tarski, A. 173
Tartakowskij, V.A. 94
Taylor, T. 58, 100, 126
Teichmüller, O. 90
Threlfall, W. 5, 9
Thurston, W.P. 46, 85, 156, 194
Tietze, H. 10f.
Tits, J. 46, 212
Tsejtin, G.S. 146
Todd, J.A. 29
Torelli, R. 203
Trofimov, V.I. 214
Turing, A.M. 143
Turner, E.C. 49, 54

Ullman, J.D. 151, 154

Vogt, E. 4f., 9, 13, 27, 29, 40, 63, 65ff., 71, 75f., 78, 82ff., 85ff., 89f., 113, 129
Volodin, I.A. 191f.

Wagon, S. 215
Wajnryb, B. 194
Waldhausen, F. 110f., 113, 188, 192, 201
Wall, C.T.C. 136, 138, 141
Whitehead, J.H.C. 48f.
Wilkie, A.J. 213
Wilkie, H.C. 75
Wolf, J.A. 212f.

Zalesskij, A.E. 220
Zassenhaus, H. 219
Zieschang, H. 4f., 9, 13, 27, 29, 40, 45, 55, 59, 63, 65ff., 71, 74ff., 78, 82ff., 85ff., 89f., 97, 105, 113f., 120ff., 126, 129, 179, 190, 197

Subject Index

Abelian group, abelianization 12
Abstract kernel 128
Accessible group 140
AC-conjecture (Andrews-Curtis) 205
Action on a graph 23, 42
– on a space 88
AG = amenable groups 224
Alexander duality theorem 114
Alexander module, polynomial 121
Alexander-Tietze deformation theorem 84
Algebraic intersection number 65, 82
Algorithm 143
Algorithmic Poincaré problem 191ff.
Almost contained in 139
– invariant subset 137
Alternating binary product 79
Amalgamated subgroup 35
Ambient isotopy 114
Amenable group 224
Anisimov's theorem 151
Annulus 64
Area of a diagram 99
Aspherical manifold 109
Augmentation ideal 219
Automatic group 156
Automaton (deterministic, finite) 151f.
Automorphism of free group 47
– of Fuchsian group 79
– geometric 54
– permutation 47
– Nielsen 47
– Whitehead 49

Baer's theorem 84, 111
Balanced presentation 106, 193
Banach-Tarski paradox 223
Base group of HNN-extension 35
Betti number of a group 6, 12
– – of a manifold 108
Bifurcation 80
Binary product 79
Bolyai-Lobachevskij plane 70, 88
Boundary irreducible manifold 110
– of a surface 62
– path 15
Braid 122
– automorphism 125
– group 9, 123, 204
Branched covering 67
Burnside group 225

Cancellation condition $C'(\lambda), C(6)$, 94, 97f.
– diagram 57, 92
Canonical automorphism group 177
– fundamental sequence 177
Cayley diagram 22, 69, 71
Central letter 32
Centre of free product 40
– of fundamental group of a knot 120
Chain 19
Chomsky normal form 155
Church's thesis 143
Circle 15
Coboundary operator (map) 130, 137
Cochain (complex), cocycle 130f.
Cohomological dimension 134
Cohomology group 131
– of braid groups 127
– of cyclic groups 133
– of 1-relator groups 60
Commutator subgroup 12

Compact surface 62
Complex 13f.
Complexity of a subgroup 53
Compressible 110
Configuration space 126
Conjugacy diagram 96
– problem 11, 185
–– for braid groups 86, 126
–– for free groups 32
–– for free products 39
–– for 1-relator groups 57
–– for small cancellation groups 97
–– for surface groups 91
–– for word hyperbolic groups 99
Context-free grammar 154f.
– language 152
Core of a graph 53
Covering 23
– of a Riemann surface 87f.
– transformation 25
Crystallographic group 75, 88
Cube with handles 187
Cut and paste 19f.
Cuts on a surface 80
Cyclic word 49
Cyclically reduced 32, 39

Decision problem (solvable) 11
–– unsolvable 143
Deficiency of a group 107
Defining relations 8
Degree 98
Dehn's algorithm 57, 91, 94, 184ff., 196, 226
Dehnsches Gruppenbild 23
Dehn's Lemma 107
Dehn-Nielsen theorem 81, 126
Dehn twist 85, 194
Dihedral group 8, 70
Disc 64, 92
Discontinuous group 88
Discrete group 88
Dodecahedral group 70
Dodecahedron space 108
Double point 114
Dyck's theorem 8

Edge 13
– group 41
Effective action of a group 78
Effectively given (fundamental sequence) 175
Efficient group 107
Elementary amenable group 224

Empty word 6
Ends of a space 136
Equation over a group (solution) 101
Equimorphism 222
Equivalence of epimorphisms 180
– of splitting homomorphisms 188, 197ff.
– of words 10
Euler characteristic 15, 93, 96, 108
– number 112
Exceptional fibre 112
Exponent of a group 98
Exponential growth 213
Extension of groups 127

Face 14
Factor set 129
– through 200
Fibred knot 120
– manifold 111
Finitely generated, presented 7
Finite support (of cochain) 137
Fixed point subgroup (of automorphism) 54
Følner sequence 224
Free group 10, 31f.
–– subgroup 27, 33f.
Free product amalgamating a subgroup 21
–– centre, finite subgroup 40
–– normal form theorem 36
Freiheitssatz 55, 57, 102f.
Fricke moduli 89
Fuchsian group 75, 88
Fundamental domain 67
– group 18
–– of Seifert manifold 112
– sequence 175

Generating solution 182
– system 6, 74
Generators, adding, deleting 10
Genus 63
Geometric generators, rank 74
– isomorphism 69
GK-conjecture (Grigorchuk-Kurchanov) 205
Graph 13
– of groups 41
Gromov's theorem 213, 215, 217
Group action, without inversion 23, 42
– discontinuous, free, action 88
– of homeomorphisms 88
– of intermediate growth 216

– of a knot 114
– ring 100
Growth exponent 225
– function 210
– rate 211
Grushko's theorem 44, 140, 156

Haken hierarchy, manifolds 110
Halting problem 146
Handlebody 106
Hauptvermutung 87, 106
Heegaard decomposition 106, 121, 187, 190f.
– genus 113, 121
– splitting 97
Hilbert's 5th problem 214
Hilbert-Poincaré series 220
HNN-extension 35
Homeomorphy problem for 4-manifolds 150
Homeotopy group 84
Homology groups 19
– , homotopy 3-sphere 190f.
Homomorphism of complexes 18
Homotopy of paths 17
Hopfian group 37, 60
Hurewicz theorem 109
Hyperbolic group 184, 226
– plane 75

Identifying 15
Inner rank of a group 104
Intermediate growth 216
Intersection form 65
– number 65, 82
Inversion, action without 23, 42
Irreducible manifold 110
Isomorphism problem 11
– – for free groups 32
– – for 1-relator groups 58
– – for 2-generated 1-relator groups w. torsion 60
– – unsolvable 149

van Kampen theorem 20f., 187
von Kerékjártó's theorem 221
Kervaire-Laudenbach problem 101
Klein bottle 30, 64
Kleinian groups 88
– surfaces 78
Knot 114
– figure eight 117
– group 114
– torus 117
– trefoil 116
– two-bridge 117f.
Kurosh theorem 40

Label (of edge) 22
Language accepted by automaton 152
Length 31, 36
– function 46
Lens space 109
Lickorish twist 85
Lies over 24
Line 16
Link 114
Locally cyclic group 58
– finite group 141
– indicable group 100
– separating point 183
Loop theorem 107
Lower central series 219
Lyndon-Hochschild-Serre spectral sequence 135

Magnus' theorem 193
Makanin's algorithm 173
Manifold 105
– Euler characteristic 108
Mapping 17
– class group 84, 193
Markov's normalisation principle 143
Mayer-Vietoris sequence 135
Measure of planar discontinuous group 70
Membership problem 39
Metabelian group 58
Metric cancellation hypothesis 94
Milnor's problem 213, 216
Minimal cyclic word 49
Möbius band 64
Modular group 9, 29, 89
– problem 87
Modulus 89

NEC-groups 75
Neighbour 62
Net, planar 66
NF = groups with no 2-generator free subgroup 225
Nielsen automorphism 47, 206
– equivalent 45, 97
– -Khmelevskij function 173
– number 182
– realisation problem 86
– reduced binary product 80
– reduced system 32ff.

– transformations 33, 44
Nilpotent semigroup 215
Non-metric cancellation hypothesis 97
Normal dissection of braids 124
– form theorem for free products 36
Number of ends 136

Obstruction 134
Octahedral group 70
One-relator group 58, 103
Order of a covering 24
Orientable, non-orientable surface 15f.
Orientation of a graph 41
– preserving (reversing) operation 66
– of a surface, complex 62
Orlicz-space 218

Parametric solution 172
Path 14, 41
PDA = pushdown automaton 152
Peak 6, 31
– reduction lemma 51
Periodic group 141
Peripheral system 109
Picard group 88
Planar discontinuous groups 67
– – – subgroup of 75f.
Planar net 66
Platonic groups 70, 88
Poincaré conjecture 108, 191, 200f.
– dodecahedron 108
– duality 105, 139
Polynomial growth 212f.
Power notation 7
Presentation of a group 7
– balanced 106
– regular 151
Primitive element (of free group) 49
Product, binary alternating 79
– of paths 17
– of words 6
Projective plane 15, 64, 85
– resolution 133
Properly embedded 110
Pushdown automaton 152

Quadratic equation (in free group) 178
– word 104
Quillen's theorem 220
Quotient complex 15

Rank of a group 6, 12
– of a free group 10, 28
Rational Euler number (Seifert manifold) 112
Razborov's theorem 177
Reading head 144
Recursively enumerable 148
Reduced product 33
– word 31
Reflection 67
Regular covering 25
– group presentation 151
Regularly filtered manifold 223
Reidemeister-Franz torsion 109
Reidemeister-Schreier theorem (method) 28, 65, 70, 75, 77, 120
Relations, relators, system of defining relations 6f.
Residually finite 60
– free group 174
Riemann-Hurwitz formula 75, 78
Riemann mapping theorem 88
– surface 87
Rotation 67
R-reduced 95
ℝ-tree 46

Scanned symbol 144
Schönflies theorem 66
Schreier property 27
Seifert manifold (fibre space) 97, 111, 121
Seifert-van Kampen theorem amalgamation form 20, 117, 120, 135
– – HNN-form 21, 135
Selberg Lemma 77
Semidirect product 129
Semigroup presentation 146
Separating point 182
Shapiro's Lemma 134
Siegel's modular group 82
Simple closed curves 83
Small cancellation 94
– – group 184f.
Soluble group 58
Solution 101, 104, 172
Source 13
Spanning tree 14, 34
Special parametric solution 179
Specialization 186
Sphere 64, 85, 88
– theorem 102, 107, 139
Splitting homomorphism 192
– over a subgroup 138
Spur 14
Stabiliser 52

Stable letter 35
Stack letter 152
Standard sphere 190
– splitting homomorphism 192
Star 62
State 144, 151
– alphabet 151
Subdivision of edge, face 20
Sufficiently complicated 112
– large 110
Surface 87
– complex 61
– group 64
– symmetry 76
Symmetrised relator 94
Symplectic matrix 202

Tailed disc 92
Tame knot 114
Tape alphabet 143, 151
Target 14
Tarski's problem 173
Teichmüller space 90
T-equivalence of epimorphisms 104
Tetrahedral group 70
Tietze operations (finite) 10
Tietze's theorem 11
Tits alternative 212
Todd-Coxeter method 29
Topological group (discrete) 88
– surface 87
Torelli group 203
Torus 21, 31, 64, 85
– knot 117
– theorem 142
Tower (lifting) 102
Transitive 29
Transversal (Schreier) 27f.
Tree 14
– product 42
Trefoil knot 116
Triangle condition 94
Trivial knot 114
Tunnel number of a knot 121
Turing machine 143ff.
Two-sided surface 110

Universal coefficient theorem 105
– covering 30
Unsolvable word problem 146

Variables 101
Vertex 13
– group 41
Virtual dimension 134
Virtually free 153f.
– nilpotent, soluble 211

Whitehead automorphism 49
– theorem 49, 83
Width of group element 183
Wirtinger presentation 115
Word 6
Word hyperbolic group 99
– problem 11, 195
– – for free groups 31
– – for free products 39, 120
– – for HNN-extensions 39
– – for 1-relator groups 55, 58
– – for small cancellation groups 94, 98
– – for surface groups 98

Zassenhaus series 219
ZG-module 130

GPSR Compliance

The European Union's (EU) General Product Safety Regulation (GPSR) is a set of rules that requires consumer products to be safe and our obligations to ensure this.

If you have any concerns about our products, you can contact us on ProductSafety@springernature.com

In case Publisher is established outside the EU, the EU authorized representative is:

Springer Nature Customer Service Center GmbH
Europaplatz 3
69115 Heidelberg, Germany

Batch number: 10682213

Printed by Printforce, the Netherlands